Industrial Wastewater Treatment by Activated Sludge

Industrial Wastewater Treatment by Activated Sludge

Derin Orhon, Fatos Germirli Babuna and Ozlem Karahan

Published by   **IWA Publishing**
Republic - Export Building
1 Clove Crescent
London E14 2BA, UK
Telephone: +44 (0)20 7654 5500
Fax: +44 (0)20 654 5555
Email: publications@iwap.co.uk
Web: www.iwaponline.com

First published 2009
© 2009 IWA Publishing

Apart from any fair dealing for the purposes of research or private study, or criticism or review, as permitted under the UK Copyright, Designs and Patents Act (1998), no part of this publication may be reproduced, stored or transmitted in any form or by any means, without the prior permission in writing of the publisher, or, in the case of photographic reproduction, in accordance with the terms of licences issued by the Copyright Licensing Agency in the UK, or in accordance with the terms of licenses issued by the appropriate reproduction rights organization outside the UK. Enquiries concerning reproduction outside the terms stated here should be sent to IWA Publishing at the address printed above.

The publisher makes no representation, express or implied, with regard to the accuracy of the information contained in this book and cannot accept any legal responsibility or liability for errors or omissions that may be made.

Disclaimer
The information provided and the opinions given in this publication are not necessarily those of IWA and should not be acted upon without independent consideration and professional advice. IWA and the Author will not accept responsibility for any loss or damage suffered by any person acting or refraining from acting upon any material contained in this publication.

British Library Cataloguing in Publication Data
A CIP catalogue record for this book is available from the British Library

Library of Congress Cataloging- in-Publication Data
A catalog record for this book is available from the Library of Congress

ISBN: 9781789065282

Contents

Foreword   xiii

1 INTRODUCTION
- 1.1 NEED FOR BIOLOGICAL TREATMENT   3
- 1.2 NEW PERSPECTIVES   5
  - 1.2.1 Hazardous wastes and micropollutants   5
  - 1.2.2 Innovative treatment   7

2 ENERGETICS AND STOICHIOMETRY OF SUBSTRATE REMOVAL
- 2.1 ENERGETICS OF SUBSTRATE REMOVAL   11
  - 2.1.1 Introduction   11
  - 2.1.2 Biosynthesis   12
  - 2.1.3 Energy generation and transformation   12
  - 2.1.4 Nutritional classification of microorganisms   14
- 2.2 THE CONCEPT OF YIELD   16
  - 2.2.1 Assessment of biomass   18
  - 2.2.2 Assessment of substrate   20
    - 2.2.2.1 The BOD parameter   20
    - 2.2.2.2 The COD parameter   21
  - 2.2.3 The definition of yield   22
  - 2.2.4 Expression of yield for different model components   26

		2.2.4.1	VSS as a biomass parameter	26
		2.2.4.2	VSS and BOD_5 as biomass and substrate parameters	27
	2.3	PROCESS STOICHIOMETRY		30
		2.3.1	System defined with biomass COD and substrate COD	30
		2.3.2	System defined with biomass VSS and substrate COD	31
		2.3.3	System defined with biomass VSS and substrate BOD_5	32
	2.4	REFERENCES		35
3	**MODELLING OF ORGANIC CARBON REMOVAL**			
	3.1	PRINCIPLES OF MODELLING		37
	3.2	MATRIX REPRESENTATION IN MODELLING		39
	3.3	MASS BALANCE		44
	3.4	CHARACTERIZATION OF ORGANIC CARBON		45
		3.4.1	Basis for COD fractionation	45
		3.4.2	Major COD fractions in wastewaters	46
	3.5	MAJOR PROCESSES FOR ORGANIC CARBON REMOVAL		49
		3.5.1	Heterotrophic growth	50
		3.5.2	Hydrolysis	52
		3.5.3	Endogenous respiration	55
		3.5.4	Generation of residual microbial products	58
			3.5.4.1 Particulate residual microbial products	58
			3.5.4.2 Soluble residual microbial products	59
	3.6	MULTI-COMPONENT MODELLING OF ORGANIC CARBON REMOVAL		61
		3.6.1	The endogenous decay model	62
		3.6.2	Death regeneration concept – ASM1	64
		3.6.3	Substrate storage concept – ASM3	65
		3.6.4	Modelling of complex industrial wastewaters – case studies	67
			3.6.4.1 Alternative models	67
			3.6.4.2 Expanded COD fractionation	69
	3.7	OPTIMUM SYSTEM DESIGN BASED ON MODELLING – A CASE STUDY		71
	3.8	REFERENCES		74

4 MODELLING OF NUTRIENT REMOVAL

4.1	INTRODUCTION		79
4.2	BIOLOGICAL NITROGEN REMOVAL		80
	4.2.1 Modelling of nitrification		81
		4.2.1.1 Major nitrogen fractions	81
		4.2.1.2 Stoichiometry of nitrification	83
		4.2.1.3 Growth of autotrophs	85
		4.2.1.4 Decay of autotrophs	86
		4.2.1.5 Conversion of organic nitrogen to ammonia	86
		4.2.1.6 Process kinetics	87
	4.2.2 Modelling of denitrification		89
		4.2.2.1 Stoichiometry of denitrification	90
		4.2.2.2 Growth of denitrifiers	93
		4.2.2.3 Hydrolysis of the slowly biodegradable COD	94
		4.2.2.4 Decay of denitrifiers	95
		4.2.2.5 Process kinetics	96
4.3	ENHANCED BIOLOGICAL PHOSPHORUS REMOVAL		96
4.4	REFERENCES		101

5 EXPERIMENTAL ASSESSMENT OF BIODEGRADATION

5.1	EXPERIMENTAL BASIS OF BIODEGRADATION		103
5.2	PRINCIPLES OF RESPIROMETRY		104
	5.2.1 Concept of oxygen uptake rate (OUR)		104
	5.2.2 Interpretation of respirometric data		109
5.3	ASSESSMENT OF BIODEGRADATION CHARACTERISTICS		112
	5.3.1 Assessment of inert fractions		113
		5.3.1.1 Estimation of inert fractions by batch experiments	115
		5.3.1.2 Estimation of inert fractions by respirometry	121
	5.3.2 Assessment of biodegradable fractions		124
5.4	EXPERIMENTAL ASSESSMENT OF MODEL COEFFICIENTS		126
	5.4.1 Assessment of yield coefficient		126
		5.4.1.1 Heterotrophic growth yield (Y_H)	126
		5.4.1.2 Substrate storage yield (Y_{STO})	128

		5.4.2	Assessment of endogenous decay coefficient (b_H)	130
		5.4.3	Assessment of maximum heterotrophic growth coefficient ($\hat{\mu}_H$)	133
		5.4.4	Assessment of k_h and K_X	134
	5.5	ASSESSMENT OF INHIBITION AND TOXICITY	135
	5.6	MODEL EVALUATION OF EXPERIMENTAL DATA	140
	5.7	REFERENCES	146
6	**MANAGEMENT OF INDUSTRIAL WASTEWATERS**			
	6.1	NEW TRENDS IN INDUSTRIAL WASTEWATER MANAGEMENT	151
	6.2	BASIC TOOLS	153
		6.2.1	Assessment at source – process and pollution profiles	153
		6.2.2	Conventional wastewater characterization	156
			6.2.2.1 Assessment of significant pollutant parameters	156
			6.2.2.2 Reliability	158
			6.2.2.3 Relationships between major pollutant parameters	161
		6.2.3	COD fractions of industrial wastewaters	165
	6.3	RESOURCE MANAGEMENT	167
		6.3.1	Material reclamation and reuse	167
		6.3.2	Auxiliary chemicals	171
		6.3.3	Xenobiotics	175
	6.4	REFERENCES	176
7	**CONTINUOUS FLOW ACTIVATED SLUDGE TECHNOLOGY**			
	7.1	INTRODUCTION	181
	7.2	BASIC PARAMETERS	182
		7.2.1	The sludge age	182
		7.2.2	The heterotrophic net yield coefficient	183
		7.2.3	The food to microorganism ratio	185
		7.2.4	The mean hydraulic retention time	186
	7.3	ASSESSMENT OF SYSTEM FUNCTIONS	186
		7.3.1	Reactor biomass	187
			7.3.1.1 Active heterotrophic biomass	188
			7.3.1.2 Inert particulate microbial products	189

		7.3.1.3	Inert particulate COD of influent origin	190
		7.3.1.4	Total biomass in the reactor	191
	7.3.2	Excess sludge production		194
	7.3.3	Effluent quality		199
	7.3.4	Oxygen consumption		201
	7.3.5	Recycle ratio		202
	7.3.6	Nutrient balance		204
		7.3.6.1	Nutrient removal	204
		7.3.6.2	Nutrient requirements	206
7.4	PROCESS DESIGN FOR ORGANIC CARBON REMOVAL			209
	7.4.1	The concept of pre-treatment		209
	7.4.2	Conceptual design procedure		211
7.5	BIOLOGICAL NITROGEN REMOVAL			219
	7.5.1	Activated sludge design for nitrification		219
		7.5.1.1	Aerobic sludge age	220
		7.5.1.2	Nitrification parameters	221
		7.5.1.3	Autotrophic biomass	222
		7.5.1.4	Autotrophic oxygen demand	222
		7.5.1.5	Alkalinity consumption	223
		7.5.1.6	Design procedure	223
	7.5.2	Activated sludge design for nitrogen removal		232
		7.5.2.1	Process configurations	232
		7.5.2.2	The overall sludge age	235
		7.5.2.3	Denitrification potential	236
		7.5.2.4	Oxygen requirement	237
		7.5.2.5	System design for pre-denitrification	238
7.6	ENHANCED BIOLOGICAL PHOSPHORUS REMOVAL			248
	7.6.1	Factors affecting EBPR		249
		7.6.1.1	Wastewater characteristics	249
		7.6.1.2	System parameters	250
		7.6.1.3	Environmental factors	250
	7.6.2	System design for EBPR without nitrogen removal		251
	7.6.3	System design for EBPR with nitrogen removal		252
7.7	REFERENCES			253

8 SEQUENCING BATCH REACTOR TECHNOLOGY

8.1	INTRODUCTION	257
	8.1.1 Historical development	258
	8.1.2 Current experience	259
	8.1.3 Unified basis for modelling and design	260
8.2	PROCESS DESCRIPTION	262
	8.2.1 Cycle frequency (m)	262
	8.2.2 Nominal hydraulic retention time (HRT)	263
	8.2.3 Duration of phases in a cycle	263
	8.2.4 Duration of periods in a process phase	264
	8.2.5 Number of tanks	264
	8.2.6 Sludge retention time (SRT)	265
	8.2.7 System modelling	265
8.3	ORGANIC CARBON REMOVAL	267
	8.3.1 Basic principles	267
	8.3.2 Specific design parameters	269
	8.3.2.1 Effective sludge age	269
	8.3.2.2 The net yield	270
	8.3.2.3 Excess sludge production and reactor biomass	270
	8.3.2.4 Reactor volume	271
	8.2.3.5 Design parameters for aeration	273
	8.3.3 Process design	275
8.4	BIOLOGICAL NITROGEN REMOVAL	280
	8.4.1 Nitrogen balance	281
	8.4.2 Selection of process options	283
	8.4.2.1 Pre-denitrification	284
	8.4.2.2 Step feeding with dual anoxic phases	284
	8.4.2.3 Intermittent aeration	286
	8.4.3 SBR design for pre-denitrification	286
8.5	ENHANCED BIOLOGICAL PHOSPHORUS REMOVAL	296
	8.5.1 EBPR without nitrogen removal	297
	8.5.2 Simultaneous nitrogen and phosphorus removal	298
8.6	REFERENCES	299

9 MANAGEMENT OF TEXTILE WASTEWATERS

9.1 GENERAL ASPECTS — 303
   9.1.1 In-plant control measures applicable to textile mills — 305
      9.1.1.1 Water conservation — 305
      9.1.1.2 Wastewater reclamation and reuse — 305
      9.1.1.3 Material reclamation — 306
      9.1.1.4 Substitution of chemicals — 307
      9.1.1.5 Process modifications — 308
   9.1.2 End-of-pipe treatment options applicable to textile mills — 309
   9.1.3 Polluting sources and characteristics — 310

9.2 A CASE ON WATER CONSERVATION AND WASTEWATER RECOVERY AND REUSE — 323
   9.2.1 Characteristics of the plant operation — 323
   9.2.2 Adopted methodology for in-plant control — 325
      9.2.2.1 Description of the processes and evaluation of the segregated effluent characteristics — 325
      9.2.2.2 Technical basis of the feasibility analysis — 332
      9.2.2.3 Appropriate treatment alternatives — 336
      9.2.2.4 Feasibility analysis — 339

9.3 REFERENCES — 345

10 MANAGEMENT OF TANNERY WASTEWATERS

10.1 GENERAL ASPECTS — 349
10.2 THE TANNING PROCESS — 350
   10.2.1 Beamhouse operations — 350
   10.2.2 Tanyard processes — 351
   10.2.3 Post-tanning operations — 352
   10.2.4 Finishing operations — 353
10.3 SUB-CATEGORIZATION IN TANNERY INDUSTRY — 353
10.4 WASTEWATER GENERATION AND CHARACTERISTICS — 355
   10.4.1 Process and pollution profiles — 356
   10.4.2 Conventional wastewater characterization — 356
10.5 IN-PLANT CONTROL MEASURES FOR TANNERY INDUSTRY — 357

	10.5.1	Material substitution and reuse	361
	10.5.2	Water use reduction	362
10.6	TREATMENT OF TANNERY EFFLUENTS		362
	10.6.1	Treatment requirements	362
	10.6.2	Treatment schemes	363
	10.6.3	Pre-treatment applications	364
		10.6.3.1 Case study-1	364
		10.6.3.2 Case study-2	364
10.7	BIODEGRADABILITY CHARACTERISTICS		366
	10.7.1	Biodegradability based characterization	366
	10.7.2	Activated sludge modelling for tannery effluents	368
10.8	REFERENCES		374

Index   377

Foreword

It gives us pleasure to present the book with a great sense of fulfilment after a long period devoted to its preparation. The idea of writing this book emerged from the work of *The Environmental Biotechnology Group* at Istanbul Technical University. The core of this group was formed in the early seventies and rapidly expanded into a productive research team focused on the biodegradation mechanisms of complex substrates and the behaviour of microbial systems. The authors have actively contributed to the scientific efforts of the group. After more than twenty years of active research, the authors felt that it was time to look back and evaluate the accumulated scientific output.

The book attempts to provide an up-to-date coverage of the treatment of industrial wastewaters by activated sludge. The main theme is essentially biodegradation, perhaps the most fascinating issue in environmental sciences and likely to remain a challenging problem for future research. Industrial effluent serves as the perfect example for investigating biodegradation processes of complex substrate mixtures.

A comprehensive coverage of the subject involves identification of the major biochemical mechanisms, review of process modelling and the necessary bridge between the design and the mechanistic understanding of the activated sludge process. The book covers these essential steps in an appropriate sequence. After the first chapter introducing the general framework of industrial wastewater control and exploring new perspectives, the energetics and stoichiometry of substrate removal is outlined in the second chapter. Chapter 3 reviews the modelling of organic carbon removal. Similarly, the principles of nutrient removal emphasizing process modelling are summarized in Chapter 4. The next chapter provides basic information on the current methodology used for the

© 2009 IWA Publishing. *Industrial Wastewater Treatment by Activated Sludge*, by Derin Orhon, Fatos Germirli Babuna and Ozlem Karahan. ISBN: 9781789065282. Published by IWA Publishing, London, UK.

experimental assessment of biodegradation. The fundamentals of industrial wastewater management are discussed in Chapter 6. The following chapter defines the essential functions for continuous flow activated sludge systems and gives a unified design methodology both for organic carbon and nutrient removal. A similar treatise of the sequencing batch reactor technology is included in Chapter 8. Chapters 9 and 10 provide detailed management analysis for two selected industrial categories, textile and leather tanning industries.

The book is primarily written for students and practitioners. The content of each chapter is organized in a way that may be used both for advanced undergraduate and graduate programs. Different parts of the text currently serve as the major reference material for three graduate courses offered at Istanbul Technical University: *Biological Treatment of Industrial Wastewaters; System Design for Aerobic Processes and Experimental Methods in Biological Treatment*. Similar key material is purposely provided in different chapters in order to complement the flow of information of each chapter so that they can be read and evaluated as separate entities. A number of examples are included in the text to illustrate significant issues discussed. Design aspects are presented to address a wider audience, in a format that can be readily used by the practitioners of the wastewater treatment industry.

The devoted assistance of our colleagues Ms. Asli S. Ciggin and Mr. Serden Basak in the final production of the manuscript is gratefully acknowledged.

DERIN ORHON
FATOS GERMIRLI BABUNA
OZLEM KARAHAN

1
Introduction

Industrial wastewaters are always a priority issue for the protection of the environment. They should be considered as the most significant components of water quality management programs. Industrial activities are incorporated into these programs at two different dimensions (i) wastewater characterization, and (ii) effluent discharge limitations. Consequently, the success of any quality management scheme in protecting the environment chiefly depends upon how thoroughly the wastewater is characterized and how effective and applicable are the effluent limitations.

Generally, effluent limitations define the quality of the wastewater to be discharged and imply appropriate treatment. Technical description of acceptable quality for discharge is done by setting numerical limits for a set of parameters in terms of concentration, load or both. This way, effluent limitations provide the technical instruments required for legal action. *Pre-treatment limitations* or *standards* control industrial wastewater discharges into public sewers or joint

© 2009 IWA Publishing. *Industrial Wastewater Treatment by Activated Sludge*, by Derin Orhon, Fatos Germirli Babuna and Ozlem Karahan. ISBN: 9781789065282. Published by IWA Publishing, London, UK.

treatment plants. Discharges to receiving waters are regulated by *direct discharge standards*.

While the major objective of a discharge limitation is protection of the environment, it should correlate properly with the characteristics of the wastewater and with the specific treatment scheme that needs to be implemented at source. In a way, a specific effluent limitation should be conceived as a fingerprint of the wastewater quality, improved by appropriate treatment. In this context, management of industrial wastewaters usually involves *categorization* and *sub-categorization* due to complexity of processes involved and the variable character of wastewaters generated. Categories and sub-categories are used to differentiate wastewater quality, appropriate treatment technology and applicable effluent limitations. Experimental surveys indicate however that striking differences may be observed in wastewater quality and quantity from one plant to another within the same sub-category, mainly because of different experiences and habits in the production scheme. Therefore, in-situ experimental information should be derived where possible for each plant on a case by case basis.

The worst approach to take in characterizing industrial wastewaters is to rely on the *end-of-pipe* inspection of effluent quality. The spot image that may be obtained with this exercise may at times be totally useless and even misleading, because wastewater characteristics are closely related to the production scheme at the industry and they are likely to exhibit significant variations with time. A much more reliable approach is to design and create a data base defining the *pollution profile* of the plant, which describes each production step and individual wastewater stream in terms of their specific properties including different flow rates, chemicals, specific pollutants, etc.

The early examples of effluent limitations did not correlate with the required treatment performance well. Then, efforts were directed towards the concept of appropriate treatment. Combinations of various treatment options have been empirically tested for different industrial wastewaters to determine the most efficient alternatives. New and advanced methods of treatment emerged from these studies which also led the way for more stringent standards and ultimately for the *zero discharge* concept. The idea of zero discharge did not have a significant practical implication due to severe economical constraints. Later, *best practicable* and *best available economically achievable* concepts were promoted and served as the technical platform for today's technology-treatment concepts based standards. Beginning in the eighties, the concept of *sustainable environment* triggered new trends such as *clean technologies, waste minimization, waste recovery and reuse, in-plant control*, etc.

Despite all efforts, the quest still continues for a rational discharge limitation system which accounts for all aspects of industrial wastewater management, including environmental protection, correlations between production and waste generation, better wastewater quality assessment, technological requirements, implementation and enforcement, and economic feasibility.

1.1 NEED FOR BIOLOGICAL TREATMENT

Industrial wastewaters involve a large spectrum of different organic compounds depending on the specific production scheme. With the exception of a few specific industrial categories, physical or chemical methods of treatment can only provide partial removal for the organic content of the wastewater. Therefore similar to domestic sewage, the majority of industrial effluents require biological treatment for effective removal of organic matter, with or without a pre-treatment step. The *activated sludge process* is by far the preferred treatment scheme for this purpose.

Assessment of the amount of substrate is the main concern for biological processes. Inorganic substrates, such as nitrogen and phosphorus fractions can be directly measured and incorporated into process stoichiometry. Assessment of the organic substrate is not so simple, both for domestic sewage and industrial wastewaters containing a great variety of different organic compounds which cannot be quantified or even identified individually on a routine basis. Consequently, as elaborated in detail in the following chapters, *the chemical oxygen demand* (COD) preferentially serves as an index value to characterize the overall organic content. Industrial wastewaters are generally stronger than domestic sewage in terms of their COD levels. Depending on the type of industry, the wastewater COD exhibits a huge variation from 500 mg/l to as high as 100,000 mg/l. COD of most industrial effluents however remains in the narrower range of 1000–2000 mg/l.

Selection of conventional parameters is also an important step for industrial wastewaters. It serves as the experimental database for the design of the appropriate treatment scheme. It also provides the necessary information for the interpretation of most operation problems. Characterization should primarily cover all parameters prescribed in related effluent discharge limitations. It should also give the relevant indication for pre-treatment, if needed. It should essentially supply statistically reliable information on the COD level, the nutrient balance (COD/N and COD/P ratios) and the buffer capacity (pH, alkalinity, etc.).

Identification of different COD fractions with different biodegradation characteristics should be regarded as the major milestone in the understanding of biological treatment of wastewaters. This scientific effort has been the driving force for new mechanistic models. The merit of process modelling for activated sludge design is now largely recognized for domestic sewage. Mechanistic models are widely used for a better interpretation of process kinetics, and stoichiometry. Resulting mass balances are readily integrated with system design for organic carbon and nutrient removal. Modelling however is generally overlooked in the biological treatment of industrial wastewaters. Performance evaluation with a *black box* approach using a few parameters and input-output functions is still very common. The related design practice is quite empirical and still depends upon simple observations and past experience. However, there is increasing agreement today on the value of modelling for the activated sludge treatment of industrial wastewaters. This new approach requires generation of specific experimental data related to biodegradation characteristics for each case.

A major drawback in the use of COD as a design parameter for organic matter is the fact that it covers both biodegradable and non biodegradable compounds. This drawback is now overcome by new experimental techniques which separately identify biodegradable and inert COD fractions. Experimental assessment of all major COD fractions is currently a major requirement both for an accurate description of wastewater character and for a rational activated sludge design based on modelling and mass balance. The total biodegradable COD is one of the key parameters in system design as it indicates the magnitude of organic carbon potentially removable through biochemical processes from industrial wastewaters.

Modelling of activated sludge systems for industrial wastewaters also requires experimental assessment of the stoichiometric and kinetic coefficients involved. Recent research efforts towards a better understanding of biodegradation characteristics of complex substrate mixtures such as industrial wastewaters not only offered the basis for comprehensive mechanistic models but also introduced respirometry as a useful instrument providing experimental support for the models. Respirometric techniques are now fully developed for the assessment of process coefficients. These coefficients should be specifically determined for each wastewater. They are also quite model specific and their values should be experimentally assessed in conformity with the structure of the model adopted for design. Extensive experimental data on characteristics and process coefficients for a wide range of industrial wastewaters is provided in the following chapters, which essentially show that activated sludge treatment of industrial wastewaters may have the same conceptual basis as domestic sewage. With sufficient information, a unified basis may be set, both applicable to domestic sewage and industrial effluents, for the design of activated sludge systems for COD and nutrient removal.

Introduction

1.2 NEW PERSPECTIVES

The promotion of COD as a collective quality and control parameter for organic compounds has been extremely useful in assessing organic fractions with different biodegradation characteristics and in modelling treatment performance. Effluent standards are now defined for COD levels achievable with biological treatment. This type of an evaluation enables, as will be also underlined in the following parts, no significant distinction between domestic sewage and industrial effluents: As schematically indicated in Figure 1.1, the biological treatment influent essentially includes biodegradable and non biodegradable COD fractions. Effective biological treatment lowers the total COD in the effluent below effluent discharge limitations by removing all biodegradable components. In other words, the effluent still contains all the non biodegradable chemicals initially present in the wastewater and an additional supplement of residual microbial products generated during treatment. However, the traditional evaluation approach based on a single total COD value finds the level of treatment in compliance with discharge regulations. While this approach may be tolerated for domestic sewage, it represents a potential environmental risk for industrial wastewaters containing a wide spectrum of chemicals likely to bypass biological treatment.

1.2.1 Hazardous wastes and micropollutants

Recognition of the ever increasing number of chemicals generated by industrial activities and discharged to the environment, with no appreciable effect from full

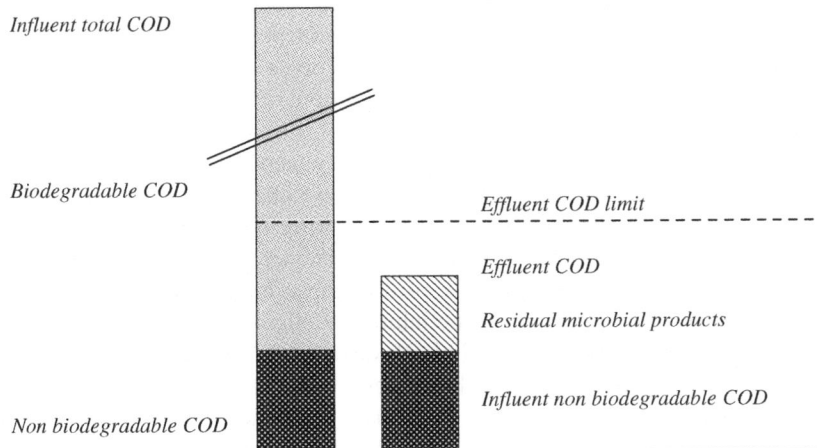

Figure 1.1. Fate of COD fractions in biological treatment

biological treatment, brought about significant conceptual developments. Major problems so far identified, while defining significant new perspectives for industrial waste management, still remain as challenging issues to be solved in the future.

The concept of *hazardous wastes* emerged this way, as a result of in-plant evaluation of polluting sources. Segregated analysis of different waste streams identified highly concentrated chemical wastes which were not treatable when combined with the rest of aqueous wastes by conventional treatment schemes. Accordingly, a new management code of practice was developed to identify hazardous waste streams in each industry for separate treatment and disposal by specific measures. Despite legal requirements, occasional discharge of hazardous wastes is still a major problem for the safe operation of treatment systems.

The in-plant analysis of industrial activities also revealed a great number of specific chemicals regularly used in the manufacturing processes, likely to cause serious adverse effects on the treatment scheme and on the environment. The most dangerous part of this discovery was the fact that such pollutants could not be accounted for by means of conventional management schemes based on traditional pollution parameters. This system deficiency led the way to a new perspective of including *priority pollutants* or *micropollutants* in the concept of industrial pollution, both as additional key parameters in the characterization of industrial wastewaters and in effluent discharge limitations.

In this context, it is easy to visualize the influent of a traditional *end-of-pipe* industrial wastewater treatment system as the melting pot for hazardous waste streams, a variety of non biodegradable, toxic chemicals together with biodegradable organics. With the new management perspectives, the solution of hazardous wastes is relatively simple as it involves identification and separate handling of these wastes. However, recognition of micropollutants is still a challenging issue and involves new problems.

The first important issue is the *inhibitory* and/or *toxic effects* of these chemicals on biological treatment. Most of these chemicals exert an inhibitory action on activated sludge, significantly impairing the treatment performance. Extensive research effort has been devoted to the measurement of inhibitory and toxic effects, with the main purpose of deriving a numerical value that would give an appropriate indication of the inhibitory effect. In practice, the common way to reach this numerical limit is through experimental observations and evaluations. Therefore, a major achievement in trying to understand and utilize the concept of inhibition would be to describe it physically and mathematically, in terms of simple parameters that could be incorporated into process rate and mass balance expressions defining the functioning of the activated sludge process. This way, the inhibitory effect may be reflected upon the stoichiometric and kinetic coefficients associated with microbial processes and become part of the mechanistic models currently in use for activated sludge systems. The merit of this approach relies upon experimental methods utilized to quantify these coefficients. In other words, it becomes useful and meaningful only if it is used in conjunction with experiments that would yield accurate and consistent results for the assessment of process kinetics.

The second important issue is *treatability*. Most micropollutants are resistant to biodegradation. Some of them can be removed to a certain degree by conventional treatment systems. Others require advanced treatment. These systems may be modified to provide enhanced removal of micropollutants. However, standardized treatment schemes for the removal of specific micropollutants are yet to be determined. Removal efficiencies for many of the micropollutants in conventional systems vary and they are often low, and mostly inadequate. The essential problem remains to be their *biodegradation* whether within the biological treatment system or in the environment. Greater research efforts are now directed towards the analysis of biodegradation profiles of these pollutants in combination with other techniques such as advanced oxidation technologies. The problem still requires extensive investigation.

The third issue is *source management*. Collection and *end-of-pipe* treatment of the combined effluent generally proves satisfactory for conventional COD removal but this approach should be avoided to the extent possible for micropollutants. A much better way for these specific pollutants is to identify their respective sources of generation and remove them where they are most concentrated. In this context, an industrial pollution profile should identify, aside from conventional parameters of pollution, hazardous wastes and to the extent possible, separate wastewater streams containing significant micropollutants.

1.2.2 Innovative treatment

Recent information on biodegradation kinetics provides a clear indication that the removal of biodegradable COD fractions only require a small fraction of hydraulic retention time and the sludge age adopted in the design of activated sludge systems. The major portion of the reactor capacity is still devoted to conditioning the biomass for upgrading its settling properties so that it may be separated from the liquid stream and re-circulated back to the reactor. Despite all the major achievements in the understanding and modelling of substrate removal, the activated sludge process still depends on an almost century old technology of gravity settling which basically controls system design. In fact gravity settling dictates the level of biomass concentration that can be maintained in the reactor, the magnitude of internal recirculation in nitrogen removal, etc. This is especially true for the biological treatment of industrial wastewaters where conservative design parameters are selected for the sole reason of ensuring stability of biomass settling. In this context, abandoning gravity settling for *membrane filtration* and traditional activated sludge systems with secondary clarifiers for the *membrane bioreactors* (MBR) should be envisaged as a major new perspective for biological treatment in the future.

The *membrane bioreactor* is now quite attractive as an innovative treatment technology. It requires less volume and a smaller footprint mainly because of its ability to separate and to filter treated effluent from biomass independently from its settling properties. This feature is commonly interpreted as a means for sustaining a higher biomass level and for operating the MBR at high sludge ages with substantially lower excess sludge production as compared to conventional activated sludge process. The current practice with the MBR raises two main questions: Do we indeed require (i) a higher sludge age and (ii) a lower excess sludge production?

The answer to the first question is very simple, especially for industrial wastewaters: We do not! In fact, exploring the possibility of maintaining a substantially higher biomass concentration with high sludge ages is closely related to empirical experiences and habits underlining the traditional practice. It is evident that once the constraint of settling is no longer a problem, the operating flexibility of the MBR should be exploited to lower the sludge age and the reactor volume to a level prescribed by biodegradation requirements alone. Furthermore, the MBR system should be operated with as low a biomass concentration as possible to facilitate and enhance membrane filtration.

The answer to the second question is not so straight forward and requires a conceptual evaluation. Wastewater characterization now differentiates soluble and particulate fractions of the biodegradable COD. The biological treatment system is essentially concerned with the soluble fraction of every pollutant. In this respect, the effluent discharge limitations only control the soluble biodegradable COD compounds of the influent, which need to be utilized and fully removed from the system before the liquid stream leaves the reactor. Experimental data indicate that the biodegradation rates of soluble COD fractions are fast enough to enable complete removal within a small fraction of the hydraulic retention time usually selected for activated sludge systems. On the other hand, particulate biodegradable COD compounds become immediately entrapped and enmeshed with the biomass so that they can be totally removed from the system with excess sludge. Traditionally, the sludge is allowed to stay in the reactor. The potential energy of the particulate biodegradable matter is burned within the reactor with the supply of excessive amounts of oxygen requiring additional energy input. This energy is practically wasted for endogenous respiration in a much larger reactor volume operated at much higher sludge ages. Especially in systems such as the MBR where biomass settling is not a problem, there is the possibility of recovering this energy from sludge under anaerobic conditions and the opportunity of using a much smaller reactor volume devoted solely to the treatment of the liquid stream. The beneficial tradeoff between liquid and solid waste streams is quite a challenging issue that deserves serious consideration in the future.

2
Energetics and stoichiometry of substrate removal

2.1 ENERGETICS OF SUBSTRATE REMOVAL

2.1.1 Introduction

Biological treatment is an ingenious system where pollutants in wastewater serve as substrate for the microbial community sustained in a reactor. Microorganisms are grown in a controlled environment at the expense of organic and inorganic pollutants in the feed stream through a complex sequence of biochemical reactions. These reactions occur as the vital steps of the *metabolic activities* of microorganisms. *Metabolism* defines the entire spectrum of intricate biochemical conversions taking place in living cells. Sound environmental engineering requires full understanding and control of these biochemical processes for optimizing treatment performance. The treatment system is called *activated sludge* when the biological reactor

© 2009 IWA Publishing. *Industrial Wastewater Treatment by Activated Sludge*, by Derin Orhon, Fatos Germirli Babuna and Ozlem Karahan. ISBN: 9781789065282. Published by IWA Publishing, London, UK.

favours *suspended growth*, mostly with the presence of dissolved oxygen. This process essentially involves an aeration tank, that is, a biological reactor where the biomass is kept in suspension by aeration. *Biofilters* however sustain *attached microbial systems* or *biofilms*.

2.1.2 Biosynthesis

The complex array of biochemical reactions related to microbial growth may be considered in two main groups: (i) *biosynthesis,* and (ii) *energy reactions*. The first group involves manufacturing of major biochemical components of microbial cells from external substrate, using specific sequences of biosynthetic pathways. These reactions are also called *assimilative* reactions or *anabolic* reactions. The essential chemical components of microbial cells, such as carbohydrates, proteins, nucleic acids, etc., are continuously synthesized through these reactions from smaller building-block molecules by the action of *enzymes*. To construct a protein molecule for example, hundreds of amino acids must be joined in the right sequence by peptide bonds. Similarly, polysaccharides are formed by hundreds of glucose molecules joined in glycosidic linkage. The simplified overall biosynthesis reaction

$$\text{building blocs} \rightarrow \text{macromolecules} + H_2O$$

is *endergonic:* it is an uphill reaction, requiring significant input of energy. The right source of energy must also be present and available in the biological reactor so that microbial growth may proceed. *Bioenergetics* is the term that defines energy transformations in living cells. It is also an essential element of biological treatment systems. A detailed treatise of bioenergetics is provided in basic textbooks (Lehninger, 1965; Racker, 1965).

2.1.3 Energy generation and transformation

The energy required by biosynthesis is supplied by *catabolic* reactions. The microbial community in the activated sludge systems require *chemical energy* like almost all the living systems of the animal kingdom. The energy generating reactions are also called *exergonic reactions* as they yield useful chemical energy. Therefore in biological treatment, substrate becomes the major ingredient of two types of reactions: (i) it may take part in *catabolic* reactions where it is converted into stable end products with energy generation, or (ii) it may be incorporated into *anabolic* reactions for the manufacturing of major macromolecules of the microbial cells. The nature of substrate utilized in these metabolic routes is directly a function of the nutritional requirements of the microbial community maintained in the system.

In catabolic reactions, only a portion of the energy release is useful energy or *free energy*, as the rest is dissipated to increase entropy according to thermodynamic laws. *Oxidation-reduction* reactions are inherently selected as the source of chemical energy in biological systems. This natural selection is quite effective as oxidation-reduction reactions yield much higher useful energy per mole, compared with other alternatives of chemical reactions. Oxidation is essentially characterized by the loss of electrons. Therefore, oxidation–reduction is basically an exchange of electrons between substrate oxidized (electron donor) and an *electron acceptor*. The flow of electron is quite significant to set the equivalence of different components of a biochemical reaction. The electron exchange in oxidation-reduction reactions is illustrated with the following typical example.

Example 2.1 Electron balance in oxidation-reduction reactions
Calculate the electron flow for the oxidation of a mole of glucose under aerobic conditions.

Assessment of the electron flow requires determination of applicable oxidation and reduction half reactions where O_2 serves as the electron acceptor.

The following steps may be suggested for an accurate formulation of an oxidation/reduction half reaction: (i) Determine the oxidized product; (ii) balance carbon; (iii) balance oxygen with H_2O; (iv) balance hydrogen with H^+; and (v) balance the charge with electrons. Glucose is fully oxidized to CO_2 under aerobic conditions. Therefore, the oxidation half reaction:

$$C_6H_{12}O_6 + 6H_2O \rightarrow 6CO_2 + 24H^+ + 24e^- + energy$$

The reduction half reaction:

$$4e^- + 4H^+ + O_2 \rightarrow 2H_2O + energy$$

Balancing the two half reactions yield the following overall oxidation reaction for glucose:

$$C_6H_{12}O_6 + 6O_2 \rightarrow 6CO_2 + 6H_2O + energy$$

Oxidation of glucose under aerobic conditions involves transfer of 24 e^-/mole. It requires six moles of oxygen acting as final electron acceptor. Therefore, the electron equivalent of one mole of glucose is six moles of oxygen.

14   Industrial Wastewater Treatment by Activated Sludge

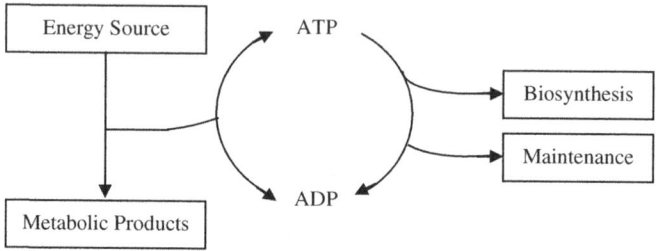

Figure 2.1. The role of ATP in the utilization of energy in microbial processes

Biochemical reactions are always coupled. This is their basic feature which differentiates them from simple chemical reactions. The coupling of biochemical reactions requires two specific mechanisms: (i) transfer of energy, and (ii) transfer of electrons. The energy released by catabolic reactions is not converted into heat or immediately utilized in biochemical work but conserved as chemical energy in the *adenosine triphosphate* molecule, commonly known as ATP. This molecule essentially serves as the coupling agent between energy-yielding and energy-requiring metabolic reactions. The energy potential acquired and stored in ATP through catabolic processes is utilized for two different purposes (i) for biosynthesis, where it is expended to reduce the carbon source to the necessary redox level for assimilation and to generate cellular macromolecules; (ii) for maintenance, which involves all the other essential functions of microbial life, aside from growth. Two major functions may be cited as requiring energy for maintenance: transport work and mechanical work. ATP is formed from ADP (adenosine diphosphate) during catabolic reactions and releases its chemical energy for the synthesis of cellular material and for maintenance requirements as schematically shown in Figure 2.1.

Energy reactions require a *final electron acceptor*. Dissolved oxygen acts as the final electron acceptor in *aerobic* systems. This is an over simplified picture for biochemical reactions as direct transfer of electron to the final electron acceptor would spontaneously release a significant amount of energy not fully utilizable in biosynthesis. Instead, the electrons are first received by *pyridine nucleotides*, NAD and NADP acting as electron carriers. In this step, these molecules become reduced to $NADH_2$ and $NADPH_2$. The electrons are then transferred to the final electron acceptor through an *electron transport chain* with a stepwise release of available energy. $NADH_2$ and $NADPH_2$ are re-oxidized for their cyclic function.

2.1.4 Nutritional classification of microorganisms

Activated sludge system is traditionally used for the removal of organic pollutants from domestic sewage. Now, it is also effectively used for removing significant inorganic pollutants like nitrogen and phosphorus. However, wastewater characteristics do not allow efficient control of all types of organic and inorganic substances by a single group of microorganisms. Consequently treatment systems need to be designed for a specific purpose. Selected operating conditions, if appropriate, will only sustain an *enrichment culture* that will outgrow other species by selective growth to achieve the targeted performance. Therefore success in the biological removal of a particular wastewater component is closely related to the nutritional requirements of microorganisms and the specific function of this component in major biochemical reactions involved.

There are a number of ways to categorize microorganisms. When this process is narrowed down to activated sludge systems, three basic factors may serve as the basis for the nutritional classification of biomass sustained in the biological reactor: *Carbon source, energy source* and *final electron acceptor*. Microorganisms sustained in the activated sludge process are *chemotrophs* as they depend on chemical energy. Organic compounds are the principal concern in biological treatment both for domestic sewage and industrial wastewaters. Microbial groups using organic compounds as their organic carbon source are called *heterotrophs*. Microorganisms that utilize CO_2 as their principal carbon source for biosynthesis are called *autotrophs*. Heterotrophic cells require complex organic carbon forms such as carbohydrates, amino acids, etc., which autotrophic cells can manufacture from CO_2, ammonia and water.

As far as the electron acceptor is concerned, activated sludge systems mostly operate under *aerobic* conditions, when the biological reactor is aerated and *dissolved oxygen* serves as the final electron acceptor. In the absence of oxygen, oxidized inorganic compounds assume the same function under *anoxic* conditions. The energy yielding process under these conditions is called *respiration*. On the other hand, fermentation defines the *anaerobic* metabolism of microorganisms where electrons are exchanged between oxidized and reduced organic compounds, yielding fermentation end products.

Activated sludge is generally an aerobic process relying on the growth of heterotrophs for the biodegradation of organic carbon. The heterotrophic biomass may be adapted to a sequence of anoxic/aerobic operation for the removal of nitrate. Autotrophic *nitrifiers*, oxidizing ammonia or nitrite as their energy source, may also be sustained together with heterotrophs under suitable growth conditions. A more detailed account of microbial classification on the basis of nutritional requirements is presented by Orhon and Artan (1994). An outline of major microbial communities for the activated sludge process is presented in Table 2.1. Similarly, Table 2.2 summarizes pollutant removal potential of activated sludge systems in terms of major metabolic functions.

Table 2.1. Significant microbial groups for activated sludge systems

Process	Microbial group	Carbon source	Energy source	Electron acceptor
Organic carbon oxidation	Heterotrophs	Organic carbon	Organic carbon	O_2
Denitrification	Heterotrophs	Organic carbon	Organic carbon	Oxidized nitrogen
Nitrification	Autotrophs	CO_2	NH_4^+	O_2

Table 2.2. Pollution removal potential of different metabolic functions

Pollutant removal	Metabolic functions
Organic carbon	Heterotrophic growth *Carbon and energy source*
Nitrogen removal	Heterotrophic & autotrophic growth *Incorporation into biomass*
Phosphorus removal	Heterotrophic & autotrophic growth *Incorporation into biomass*
Ammonia oxidation	Autotrophic growth *Nitrification*
Oxidized nitrogen removal	Anoxic growth of heterotrophs *Denitrification*
Enhanced biological phosphorus removal	Growth of PAOs* *Internal storage of polyphosphate*

* PAOs = Phosphorus accumulating organisms

2.2 THE CONCEPT OF YIELD

In biological treatment, substrate removal takes place by conversion into living matter – *microbial cell* – through a complex sequence of biochemical reactions. The conglomerate of different microbial species sustained in a biological reactor is commonly called *biomass*. Removal of organic substrate, the major pollutant of concern in domestic sewage and industrial wastewater, is therefore coupled with the formation of biomass and consumption of dissolved oxygen, which serves as the final electron acceptor under aerobic conditions. Inorganic substrate such as nitrogen and phosphorus is also incorporated into biomass to fulfill the nutritional requirements of microorganisms. Accurate understanding of these processes is essential not only for scientific evaluations but also for the practical issues of treatment system operation. Steady-state conditions of biological reactor are maintained by regularly discharging the biomass generated by substrate removal, as *excess sludge* from the reactor. Treatment and disposal of *biological sludge* is at least as important and as costly an operation as substrate removal. On the other hand, supply of necessary oxygen to the biological reactor is the major component of operational costs. Therefore optimum system operation requires correct assessment of *excess biological sludge* produced and *oxygen* consumed. The *yield* parameter plays a major role in this assessment.

The concept of yield basically establishes and reflects the balance between substrate removed, biomass generated and dissolved oxygen consumed. This balance may best be expressed in terms of electron equivalence or, for practical purposes, as a mass balance. The definition of yield is closely related to the way in which basic biochemical processes are interpreted. Within a simplified context, *microbial growth* and *endogenous respiration* may be considered as two basic processes directly affecting the concept of yield. The commonly adopted modelling approach today conveniently assumes that the two processes occur simultaneously but in successive steps. As schematically illustrated in Figure 2.2, substrate is first utilized by means of *energy* and *biosynthesis* reactions, and converted into biomass. In the modelling jargon, this step is called the *growth* process. Then, biomass generated through growth undergoes oxidation and generates *energy for maintenance*. This step is defined as *endogenous respiration* mainly because it also requires oxygen as an electron acceptor. Experimental observation and evaluation of the two processes individually is not possible as they occur simultaneously.

In this context, process modelling and system design involves two different yield coefficients. The first one is the *true yield coefficient* Y, which is associated with the growth process. This coefficient defines the numerical relationship between the

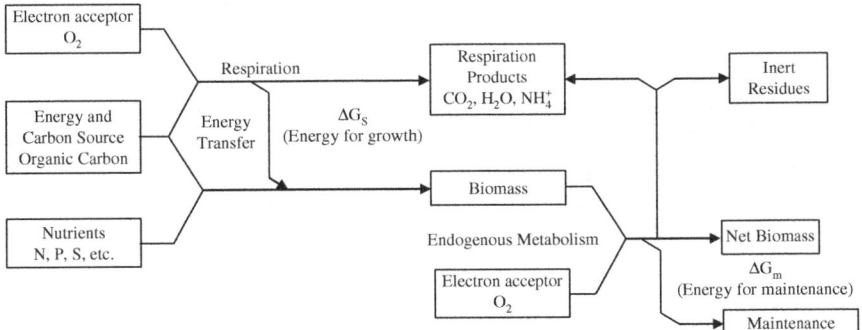

Figure 2.2. Substrate and energy utilization of heterotrophic microorganisms under aerobic conditions (Orhon and Artan, 1994)

18   Industrial Wastewater Treatment by Activated Sludge

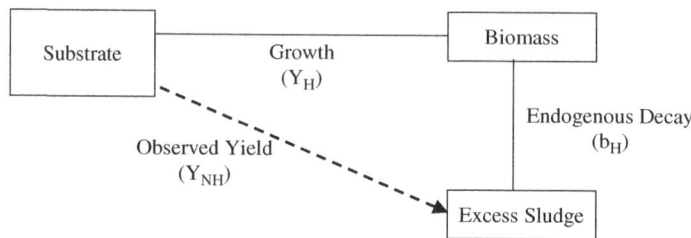

Figure 2.3. Relationship between the true yield coefficient and the net yield coefficient for heterotrophic biomass

amount of biomass produced per unit amount of substrate utilized. This coefficient is considered as the major parameter defining the stoichiometry of biological systems mainly because it is not affected by reaction rates. The true yield coefficient, Y can be experimentally calculated and adopted as a constant for a given substrate under specific operating conditions. It may assume different values as the type of substrate changes, due to different energy conversion and utilization mechanisms involved.

In biological reactors, the amount of biomass remains always under the level that may be computed using the true yield coefficient. The observed biomass concentration becomes reduced in proportion to the relative magnitude of endogenous respiration as compared to growth. As schematically plotted in Figure 2.3, the observed amount of biomass generated per unit amount of substrate utilized is defined by means of the *net yield coefficient* Y_N. As opposed to the true yield, this coefficient is not a constant even for a given substrate but varies, as will be mentioned in the following sections, in a way that is inversely proportional with the *sludge age*, under the selected operating conditions. The observed yield coefficient serves as an important and useful design parameter in the assessment of major operating variables such as excess sludge, oxygen consumption and nutrient requirements.

Biological treatment of wastewater involves the interaction of a complex mixture of substrate and a mixed culture of microorganisms enriched for selected operating conditions of the system. In such systems, the relevance and the restrictions of analytical procedures commonly used for defining *biomass* and *substrate* should be well understood. This understanding is essential for an accurate determination of related stoichiometric relationships and mass balance expressions.

2.2.1 Assessment of biomass

The term *activated sludge* is now universally recognized for defining the mixed microbial community sustained in the aeration tank of biological treatment systems. The content of the biological reactor is called *mixed liquor*. Interpretation of activated sludge behaviour using stoichiometric and kinetic expressions essentially requires (i) an analytical procedure defining the amount or the concentration of biomass in the mixed liquor, and sometimes (ii) an empirical formula approximating activated sludge composition. Both are very difficult to assess and the approach selected may at times be a major drawback itself for an acceptable evaluation of related biochemical processes.

At this time, no acceptable method has been proposed to quantify the total *active biomass* in the biological reactor, let alone biomass of separate communities with different metabolic functions, i.e. heterotrophs, nitrifiers, etc. Active biomass of different activated sludge groups can only be assessed indirectly, by model calibration of relevant experimental data. The only analytical procedure so far available for this purpose is the historical *volatile suspended solids* (VSS), measurement or in other words, the VSS content of the mixed liquor (MLVSS). VSS is a collective parameter basically defining the particulate organic fraction in the reactor. This fraction is commonly differentiated from its soluble counterpart by the filtration size in the analytical procedure and 0.45 μm (membrane filters) or 1.2–1.6 μm (glass fiber filters) are commonly adopted for filtration of the mixed liquor. Inevitably, VSS measurements include aside from active biomass, inert organics in the influent, the remaining portion of particulate biodegradable substrate entrapped in the activated sludge flocs and particulate residual metabolic products, i.e. non biodegradable particulate residues of endogenous respiration. Similarly, *suspended solids* (MLSS) measurement in the mixed liquor reflects the total suspended solids (TSS) content of the activated sludge above the same particle size as VSS. This parameter additionally includes inorganic (*fixed*) particulate matter either present in the influent or generated through cellular decay. The VSS/TSS ratio is commonly used as an empirical design parameter for the estimation of the amount of excess sludge generated by biological treatment. In recent years, the chemical oxygen demand (COD) has been proposed as a model component to also identify active biomass – *cell COD* – and other particulate biomass fractions in multi-component mechanistic models (Henze *et al.*, 1987; Orhon and Artan, 1994).

A specific chemical formula characterizing activated sludge composition, although useful in establishing stoichiometric relationships for biomass, is bound to be a very crude approximation. First, a large variety of microorganisms from different trophic levels may be sustained with the substrate mixture in wastewaters. Second, even a single microbial cell is composed of different macromolecules with different chemical compositions. Therefore, the chemical description of activated sludge is usually made on the basis of elementary composition of microbial cells. On a dry weight basis, around 90% of biomass is organic and the remaining 10% is inorganic. The inorganic constituents may be listed as iron, calcium, magnesium, sodium and trace amounts of other elements. The organic fraction of biomass includes around 50–55% carbon, 25–30% oxygen, 10–15% nitrogen, 6–10% hydrogen, 1–3% phosphorus and 0.5–1.5% sulphur (Pitter and Chudoba, 1990; Orhon and Artan, 1994). Among many alternatives, $C_5H_7NO_2$ proposed by Porges *et al.* (1956) has survived over the years as the most widely used overall biomass formula used in the activated sludge stoichiometry. This formula defines a composition with 53% carbon and 12% nitrogen. It does not include a phosphorus fraction, mainly because of the concern for keeping the formula as simple as possible. More elaborate empirical formulas are also available for the stoichiometry of phosphorus removal (McCarty, 1970).

2.2.2 Assessment of substrate

Identification of the organic content of wastewaters has always been the major drawback for appropriate evaluation of biological treatment systems. While inorganic substrates such as nitrogen and phosphorus can be measured directly and incorporated into stoichiometric expressions, the same is not possible for organic carbon. Since their discovery at the beginning of the twentieth century, activated sludge systems, have been mostly used for organic carbon removal, especially for industrial wastewaters. In these systems, the large spectrum of different organic compounds in the influent could only be characterized by means of collective parameters. *Biochemical oxygen demand* (BOD) and *chemical oxygen demand* (COD) should be recognized as the most widely used parameters of organic carbon for wastewater characterization and activated sludge modelling. *Total organic carbon* (TOC) is also used for the same purpose, mainly in experimental biodegradation studies.

2.2.2.1 The BOD parameter

The BOD parameter, although still widely used in the assessment of biological treatment performance and effluent quality, is quite unreliable in providing an accurate reflection of the amount and the fate of organic carbon. It is measured in a very poorly defined reactor – a BOD bottle. The sample is highly diluted most of the time, especially for strong industrial wastewaters, so that the oxygen consumption during the test remains lower than the limited amount that is available in the bottle. Alternately, manometric procedures may require less dilute samples but use the same basic concept. The test is started with a small

Energetics and stoichiometry

amount of biomass seeding and stopped, usually after five days, before the completion of the poorly defined set of biochemical reactions. Therefore, the *five day biochemical oxygen demand* (BOD_5) merely indicates a portion of the organic carbon in the sample. It only provides an *iceberg image* of the *total (ultimate) oxygen demand* (L) which is likely to be the stoichiometric equivalent of the biodegradable organic carbon content. Furthermore, the ultimate oxygen demand cannot be measured directly but correlated with BOD_5:

$$BOD_5 = f \times L \tag{2.1}$$

The correlation factor f is by no means a constant. It was observed to vary within a very wide range, from 0.2 to 0.8 for different types of industrial wastewaters (Orhon *et al.*, 1992; 1995). Therefore BOD_5 can hardly serve as a model or a control parameter for substrate because its incorporation into stoichiometric and kinetic expressions may yield questionable and most of the time, misleading results. This issue will be further elaborated in the following chapters.

2.2.2.2 The COD parameter

COD is measured under a strong oxidation environment, as the oxygen equivalent of organic matter. As ammonia is not oxidized under testing conditions, it may be regarded as a reliable substrate parameter only defining the organic carbon balance in biochemical reactions. COD covers all organic compounds likely to serve as the carbon or the energy source in *heterotrophic growth*, with the exception of a few aromatic compounds like, toluene, pyridine, etc., which are not oxidized in the test. As the relative magnitude of these aromatic compounds remains negligible in most industrial wastewaters, COD, unlike BOD_5, is now used as a stoichiometric equivalent of organic matter. It cannot differentiate however biodegradable organics from the non biodegradable compounds, which are also oxidized in the test. This is no longer a major drawback since new techniques now enable separate experimental identification of biodegradable and non biodegradable fractions (Germirli *et al.*, 1991; Orhon *et al.*, 1999a; Orhon and Okutman, 2003). These techniques will be described in detail in the following chapters. It should be noted that if BOD_5 and COD are measured together for a given wastewater, the value of the total biodegradable COD concentration, C_S, may be used as the ultimate biochemical oxygen demand, L and if necessary, the correlation factor f for BOD_5, an important parameter for stoichiometric relationships, may be estimated as follows:

$$f = \frac{BOD_5}{C_S} \tag{2.2}$$

Although a more precise parameter for scientific experiments, TOC has the same deficiency of not differentiating the biodegradable and non biodegradable fractions of organic matter. Furthermore, it does not indicate the oxidation state of substrate and for this reason it cannot be used to set a direct balance between substrate utilized, biomass generated and oxygen consumed. TOC parameter cannot always be used on a routine basis because it requires sophisticated and costly analytical instrumentation.

2.2.3 The definition of yield

In the light of recent developments concerning understanding and modelling of the activated sludge process, the yield coefficient is best defined in terms of electron flow and balance between substrate utilized and biomass generated. In fact, energy and electron balance are the main ingredients for the stoichiometry of all biochemical reactions. This concept is applicable both for inorganic or organic substrates. In the *heterotrophic growth* process, organic substrate serves both as energy and carbon sources. In the *energy reactions,* a part of the organic substrate is oxidized and the corresponding fraction of the total electron potential is transferred to the final electron acceptor, dissolved oxygen. Remaining organic substrate is converted into biomass through *biosynthesis reaction*, the second step of the growth process.

In terms of electron balance, the total electron potential of substrate is calculated as if all substrate is oxidized to CO_2. The yield coefficient identifies the part of substrate converted to the overall oxidation state of the biomass. The electron equivalent of this substrate fraction used in biosynthesis may be determined from the amount of electrons liberated when the generated biomass is also assumed to be totally oxidized to CO_2. With this approach, the yield coefficient for heterotrophs, Y_H may be defined as:

$$Y_H = \frac{\text{electron potential of generated biomass}}{\text{total electron potential of substrate utilized}}$$

For *heterotrophic growth,* which is the main biochemical process for the removal of organic pollutants by activated sludge, the heterotrophic yield, Y_H may be also defined in terms of oxygen equivalence, using the same principle of electron balance:

$$Y_H = \frac{\text{oxygen equivalent of generated biomass}}{\text{oxygen equivalent of organic substrate utilized}}$$

Energetics and stoichiometry

As oxygen equivalence may be conveniently expressed using the COD parameter, Y_H is now commonly defined in terms of COD for heterotrophic growth:

$$Y_H = \frac{\text{cell COD of generated biomass}}{\text{COD of substrate utilized}}$$

For practical applications, the above expression is not as simple as it looks and Y_H cannot be readily calculated by direct measurements of biomass COD and substrate COD. First of all, the expression requires COD of the active heterotrophic biomass; however, direct COD measurement of particulate matter in the reactor includes aside from active biomass, a number of other particulate COD fractions entrapped into microbial flocs. Furthermore, the difference between influent and effluent COD concentrations in a biological treatment system, ΔCOD does not accurately reflect the magnitude of biodegradable COD utilized, as the effluent COD is likely to include soluble residual microbial products at levels which may be quite significant for industrial wastewaters (Orhon *et al.*, 1989; 1999a). The problems involved in the numerical assessment of Y_H are now largely solved by the promotion of respirometric techniques and model calibration of the experimental data (Sollfrank and Gujer, 1991; Cokgor *et al.*, 1998). These methods will be described in detail in the following chapters. The example below intends to provide additional clarification to the concept of yield and electron equivalence presented in this section.

Example 2.2 Heterotrophic yield and electron balance
The stoichiometry of heterotrophic growth in an activated sludge process fed with organic substrate is characterized by the following equation:

$$8CH_2O + 3O_2 + NH_3 \rightarrow C_5H_7NO_2 + 3CO_2 + 6H_2O$$

where CH_2O *denotes organic substrate and* $C_5H_7NO_2$ *active heterotrophic biomass.*

(a) *Identify the stoichiometry of applicable energy and biosynthesis reactions*
(b) *Establish the electron balance between substrate utilized, biomass generated and dissolved oxygen consumed*
(c) *Calculate the heterotrophic yield,* Y_H

(a) As the simplest formula for carbohydrate, CH_2O provides an overall stoichiometric description of the variety of organic compounds taking part in the influent organic substrate:

The solution of the problem should start with the identification of the biosynthesis reaction. The empirical formula adopted for biomass has 5 carbon atoms and therefore the corresponding stoichiometric equation should include 5 CH_2O as given below:

$$5CH_2O + NH_3 \rightarrow C_5H_7NO_2 + 3H_2O$$

In essence, the energy reaction can only be formulated on the basis of an overall energy balance. The fundamentals of this approach are provided in detail by McCarty (1970) and Orhon and Artan (1994). Since the total growth stoichiometry is provided in this example, the biosynthesis reaction developed above indicates that 3 CH_2O remains for the energy requirements: Accordingly, the overall substrate oxidation reaction for energy may be expressed as:

$$3CH_2O + 3O_2 \rightarrow 3CO_2 + 3H_2O$$

(b) The amount of organic substrate used for the overall heterotrophic growth process is given as 8 CH_2O. The total electron potential of substrate utilized can be derived from the following oxidation/reduction reactions:

$$8CH_2O + 8H_2O \rightarrow 8CO_2 + 32H^+ + 32e^-$$

$$32e^- + 32H^+ + 8O_2 \rightarrow 16H_2O$$

These equations show a flow of 32 electrons with total oxidation of organic substrate to CO_2. The resulting COD expression from these equations may be written as:

$$8CH_2O + 8O_2 \rightarrow 8CO_2 + 8H_2O$$

As shown in this equation the oxygen equivalent of organic substrate is 8 O_2 corresponding to 256 mg/l of biodegradable COD.

In the second step, the electron equivalent of the generated biomass is calculated by means of the following oxidation/reduction equations:

$$C_5H_7NO_2 + 8H_2O \rightarrow 5CO_2 + NH_3 + 20H^+ + 20e^-$$

$$20e^- + 20H^+ + 5O_2 \rightarrow 10H_2O$$

These equations indicate that the generated biomass incorporated 20 electrons from the total electron potential of the organic substrate. They yield the

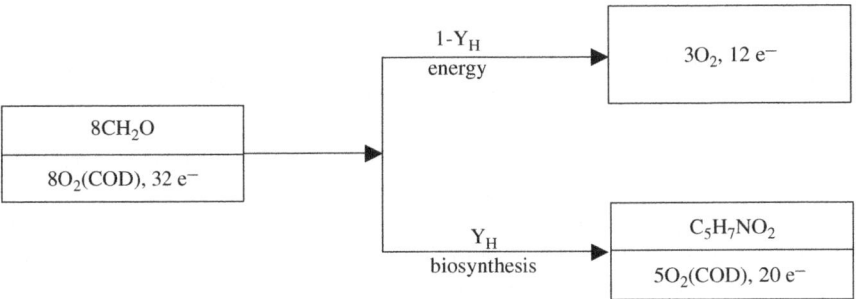

Figure 2.4. Electron and COD balance between substrate utilized, biomass generated and oxygen consumed

following overall stoichiometry for the oxidation of biomass, with an oxygen equivalent of 5 O_2 and a cell COD of 160 mg/l:

$$C_5H_7NO_2 + 5O_2 \rightarrow 5CO_2 + NH_3 + 2H_2O$$

The overall electron balance is confirmed by means of the following oxidation/reduction equations expressed for substrate oxidized for energy:

$$3CH_2O + 3H_2O \rightarrow 3CO_2 + 12H^+ + 12e^-$$

$$12e^- + 12H^+ + 3O_2 \rightarrow 6H_2O$$

The electron and COD balance established between substrate, biomass and oxygen is schematically displayed in Figure 2.4.

From the results obtained the heterotrophic yield may be computed in two different ways using the electron data and the COD data:

$$Y_H = \frac{20e^-}{32e^-} = 0.625 \ e^-. \text{ equivalent biomass}/e^-.\text{equivalent substrate}$$

$$Y_H = \frac{160}{256} = 0.625 \text{ mg cell COD/mg COD}$$

In a number of studies conducted on domestic sewage and different industrial wastewaters, the value of Y_H is determined in the narrow range of 0.60 − 0.66 mg cell COD/mg COD (Orhon *et al.*, 1999b). If no other reliable information is available, the default value of 0.64 mg cell COD/mg COD suggested as part of Activated Sludge Model No.2, (ASM2) may be adopted for design and similar performance calculations (Henze *et al.*, 1995).

2.2.4 Expression of yield for different model components

In all the modelling studies and computer-aided design applications adopted today, the COD parameter is used to characterize both substrate and biomass. The specific feature of COD that makes it attractive as a substrate and biomass parameter is the ability to set an electron balance between substrate, biomass and dissolved oxygen, as illustrated in Example 2.2. In fact, if Y_H is expressed as mg cell COD/mg COD, for each mg of substrate utilized, Y_H mg of biomass is generated and $(1-Y_H)$ mg of dissolved oxygen is consumed when the microbial growth process is considered alone. For the actual operating conditions of the activated sludge system, the effect of endogenous respiration will induce a lower net yield value, Y_{NH}, lowering the amount of biomass and increasing dissolved oxygen consumption. Under any selected operating condition however, the electron balance will still hold true in the sense that for each mg of substrate utilization the biomass generation will drop to Y_{NH} mg and the demand for dissolved oxygen will increase to $(1-Y_{NH})$ mg. Such an electron balance is not directly possible when VSS is used as a biomass parameter or BOD_5 as a substrate parameter. This issue is not well understood among scientists and designers.

In case COD is not adopted as a joint parameter for substrate and biomass, values of Y_H or Y_{NH} can no longer be justified based on energy balance and they become coefficients. Arbitrary selection of these coefficients is not allowed and may lead to dramatic errors in estimating process behaviour and performance.

2.2.4.1 VSS as a biomass parameter

Volatile suspended solids (VSS) are the most convenient parameter yielding direct estimation of biomass concentration. It can be easily converted into total suspended solids (TSS), for assessing the amount of excess biological sludge. In this context in design applications and similar calculations, it may be desirable to express Y_H or Y_{NH} in terms of g VSS/g COD. Accurate definition of Y_H using VSS as a biomass parameter requires inclusion of an f_x coefficient, which gives the COD equivalent of biomass VSS, for necessary conversions. This coefficient is calculated by means of the following total oxidation reaction reflecting endogenous respiration of biomass:

$$C_5H_7NO_2 + 5O_2 \rightarrow 5CO_2 + NH_3 + 6H_2O$$

This equation yields the COD equivalent of 113 g of biomass VSS ($C_5H_7NO_2 = 113$ g) as 160 g ($5O_2 = 160$ g). Using these values, f_x is calculated as:

Energetics and stoichiometry

$$f_x = \frac{160}{113} = 1.42 \text{ g cell COD/g VSS}$$

This value is now commonly adopted as the COD equivalent of biomass VSS in all similar evaluations. This way, the following relationship allows calculation of the value of $Y_{H,VSS}$ on the basis of Y_H (g cell COD/g COD):

$$Y_{H,VSS} = \frac{Y_H}{f_x} (\text{g VSS/g COD}) \qquad (2.3)$$

As previously mentioned, the default value Y_H is now commonly accepted as 0.64 g cell COD/g COD both for domestic sewage and industrial wastewaters. The corresponding $Y_{H,VSS}$ of 0.45 g VSS/g COD may also be adopted as a default value and used in practical applications, unless determined otherwise in specific case studies.

2.2.4.2 VSS and BOD₅ as biomass and substrate parameters

Despite major problems involved in setting accurate stoichiometric relationships, BOD_5 is still used in many instances as a substrate parameter in the design and performance evaluation of biological treatment systems. In this case, the common approach is to adopt VSS as the biomass parameter and select an arbitrary value for the corresponding yield coefficient, Y_h expressed in terms of g VSS/g BOD_5. While this approach may somehow be tolerated for domestic sewage where the BOD_5/COD ratio remains most of the time around 0.50–0.55, it may be totally unacceptable for industrial wastewaters. The best would be to replace COD for BOD_5. In case there is administrative pressure or strong preference for BOD_5 as a substrate parameter, then the corresponding yield coefficient Y_h (g VSS/g BOD_5) should not be arbitrarily selected but calculated, based on Y_H (g cell COD/g COD) using two conversion coefficients, namely, f_x and f. As previously defined in equation (2.2), f may be conveniently expressed as the ratio of BOD_5 to the total biodegradable COD, C_S. For conventional, medium-strength domestic sewage, f generally assumes a value around 0.50. For Istanbul domestic sewage, the average BOD_5 was calculated as 200 mg/l and total COD as 430 mg/l, with a biodegradable fraction of 87%. Then the corresponding f value could be computed as 0.53 (Orhon *et al.*, 1994; Orhon and Ubay Cokgor, 1997). Characteristics of industrial wastewaters greatly vary in terms of their COD and BOD_5 contents and this variation reflects on the corresponding f ratio. Table 2.3 outlines values of experimentally determined f ratios for selected industrial wastewaters.

Table 2.3. Wastewater characterization and f ratio for selected industrial wastewaters

Wastewater	Total COD C_{TI} (mg/l)	Biodegradable COD C_{S1} (mg/l)	BOD_5 (mg/l)	$f = \frac{BOD_5}{C_{SI}}$	Reference
Confectionary					
Plant A	8085	8040	6390	0.79	Orhon et al., 1995
Plant B	1800	1695	1170	0.69	Orhon et al., 1995
Meat processing					
Slaughterhouse	7230	6800	3180	0.47	del Pozo et al., 2003
Poultry processing	2700	2400	1600	0.66	Eremektar et al., 1999
Integrated dairy processing	2395	2085	1173	0.56	Andreottola et al., 2002
Tannery					
Plain-settled wastewater	2460	1940	1030	0.53	Orhon et al., 1999c
Textile					
Organized industrial district predominantly textile	930	884	490	0.55	Orhon et al., 2002
Cotton knit fabric	981	791	166	0.21	Orhon et al., 1992

Energetics and stoichiometry

Expressing Y_h using the default value of Y_H involves conversion of the biomass COD into VSS by means of f_x and substrate COD into BOD$_5$ by means of f, as shown in the following equation:

$$Y_h = \frac{Y_H}{f_x f} \text{ (g VSS/g BOD}_5\text{)} \tag{2.4}$$

Calculation of the yield coefficient in terms of different parameters selected for substrate and biomass is illustrated in the following example.

Example 2.3 Y_H calculation for different parameters

Calculate the yield coefficient for the stoichiometric equation given in Example 2.1 with (a) g VSS/g COD, and (b) g VSS/g BOD$_5$ units.

Assume $f = 0.50$ g BOD$_5$/g COD and $f_x = 1.42$ g cell COD/g VSS for the calculations.

(a) In example 2.1, the yield coefficient, Y_H was computed as 0.62 g cell COD/g COD. Using equation (2.3):

$$Y_{H,VSS} = \frac{Y_H}{f_x} = \frac{0.62}{1.42} = 0.44 \text{ g VSS/g COD}$$

$Y_{H,VSS}$ can also be calculated directly from the stoichiometric equation given in example 2.1:

$$8CH_2O + 3O_2 + NH_3 \rightarrow C_5H_7NO_2 + 3CO_2 + 6H_2O$$

In this equation $C_5H_7NO_2$ (biomass) is generated from 8 CH_2O (organic substrate). The molecular weight of $C_5H_7NO_2$ is 113 g (VSS). In example 2.2, the COD equivalent of $8CH_2O$ was computed as 256 g COD. With these values,

$$Y_{H,VSS} = \frac{113 \text{ g VSS}}{256 \text{ g COD}} = 0.44 \text{ g VSS/g COD}$$

(b) One of the following relationships may be used for the calculation of Y_h where biomass is expressed in terms of VSS and substrate in terms of BOD$_5$:

$$Y_h = \frac{Y_{H,VSS}}{f} = \frac{0.44}{0.5} = 0.88 \text{ g VSS/g BOD}_5$$

or,

$$Y_h = \frac{Y_H}{f_x f} = \frac{0.62}{0.5 \times 1.42} = 0.88 \text{ g VSS/g BOD}_5$$

2.3 PROCESS STOICHIOMETRY

Stoichiometry is one of the most useful tools for the understanding and interpretation of complex biochemical reactions describing the behaviour of activated sludge. It may be used for a selected process or for an overall reaction that may be derived from a number of simultaneously occurring processes, to reflect system performance under preset operating conditions.

When used for a selected process, stoichiometry defines appropriate coefficients which convert and adapt the general process rate expression for all model components taking part in the process. For example, if a rate expression is specifically defined for microbial growth in activated sludge, coefficients may be calculated using stoichiometric concepts for converting this rate expression into substrate removal rate or oxygen uptake rate, *etc*. This approach, when applied to real operating conditions may yield numerical values for basic system requirements (daily oxygen demand) or overall system outputs (excess biological sludge). Process stoichiometry should be established on the basis of parameters selected for biomass and substrate.

2.3.1 System defined with biomass COD and substrate COD

In the modelling and design of activated sludge systems where both substrate and biomass are expressed in terms of COD, the stoichiometric relationship between the main model components, namely substrate, active biomass and dissolved oxygen may be expressed as shown in Figure 2.5.

The basic backbone of this relationship is the heterotrophic yield coefficient Y_H or Y_{NH}. A rate expression for substrate (dS/dt) or a given amount of substrate is converted into active biomass by multiplying it with Y_H. For dissolved oxygen, the applicable stoichiometric coefficient becomes $(1-Y_H)$. Conversely, the coefficient $1/Y_H$ converts a rate expression for active biomass or a given amount of active biomass into organic substrate, and $(1-Y_H)/Y_H$ into dissolved oxygen. Obviously, a rate expression directly applies to the model component for which it is expressed (the stoichiometric coefficient is 1). The algebraic sum of the three coefficients indicates that only the COD can establish a direct balance between substrate, biomass and dissolved oxygen.

The stoichiometry outlined above and illustrated in Figure 2.5 also applies to other parameters. For example, nitrogen removal during the biodegradation of organic carbon is accomplished as part of biosynthesis, through incorporation into active biomass. If the coefficient i_{XN} is used to define the nitrogen content of biomass (mg N/mg cell COD), nitrogen removal could be expressed in terms of substrate utilization using the stoichiometric coefficients $i_{XN} Y_H$ or $i_{XN} Y_{NH}$.

Energetics and stoichiometry 31

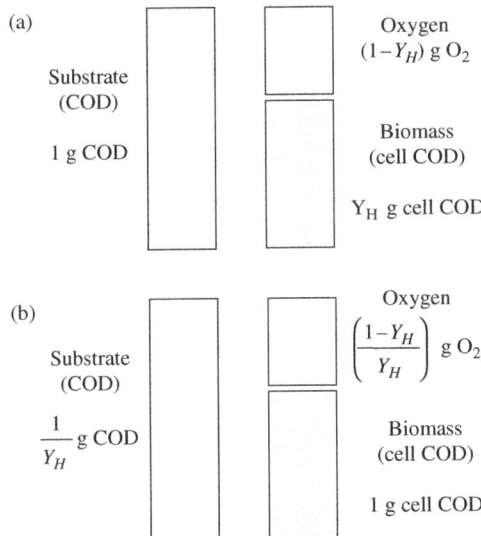

Figure 2.5. Stoichiometric relationships for a system defined with biomass COD and substrate COD, when the process rate expression is developed for (a) substrate, and (b) active biomass

2.3.2 System defined with biomass VSS and substrate COD

In an activated sludge system where volatile suspended solids parameter defines active biomass instead of cell COD, the yield coefficient $Y_{H,VSS}$ should be expressed as mg VSS/mg COD and it should be incorporated into different mass balance equations with this unit. In this case $Y_{H,VSS}$ can also be used for direct conversion of a given substrate entity into active biomass. However, the same does not hold true for dissolved oxygen. In other words, the term $(1-Y_{H,VSS})$ does not yield dissolved oxygen consumption using biodegradation of organic substrate. An accurate stoichiometric balance is only established when active biomass is re-expressed as a COD equivalent as schematically plotted in Figure 2.6. Therefore, a factor of $(f_x\ Y_{H,VSS})$ should be used for adjusting the corresponding biomass and the applicable coefficient for dissolved oxygen should be computed as $(1-f_x\ Y_{H,VSS})$ per unit amount of substrate COD utilized. This way, the balance in Figure 2.6 becomes exactly the same as the one in Figure 2.5. Conversely, a process rate expression for active biomass becomes substrate removal rate when multiplied by $1/Y_{H,VSS}$. The same rate is converted into oxygen consumption rate when multiplied by $(1-f_x\ Y_{H,VSS})/Y_{H,VSS}$.

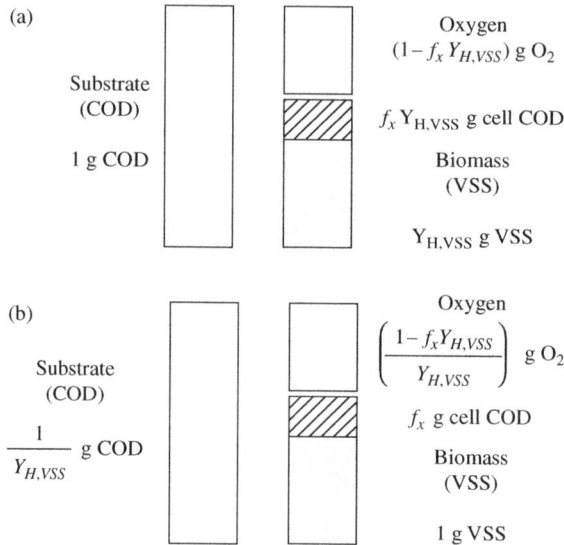

Figure 2.6. Stoichiometric relationships for a system defined with biomass VSS and substrate COD, when the process rate expression is developed for (a) substrate, and (b) active biomass

2.3.3 System defined with biomass VSS and substrate BOD₅

In this modelling approach using BOD_5 as substrate and VSS as biomass, it should be born in mind that BOD_5 only represents an unknown portion of biodegradable substrate. This does not pose a problem if an accurate value for Y_h is calculated and used to assess the amount of active biomass generated per unit amount of BOD_5 utilized. However, calculation of oxygen consumption is only possible when both substrate and biomass are expressed in terms of their COD equivalents. Therefore substrate has to be corrected first with a coefficient $1/f$ and then biomass with $f_x Y_h$. This way, the stoichiometric coefficient for dissolved oxygen may be calculated as $(1/f - f_x Y_h)$. Similar to other applications $1/Y_h$ converts a rate expression defined for active biomass into BOD_5 removal rate, and $(1 - f\, f_x\, Y_h)/f\, Y_h$ into oxygen consumption rate. The corresponding stoichiometric balance is schematically given in Figure 2.7.

The following example illustrates the way in which significant performance outputs of activated sludge operation are calculated using different substrate and biomass parameters.

Energetics and stoichiometry 33

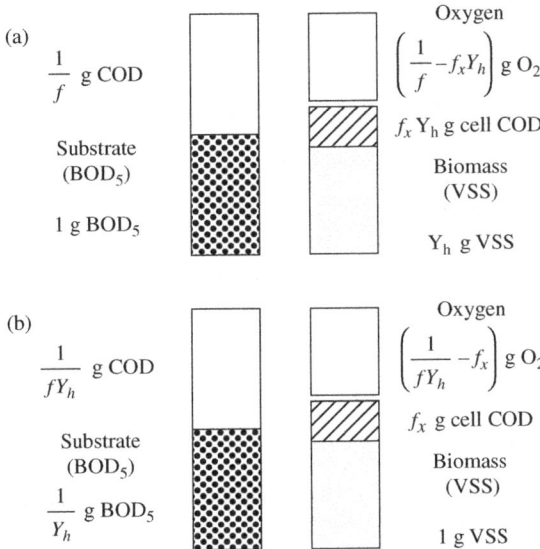

Figure 2.7. Stoichiometric relationships for a system defined with biomass VSS and substrate BOD_5, when the process rate expression is developed for (a) substrate, and (b) active biomass

---

Example 2.4 Process stoichiometry in terms of different parameters

An activated sludge system has been designed for a dairy wastewater with a flow rate of $Q = 100$ m³/day. The characteristics of the dairy wastewater have been outlined below (Orhon and Ubay Cokgor, 1997):

Total COD, $C_{TI} = 1410$ mg/l
Biodegradable COD fraction: 92%
$BOD_5 = 930$ mg/l
$Y_{NH} = 0.44$ g cell COD/g COD
(Calculated for the sludge age selected for operational conditions)
$f_x = 1.42$ g cell COD/g VSS

Calculate the daily biological excess sludge and the daily oxygen consumption of the activated sludge system when substrate and biomass parameters are expressed in terms of (a) cell COD/COD; (b) VSS/COD;, and (c) VSS/BOD_5. Assume all available biodegradable substrate is totally removed in the treatment system.

(a) Biodegradable COD concentration, C_{S1}:

$$C_{S1} = 1410 \times 0.92 = 1297\ mg/l = 1.297\ kg/m^3$$

The daily amount of COD removed, ΔCOD:

$$\Delta COD = Q\ C_{S1} = 100 \times 1.297 = 129.7\ kg\ COD/day$$

When COD is used as a common parameter for both substrate and biomass, for a unit removal of 1 kg of COD, Y_{NH} kg of excess biomass will be generated and $(1-Y_{NH})$ kg of oxygen will be consumed, as schematically indicated in Figure 2.5. Then, the daily amount of excess biological sludge, P_{XH}, may be calculated as:

$$P_{XH} = Y_{NH}\ \Delta COD = 0.44 \times 129.7 = 57.1\ kg\ cell\ COD/day$$

The daily oxygen consumption, ΔO_2:

$$\Delta O_2 = (1 - Y_{NH})\ \Delta COD = (1 - 0.44) \times 129.7 = 72.6\ kg\ O_2/day$$

This evaluation system sets, as explained before, equivalence between substrate, biomass and oxygen. In fact, the amount substrate COD removed is equal to the sum of biomass COD generated and oxygen (*negative COD*) consumed.

(b) When VSS is used as biomass parameter and COD as substrate parameter, the corresponding net yield parameter, $Y_{NH,VSS}$ can be calculated as:

$$Y_{NH,VSS} = \frac{Y_{NH}}{f_x} = \frac{0.44}{1.42} = 0.31\ g\ VSS/gCOD$$

The amount of excess sludge, P_{VSS}:

$$P_{VSS} = Y_{NH,VSS}\ \Delta COD = 0.31 \times 129.7 = 40.2\ kg\ VSS/day$$

The daily oxygen consumption, ΔO_2:

$$\Delta O_2 = (1 - f_x Y_{NH,VSS})\ \Delta COD = (1 - 1.42 \times 0.31) \times 129.7$$

$$\Delta O_2 = 72.6\ kg\ O_2/day$$

(c) When VSS is used as biomass parameter and BOD_5 as substrate parameter, the corresponding f ratio and the net yield parameter, Y_{Nh} can be calculated as:

$$f = \frac{BOD_5}{C_S} = \frac{930}{1297} = 0.717$$

$$Y_{Nh} = \frac{Y_{NH}}{f_x f} = \frac{0.44}{1.42 \times 0.717} = 0.432\ g\ VSS/g\ BOD_5$$

The daily amount of BOD$_5$ removed, ΔBOD_5:

$$\Delta BOD_5 = 100 \times 0.930 = 93 \ kg \ BOD_5/day$$

The amount of excess sludge, P$_{VSS}$:

$$P_{vss} = Y_{Nh} \Delta BOD_5 = 0.432 \times 93 = 40.2 \ kg \ VSS/day$$

The daily oxygen consumption, ΔO_2:

$$\Delta O_2 = \left(\frac{1}{f} - f_x Y_{Nh}\right) \Delta BOD_5 = \left(\frac{1}{0.717} - 1.42 \times 0.432\right) \times 93$$

$$\Delta O_2 = 72.6 \ kg \ O_2/day$$

As expected, the amount of daily excess sludge and the daily oxygen consumption are not affected by different choices for biomass and substrate parameters, owing to the accurate conversion and calculation of the corresponding net yield coefficient.

2.4 REFERENCES

Andreottola, G., Foladori, P., Ragazzi, M. and Villa, R. (2002) Dairy wastewater treatment in a moving bed biofilm reactor. *Water Sci. Technol.* **45**(12), 321–328.

Cokgor, E.U., Sozen, S., Orhon, D. and Henze, M. (1998) Respirometric assessment of activated sludge behaviour I. Assessment of readily biodegradable substrate. *Water Res.* **32**(2), 461–475.

del Pozo, R., Tas, D.O., Dulkadiroglu, H., Orhon, D. and Diez, V. (2003). Biodegradability of slaughterhouse wastewater with high blood content under anaerobic and aerobic conditions. *J. Chem. Technol. Biot.* **78**(4), 384–391.

Eremektar, G., Ubay Cokgor, E., Ovez, S., Germirli Babuna, F. and Orhon, D. (1999). Biological treatability of poultry processing plant effluent: a case study. *Water Sci. Technol.* **40**(1), 323–329.

Germirli, F., Orhon, D. and Artan, N. (1991) Assessment of the initial inert soluble COD in industrial wastewaters. *Water Sci. Technol.* **23**(4–6), 1077–1086.

Henze, M., Grady, C.P.L. Jr., Gujer, W., Marais, G.v.R. and Matsuo, T. (1987) *Activated Sludge Model No.1*, IAWPRC Scientific and Technical Report No.1, London: IAWPRC.

Henze, M., Gujer, W., Mino, T., Matsuo, T., Wentzel, M.C. and Marais, G.v.R. (1995) *Activated Sludge Model No.2*, IAWPRC Scientific and Technical Report No.2, London, IAWQ.

Lehninger, A.L. (1965). *Bioenergetics.* 2nd edn, W.A. Benjamin Inc., New York, N.Y.

McCarthy, P.L. (1970) *Proceedings of the Wastewater Reclamation and Reuse Workshop.* June 25–27, Lake Tahoe, California, pp: 226–251.

Orhon, D., Artan, N. and Cimsit, Y. (1989) The concept of Soluble Residual Product Formation in the Modelling of Activated Sludge – Reply to Discussion. *Water Sci. Technol.* **21**(12), 1566.

Orhon, D., Artan, N., Buyukmurat, M. and Gorgun, E. (1992) The effect of residual COD on the biological treatability of textile wastewater. *Water Sci. Technol.* **26**(3–4) 815–825.

Orhon, D. and Artan, N. (1994) *Modelling of Activated Sludge Systems,* Technomic Publishing Co. Inc., Lancaster, PA.

Orhon, D., Sozen, S. and Ubay, E. (1994) Assessment of nitrification-denitrification potential of Istanbul wastewaters. *Water Sci. Technol.* **30**(6), 21–30.

Orhon, D., Yildiz, G., Ubay Cokgor, E. and Sozen, S. (1995). Respirometric evaluation of the biodegradability of confectionary wastewaters. *Water Sci. Technol.* **32**(12), 11–19.

Orhon, D. and Ubay Cokgor, E. (1997). Fractionation in wastewater characterization – the state of the art. *J. Chem. Technol. Biot.* **68**, 283–293.

Orhon, D., Karahan, O. and Sozen, S. (1999a) The Effect of Residual Microbial Products on the Experimental Assessment of the Particulate Inert COD in Wastewaters. *Water Res.* **33**(14), 3191–3203.

Orhon, D., Taslı, R. and Sozen, S. (1999b) Experimental basis of activated sludge treatment for industrial wastewaters – the state of the art. *Water Sci. Technol.* **40**(1), 1–11.

Orhon, D., Ates Genceli, E. and Ubay Cokgor, E. (1999c) Characterization and modeling of activated sludge for tannery wastewater. *Water Environ. Res.* **71**(1), 50–63.

Orhon, D., Kabdasli, I., Ubay Cokgor, E., Tunay, O. and Artan, N. (2002). Experimental assessment of optimum operation strategy for large industrial wastewater treatment plants – A case study. *Enviro. Eng. Sci.* **19**(1), 47–58.

Orhon, D. and Okutman, D. (2003) Respirometric assessment of residual organic matter for domestic sewage. *Enzyme Microb. Tech.* **32**(5), 560–566.

Pitter, P. and Chudoba, J. (1990) *Biodegradability of Organic Substances in the Aquatic Environment.* Boca Raton, FL: CRC Press, Inc.

Porges, N., Jasewicz, L. and Hoover, S.R. (1956) Principles of Biological Oxidation. In *Biological Treatment of Sewage and Industrial Wastes.* Vol.1, J. McCabe and W.W. Eckenfelder, Jr., eds., pp. 35–48, Reinhold Publishing Co., New York, NY.

Racker, E. (1965) *Mechanisms in Bioenergetics.* Academic Press, NewYork, NY.

Sollfrank, U. and Gujer, W. (1991) Characterization of domestic wastewater for mathematical-modeling of the activated-sludge process. *Water Sci. Technol.* **23**(4–6), 1057–1066.

3
Modelling of organic carbon removal

3.1 PRINCIPLES OF MODELLING

The merit of modelling of the activated sludge process, while recognized and significantly improved for domestic sewage, is largely overlooked for industrial wastewaters. Practice has so far managed to keep away from modelling, generally with some disdain and the related experience remained mostly empirical, depending upon practical observation and experience. While such empirical techniques, when successful, have set the basis of current practice, the complexity of industrial wastewaters require deeper understanding of biodegradation processes using all the tools of mechanistic modelling, now available for domestic sewage. This first requires a better grasp of what a model is and how it can be useful. It also requires a new approach concerning the experimental data that should be specifically generated for the wastewater

© 2009 IWA Publishing. *Industrial Wastewater Treatment by Activated Sludge*, by Derin Orhon, Fatos Germirli Babuna and Ozlem Karahan. ISBN: 9781789065282. Published by IWA Publishing, London, UK.

studied. The nature and the extent of the experimental basis for modelling is not fully understood and explored for industrial wastewaters.

Appropriate design of a biological treatment plant should ensure desired effluent quality under variable operating conditions. Then, the designer should be able to define, handle and manipulate effective engineering tools for accurate performance prediction. A model is needed for this purpose. The model adopted may simply be structured upon personal experience and expertise as it has been common practice in the past. This approach however, often fails in the face of unusual conditions, unexpected problems or new requirements. Then, a *physical model* such as a lab-scale reactor or a pilot plant may be used for the purpose of deriving the required information for appropriate design. The physical model, although potentially useful to test and replicate actual plant characteristics, usually remains site-specific and does not provide alone full understanding of different phenomena controlling treatment processes. This type of an insight can only be acquired and improved by means of *mechanistic models* that would provide a reasonably acceptable description of major biochemical processes in terms of microbial kinetics and stoichiometric relationships among significant parameters. These models, when reliable, may then be used for the identification of potentially feasible engineering solutions.

Modelling should begin with the understanding that models developed can only provide a simplified approximation of complex biochemical reactions taking place in activated sludge systems. Therefore conceptual insight is much more valuable than the mathematics involved, however complicated they may be. Furthermore, the merit of the model chiefly depends upon the quality of the information it provides for system design and operation. This information should be readily applicable, sufficient and reliable. In this context, three important factors should be mentioned as a preliminary quality check: (i) the model should be supplied with adequate and accurate information about wastewater characterization and reactor hydraulics; (ii) the model structure should include the necessary model components and processes for the evaluation. Otherwise the stoichiometric and kinetic data generated may be distorted in order to fit the adopted structure; (iii) model calibration should be verified with sufficient experimental results reflecting system operation under different operating conditions. This is the only way to ensure the validity of the stoichiometric and kinetic coefficients determined at the calibration stage.

An activated sludge model describes *reactor kinetics*, or in simpler terms, changes that take place in the concentration of the selected model components with time. However, this information cannot be directly converted into system efficiency without considering *reactor hydraulics,* which defines the mixing characteristics of the reactor where biochemical reactions take place. In other words, *removal rate* is not *reaction rate*. The latter should be incorporated into a mass balance expression together with appropriate description of reactor hydraulics in order to define *removal rate*, i.e. the fate of a model component in a biological reactor.

Hydraulic properties identify two general categories of biological reactors: *batch reactors* and *continuous flow reactors*. A batch reactor is initially fed with wastewater and biochemical reactions proceed in a constant reactor volume without further feeding. After a selected retention time which secures the desired removal rate, the operation is stopped and the treated effluent is discharged. In practice, *fill and draw* systems applicable to industrial effluents with low flow rates are typical examples of batch biological reactors. *Sequencing batch reactor,* (SBR), may also be cited as the modified version of batch biological reactors. Conventional activated sludge system is usually a *continuous flow biological reactor*, as it is operated with a continuous wastewater feeding.

In a continuous flow biological reactor, an approximation of the residence time distribution in mathematical terms is not only very difficult but also quite useless, as this information cannot be properly incorporated into mass balance equations. This is usually done by means of *ideal reactors*. For continuous systems, three different types of ideal reactors may be envisaged: (i) completely mixed reactor, (CSTR); (ii) plug flow reactor, and (iii) dispersed plug flow reactor. The most convenient and acceptable approach in process modelling is to approximate the hydraulics of a real reactor by a series of completely mixed ideal reactors. Detailed descriptions of the properties of ideal reactors is beyond the scope of this book and can be found elsewhere (Levenspiel, 1972; Grady *et al.,* 1999; Orhon and Artan, 1994).

3.2 MATRIX REPRESENTATION IN MODELLING

As a common feature, mechanistic models describing biochemical processes become more difficult to follow as they improve. Failure to grasp the ever increasing complexity in interactions between system components often leads to serious misinterpretations. As previously mentioned, the mass balance equation for each parameter should include an overall conversion rate. This rate needs to be correlated with the process rate of a biochemical reaction by means of a specific stoichiometric coefficient for this parameter. An error in the selection of applicable coefficient leads to significant problems in the accurate interpretation of mass balance. In order to overcome this problem and simplify the understanding of the model, a matrix representation was first suggested by Petersen (1975). Henze *et al.* (1987) adopted the same matrix format for a

Table 3.1. Matrix representation of the activated sludge model using COD both for substrate and biomass

Component i	1	2	3	Process rate, ρ_j
j Process	X_H	S_S	S_O	$ML^{-3}T^{-1}$
1 Growth	1	$-\dfrac{1}{Y_H}$	$-\left[\dfrac{1-Y_H}{Y_H}\right]$	$\mu_H X_H$
2 Decay	-1		-1	$b_H X_H$
Observed removal rate, $ML^{-3}T^{-1}$		$r_i = \sum_j r_{ij} = \sum_j v_{ij}\rho_j$		
Parameters, ML^{-3}	Cell COD	COD	O_2	

simpler presentation of the activated sludge models. The first step in this approach is to identify the *model components* and the *processes* constituting the basic backbone of the selected model. Table 3.1 provides an example for the matrix representation of a simple model defined for the biodegradation of a simple soluble organic compound. The model involves only *active biomass*, X_H, *substrate*, S_S and *dissolved oxygen*, S_O, as model components and, *growth* and *endogenous respiration* as processes. The matrix format lists the two selected processes down the left side of the table and the three model components across the top by symbol and across the bottom by definition. The stoichiometric coefficients, v_{ij} are placed in specific boxes inside the matrix, relating the reaction rate of model component i, r_{ij} to the process rate ρ_j:

$$r_{ij} = v_{ij}\rho_j \tag{3.1}$$

The observed overall conversion rate, r_i for model component i is then defined as the algebraic sum of all the reaction rates, r_{ij} for all the processes related to this component:

$$r_i = \sum_j r_{ij} = \sum_j v_{ij}\rho_j \tag{3.2}$$

In this example, the first process is *microbial growth* which directly relates to active biomass, X_H, the first model component in the matrix, so that the corresponding stoichiometric coefficient is defined as $v_{11} = 1$. Similarly the process rate for microbial growth is converted to substrate removal by means of $v_{21} = -1/Y_H$ and to oxygen consumption by means of $v_{31} = -(1-Y_H)/Y_H$ as explained in detail in the preceding chapter. This approach also provides mass balance equilibrium by moving across the same line in the matrix indicating that the algebraic sum of the stoichiometric coefficients would be zero if both X_H and S_S are expressed in terms of COD as in the example provided in Table 3.1. It should be noted that dissolved oxygen is negative COD and the sign before S_O must be changed before the continuity check.

The overall conversion rate derived from the matrix can be directly incorporated into corresponding mass balance equations. In this context, the model presented in Table 3.1 yields the following overall rate expressions:

$$r_1 = r_{XH} = \mu_H X_H - b_H X_H \tag{3.3}$$

$$r_2 = r_{SS} = -\frac{1}{Y_H}\mu_H X_H \tag{3.4}$$

$$r_3 = r_{SO} = -\frac{(1-Y_H)}{Y_H}\mu_H X_H - b_H X_H \tag{3.5}$$

The observed yield, Y_{NH}, a very useful parameter for the modelling and design of activated sludge systems, defines in essence the ratio between r_{XH} and r_{SS}:

$$r_{XH} = -Y_{NH}\, r_{SS} \tag{3.6}$$

Although the use of COD directly provides equivalence between major parameters in mass balance, there may be cases where VSS and BOD_5 are adopted for biomass and substrate parameters due to familiarity and past experience. In such cases, matrix representation of the model for selected parameters becomes very useful in the formulation of expressions related to excess sludge production and oxygen consumption, provided that values of applicable yield coefficients are accurately computed. Matrix representation of reaction kinetics for VSS and BOD_5 parameters is outlined in the following example.

Example 3.1 Matrix representation of models using VSS and BOD$_5$

Set the matrix for the model presented in Table 3.1 and derive the corresponding overall oxygen consumption rate expression when (a) VSS is used as the model component for biomass, and (b) VSS and BOD$_5$ are used for biomass and substrate parameters.

(a) In this modelling approach, the parameter for defining the active biomass is changed from *cell COD* into *VSS*. Substrate is still defined in terms of *COD*. Consequently, the yield coefficient $Y_{H,VSS}$ should be defined as g VSS/g COD:

$$Y_{H,VSS} = \frac{Y_H}{f_x}$$

where, f_x = g cell COD/g VSS. Then,

$$v_{21} = -\frac{1}{Y_{H,VSS}} \quad \text{and} \quad v_{32} = -f_x$$

The stoichiometric coefficient, v_{31} for dissolved oxygen can be derived from the difference between the rate of substrate utilization and the rate of biomass generation when both are expressed in terms of COD. Consequently,

$$r_{21,COD} = \frac{1}{Y_{H,VSS}} \mu_H X_H$$

$$r_{11,COD} = f_x \mu_H X_H$$

Then, $\quad r_{21,COD} - r_{11,COD} = \left(\frac{1}{Y_{H,VSS}} - f_x\right) \mu_H X_H$

and $\quad v_{31} = \frac{(1 - f_x Y_{H,VSS})}{Y_{H,VSS}}$

The matrix representation of the model where biomass is expressed in terms of VSS is given in Table 3.2.

Table 3.2. Matrix representation of the activated sludge model using VSS for biomass and COD for substrate

Component	i	1	2	3	Process rate, ρ_j
j	Process	X_H	S_S	S_O	$ML^{-3}T^{-1}$
1	Growth	1	$-\dfrac{1}{Y_{H,VSS}}$	$-\left[\dfrac{1-f_x Y_{H,VSS}}{Y_{H,VSS}}\right]$	$\mu_H X_H$
2	Decay	−1		$-f_x$	$b_H X_H$
Observed removal rate, $ML^{-3}T^{-1}$		\multicolumn{3}{c}{$r_i = \sum_j r_{ij} = \sum_j v_{ij}\rho_j$}	
Parameters, ML^{-3}		VSS	COD	O_2	

Organic carbon removal

This matrix defines the overall oxygen consumption rate as:

$$\frac{dS_O}{dt} = r_3 = -\frac{(1-f_x Y_{H,VSS})}{Y_{H,VSS}} \mu_H X_H - f_x b_H X_H$$

(b) Similarly, the yield coefficient Y_h should be defined in terms of g VSS/g BOD_5:

$$Y_h = \frac{Y_H}{f f_x}$$

where, f = g BOD_5/g biodegradable COD for the substrate in the system. Then,

$$v_{21} = -\frac{1}{Y_h} \text{ and } v_{32} = -f_x$$

$$\text{also, } r_{21,COD} = \frac{1}{f Y_h} \mu_H X_H$$

$$r_{11,COD} = f_x \mu_H X_H$$

$$r_{21,COD} - r_{11,COD} = \left(\frac{1}{fY_h} - f_x\right) \mu_H X_H$$

Table 3.3 shows the matrix representation of the model where VSS defines biomass and BOD_5 defines substrate. The overall oxygen consumption rate is derived as follows from the matrix given in Table 3.3:

$$\frac{dS_O}{dt} = r_3 = -\frac{(1-f f_x Y_h)}{Y_h} \mu_H X_H - f_x b_H X_H$$

Table 3.3. Matrix representation of the activated sludge model using VSS for biomass and BOD_5 for substrate

Component i	1	2	3	Process rate, ρ_j
j Process	X_H	S_S	S_O	$ML^{-3}T^{-1}$
1 Growth	1	$-\frac{1}{Y_h}$	$-\left[\frac{1-f f_x Y_h}{f Y_h}\right]$	$\mu_H X_H$
2 Decay	−1		$-f_x$	$b_H X_H$
Observed removal rate, $ML^{-3}T^{-1}$	\multicolumn{3}{c}{$r_i = \sum_j r_{ij} = \sum_j v_{ij}\rho_j$}	
Parameters, ML^{-3}	VSS	BOD	O_2	

3.3 MASS BALANCE

The *reaction kinetics* expressed in the adopted model becomes only meaningful when applied to the selected reactor. Mass balance provides all relevant information about *reaction kinetics* and *reactor hydraulics* for each model component and defines *reactor kinetics*. It is set around a selected *control volume* on the basis of conservation of mass. The control volume is usually identified as the entire reactor volume for real continuous flow and batch reactors and completely mixed ideal reactors. It may also be assumed as a differential volume dV for computational purposes within certain ideal reactors. It may be neglected (set to zero) if the reactor is not associated with any reaction, as in the case of secondary settling tanks. A mass balance equation expressed for a given model component should always incorporate the following terms:

$$\left\{\begin{array}{c}\text{Rate of}\\ \text{accumaltion in}\\ \text{the control volume}\end{array}\right\} = \left\{\begin{array}{c}\text{Rate of mass}\\ \text{flow into the}\\ \text{control volume}\end{array}\right\} - \left\{\begin{array}{c}\text{Rate of mass}\\ \text{flow out of the}\\ \text{control volume}\end{array}\right\}$$
$$+ \left\{\begin{array}{c}\text{Rate of conversion}\\ \text{of mass by reaction}\\ \text{in the control volume}\end{array}\right\}$$

The following mathematical expression defines this balance:

$$\frac{dm_i}{dt} = G_{i1} - G_i + r_i V_r \tag{3.7}$$

or

$$\frac{dm_i}{dt} = -\Delta G_i + r_i V_r \tag{3.8}$$

where, m_i = mass of model component i in the control volume, [M]
- ΔG_i = net influx of model component i in the control volume, [M/T]
  r_i = observed conversion rate of model component i in the control volume, [M/L^{3}T]
  V_r = control volume, [L^3]

In a continuous flow activated sludge reactor with a constant volume, V_r and a steady wastewater flow, Q, m_i and G_i may be defined as follows:

$$m_i = V_r C_i \tag{3.9}$$

and,

$$G_i = Q C_i \tag{3.10}$$

Substituting the values of m_i and G_i the continuity equation (3.8) becomes:

$$\frac{dC_i}{dt} + \frac{1}{\theta_h}\Delta C_i = r_i \tag{3.11}$$

where, the average hydraulic detention time, $\theta_h = V_r/Q$

An ideally mixed *batch reactor* does not involve the net inflow term and the continuity equation (3.11) is reduced to:

$$\frac{dC_i}{dt} = r_i \tag{3.12}$$

In a completely mixed continuous flow reactor, where the concentration of a model component in the effluent is the same as in the reactor volume, the mass balance equation is expressed as:

$$\frac{dC_i}{dt} = \frac{1}{\theta_h}(C_{i1} - C_i) + r_i \tag{3.13}$$

where, C_{i1} = influent concentration of model component i, $[M/L^3]$
C_i = reactor concentration of model component i, $[M/L^3]$

At steady state,

$$\frac{dC_i}{dt} = 0$$

and the corresponding mass balance equation becomes:

$$C_{i1} - C_i + \theta_h r_i = 0 \tag{3.14}$$

The hydraulics of full-scale activated sludge systems may conveniently be approximated by a series of appropriate number of completely mixed ideal reactors and the resulting flow scheme may be readily incorporated into available software programs for model simulations.

3.4 CHARACTERIZATION OF ORGANIC CARBON

3.4.1 Basis for COD fractionation

Wastewater characterization with respect to organic carbon is a key issue in dealing with basic process design. Appropriate characterization is a prerequisite for a better understanding of fundamental processes and may define the way in which they should be incorporated into design. A detailed understanding of wastewater composition significantly improves efforts to evaluate and upgrade the performance of the activated sludge process. The level of characterization required depends upon the specific objective for modelling and design. It should be flexible for each particular case.

A significant development in the mechanistic understanding of the activated sludge process is the adoption of COD as the model component for organic substrate and the ensuing concept of electron equivalence between substrate, active biomass and oxygen. This may be regarded as a turning point in ending the empirical guesswork for substrate utilization, excess sludge generation and oxygen requirement calculations based on BOD_5. This step has also been very important in encouraging efforts for a better understanding of the nature and composition of organic substrate, leading to the concept of COD fractionation. Biodegradable COD fraction was introduced based upon the early observations that COD also accounted for non biodegradable compounds (Eckhoff and Jenkins, 1967). Later, the bi-substrate model proposed by Dold *et al.* (1980) recognized the fact that the wide array of organics in wastewaters may be evaluated in two broad groups represented by markedly different rates of biodegradation. This approach was further elaborated for the identification and the mechanistic description of readily biodegradable and slowly biodegradable COD components (Henze *et al.*, 1987).

3.4.2 Major COD fractions in wastewaters

The COD is now considered not as a single overall substrate parameter, but a spectrum of organic carbon fractions that require further differentiation in terms of their biodegradation rates. The use of COD should first involve identification of its biodegradable portion. In this context, the *total COD*, C_{T1}, in wastewater is evaluated in terms of two major components: The *total non biodegradable* or *inert COD*, C_{I1} and the *total biodegradable COD*, C_{S1}:

$$C_{T1} = C_{S1} + C_{I1} \tag{3.15}$$

The inert COD should be analyzed in two subgroups identifying soluble and particulate fractions as *soluble inert COD*, S_{I1} and *particulate inert COD*, X_{I1}:

$$C_{I1} = S_{I1} + X_{I1} \tag{3.16}$$

These fractions can also be defined as fractions of the total COD in wastewater:

$$S_{I1} = f_{SI} C_{TI} \tag{3.17}$$

$$X_{I1} = f_{XI} C_{TI} \tag{3.18}$$

Organic carbon removal

and,
$$C_{II} = (f_{SI} + f_{XI})C_{TI} = f_I C_{TI} \tag{3.19}$$

where,
- f_{SI} = soluble inert fraction of the influent COD [M(COD)/M(COD)]
- f_{XI} = particulate inert fraction of the influent COD [M(COD)/M(COD)]

and,
- f_I = total inert fraction of the influent COD [M(COD)/M(COD)]

The *total biodegradable COD*, C_{S1} similarly embodies two major fractions conveniently differentiated as *readily biodegradable COD*, S_{S1} and slowly biodegradable COD, X_{S1}.

$$C_{SI} = S_{SI} + X_{SI} \tag{3.20}$$

Similarly, they can be expressed as fractions of the total COD:

$$S_{SI} = f_{SS} C_{TI} \tag{3.21}$$

$$X_{SI} = f_{XS} C_{TI} \tag{3.22}$$

and,
$$C_{SI} = (f_{SS} + f_{XS})C_{TI} = f_S C_{TI} \tag{3.23}$$

where,
- f_{SS} = readily biodegradable fraction of the influent COD [M(COD)/M(COD)]
- f_{XS} = slowly biodegradable fraction of the influent COD [M(COD)/M(COD)] and,
- f_S = total biodegradable fraction of the influent COD [M(COD)/M(COD)]

The differentiation is primarily based upon experimental observations showing a significant difference of approximately an order of magnitude between the rates of biodegradation of the two fractions (Dold and Marais, 1986). Evidently each fraction contains a number of compounds associated with a range of biodegradation rates, but this range is relatively small compared with the rate difference between the two groups.

Recently, the readily biodegradable COD was further subdivided into *fermentable readily biodegradable COD*, S_{F1}, and *fermentation products*, S_{A1}, considered being acetate, but in fact covering a whole range of other fermentation products (Henze et al., 1995). This new subdivision was mainly proposed for mathematical modelling of biological phosphorus removal.

The slowly biodegradable fraction, originally defined as particulate organics in the model proposed by Dold and Marais (1986), is now found to cover a wide range of particle size distribution from soluble to colloidal and larger organic particles of complex structure. The common feature of the particulate organics is that they cannot pass through the cell wall and need to undergo extra cellular hydrolysis prior to adsorption. Hydrolysis is advocated as the rate-limiting step for the utilization rate of the slowly biodegradable organic matter. The slowly biodegradable COD was reported to account from an average of 470 mg/l COD in domestic sewage to around 3000 mg/l in confectionery wastewaters (Orhon et al., 1999a). It is then conceivable that it may be difficult and sometimes misleading to characterize this major fraction by a single hydrolysis rate, as it is likely to cover a wide array of compounds with different composition and biodegradation pattern. This argument provided the basis of the recent approach to subdivide this group into further fractions, namely *rapidly hydrolysable COD*, S_{H1}, and *slowly hydrolysable COD*, X_{S1} (Henze et al., 1995). S_{H1} is conveniently differentiated as the soluble portion of the slowly biodegradable COD. Experimental studies also indicated a significantly slower biodegradation mechanism for the settled portion of X_{S1} (Orhon et al. 2002a). *Heterotrophic biomass*, X_{H1}, although reported as part of the settleable COD in domestic sewage (Henze et al., 1995), may not be considered as a significant parameter due to source and nature of most industrial wastewaters. Figure 3.1 gives a schematic description of different wastewater COD fractions. A simple differentiation of the soluble and particulate portions of the COD in the biological reactor is required for model evaluation. For this purpose, filtration is conveniently used as a practical index of solubility. This differentiation also

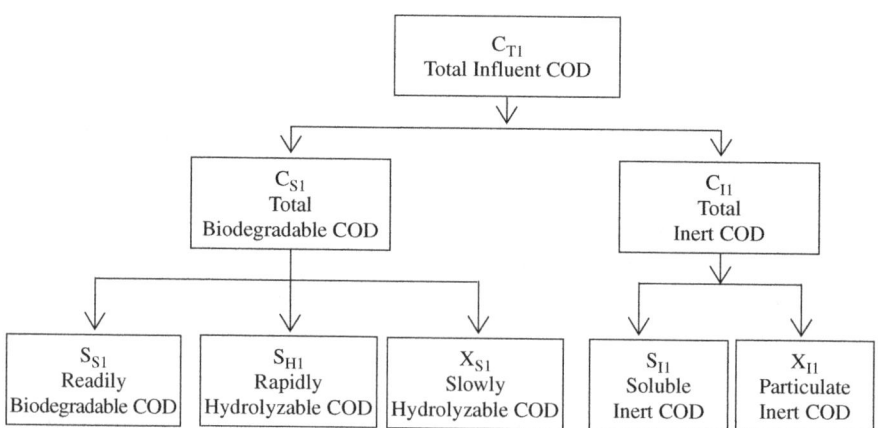

Figure 3.1. Distribution of major COD fractions in wastewater

Organic carbon removal 49

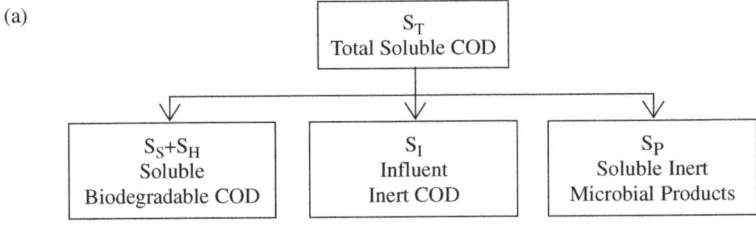

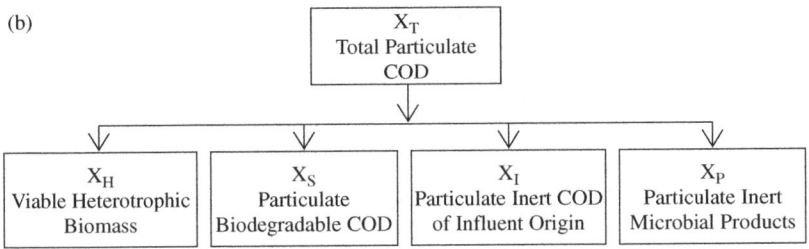

Figure 3.2. Major soluble and particulate COD components of mechanistic models

defines major process components in currently used mechanistic models. Major soluble and particulate model components defined in terms of COD fractions are indicated in Figure 3.2.

3.5 MAJOR PROCESSES FOR ORGANIC CARBON REMOVAL

An engineering evaluation of activated system behaviour should involve a reasonable understanding and a realistic description of the most important processes taking place within the system. In this evaluation, the term *process* is used to define a distinct event affecting the fate of one or more process components. Processes act as vital building blocks for mechanistic models. They should be selected and defined in a way to reflect appropriate kinetic expressions and stoichiometric relationships.

Current mechanistic models for organic carbon removal include three major processes: *growth* of heterotrophic biomass; *endogenous decay* and *hydrolysis* of slowly biodegradable organics.

3.5.1 Heterotrophic growth

It is now commonly accepted that heterotrophic growth only occurs at the expense of readily biodegradable substrate. This is in contrast with the classical concept that relates aerobic growth with the overall carbonaceous substrate in the reactor (Pearson, 1968; Jenkins and Garrison, 1968). Both readily biodegradable COD and dissolved oxygen concentrations are assumed to exert a rate limiting function in the kinetics of the aerobic growth of heterotrophic biomass. The effect of each model component is conveniently modelled with a Monod-type saturation function:

$$\frac{dX_H}{dt} = \hat{\mu}_H \frac{S_S}{K_S + S_S} \frac{S_O}{K_{OH} + S_O} X_H \qquad (3.24)$$

The rate expression includes three kinetic coefficients: *The maximum heterotrophic growth rate*, $\hat{\mu}_H$; *the half saturation coefficient for heterotrophic biomass*, K_S and *the oxygen half saturation coefficient for heterotrophic biomass*, K_{OH}. The oxygen term is primarily inserted as a switching function that would stop aerobic growth at low dissolved oxygen concentrations and thus a relatively small default value of 0.2 mg O_2/l is suggested for K_{OH} (Henze *et al.*, 1987).

This expression, although similar to the Monod equation used in traditional models, has a totally different rate implication in the sense that it relates to readily biodegradable COD, S_S, a small fraction of the total COD content, and to viable heterotrophic biomass, X_H. This version of a dependency between the process rate and the process components significantly affects the magnitude of rate coefficients. It is only satisfied for higher $\hat{\mu}_H$ values and lower K_S values. Experimental assessment of these coefficients reported in the literature for domestic sewage and a number of industrial wastewaters are outlined in Table 3.4 and Table 3.5. It is important to note that values of $\hat{\mu}_H$ and K_S displayed in these tables are generally compatible with the ones characterizing domestic sewage. This observation is justifiable by the fact that microbial

Table 3.4. Kinetic coefficients of heterotrophic growth for domestic sewage

Wastewater type	Y_H $\frac{g\,cell\,COD}{g\,COD}$	$\hat{\mu}_H$ day^{-1}	K_S mg COD/l	References
ASM2	0.63	6.0	20	Henze *et al.*, 1995
Plant 1	0.66	4.8	10	Ubay Cokgor, 1997
Plant 2	0.67	6.0	20	Orhon *et al.*, 1999b
Plant 3	0.62	3.9	10	Orhon *et al.*, 2002b

Table 3.5. Kinetic coefficients of heterotrophic growth for different industrial wastewaters

Wastewater type	Y_H $\frac{g\,cell\,COD}{g\,COD}$	$\hat{\mu}_H$ day^{-1}	K_S mg COD/l	References
Textile				
Organized industrial district predominantly textile	0.62	3.9	10	Ubay Cokgor, 1997
Cotton knit fabric	0.69	4.1	5	Germirli Babuna et al., 1998
Cotton knit fabric	0.62	3.2	13	Germirli Babuna et al., 1999
Cotton polyester mixture	0.69	5.3	5	Germirli Babuna et al., 1998
Polyester finishing	0.69	5.3	25	Germirli Babuna et al., 1998
Acrylic processing	0.60	3.9	10	Germirli Babuna et al., 1999
Denim processing	0.68	3.2	20	Orhon et al., 2001
Acrylic carpet finishing	0.66	1.7	1	Yildiz et al., 2007
Polyamide carpet finishing	0.66	6.0	1	Yildiz et al., 2008
Tannery				
Plain settling effluent	0.64	2.2	28	Orhon et al., 1999a
Plain settling effluent	0.67	2.55	5	Ubay Cokgor et al., 2008a
Chemical settling effluent	0.64	4.3	26	Orhon et al., 1999a
Integrated dairy processing	0.60	3.1	35	Orhon et al., 1993
Confectionary				
Plant I	0.69	5.1	75	Orhon et al., 1995
Plant II	0.64	4.4	150	Orhon et al., 1995

Table 3.5. *Continued*

Wastewater type	Y_H $\frac{g\,cell\,COD}{g\,COD}$	$\hat{\mu}_H$ day^{-1}	K_S mg COD/l	References
Meat processing				
Integrated plant	0.68	4.2	30	Gorgun *et al.*, 1995
Poultry processing	0.68	3.9	14	Eremektar *et al.*, 1999
Slaughterhouse	–	4.5	3	del Pozo *et al.*, 2003
Pharmaceutical	0.55	2.5	9	Ubay Cokgor *et al.*, 2008a

growth is considered to be proportional to S_S as previously mentioned, and the nature of S_S should not be expected to show significant variations from one industrial effluent to the other, because by definition, it is likely to exhibit a very similar composition for most wastewaters (simple carbohydrates, fatty acids, etc.). Slightly lower $\hat{\mu}_H$ and higher K_S levels should be attributed to inhibitory effects of various adverse compounds in industrial wastewaters.

3.5.2 Hydrolysis

Experimental observations indicate that the slowly biodegradable COD accounts for the major part of the organic content of industrial wastewaters. This is also the case for domestic sewage. In reality, the vast array of organics lumped in this fraction undergo, depending upon their nature and size, a sequence of complex reactions ranging from physical entrapment and adsorption to enzymatic breakdown and storage prior to biochemical oxidation and synthesis. As separate and detailed identification and description of these reactions is practically impossible in activated sludge systems, a single hydrolysis step is conveniently advocated to reflect the cumulative effect of all these complex interactions, where this COD fraction as a whole, is assumed to be enmeshed in sludge and solubilised to readily biodegradable substrate.

From a process standpoint, hydrolysis is assumed to be the rate limiting step for the utilization of the slowly biodegradable COD, X_S and it is commonly defined by means of a surface-limited reaction kinetics (Dold *et al.*, 1980; Henze *et al.*, 1987):

$$\frac{dX_S}{dt} = -k_h \frac{(X_S/X_H)}{K_X + (X_S/X_H)} \frac{S_O}{K_O + S_O} X_H \qquad (3.25)$$

Table 3.6. Kinetic coefficients of hydrolysis for different industrial wastewaters and domestic sewage

Wastewater type	k_h day^{-1}	K_X g cell COD / g COD	References
Textile			
Organized industrial district predominantly textile	2.0	0.38	Ubay Cokgor, 1997
Cotton knit fabric	0.8	0.7	Germirli Babuna et al., 1999
Polyester finishing	3.8	0.65	Germirli Babuna et al., 1998
Acrylic processing	1.6	0.7	Germirli Babuna et al., 1999
Denim processing	1.0	0.16	Orhon et al., 2001
Acrylic carpet finishing	1.4	0.02	Yildiz et al., 2007
Polyamide carpet finishing	3.5	0.04	Yildiz et al., 2008
Tannery			
Plain settling effluent	1.1	0.05	Murat et al., 2002
Plain settling effluent	2.0	0.1	Ubay Cokgor et al., 2008a
Chemical settling effluent	2.0	0.08	Orhon et al., 1999a
Integrated dairy processing	1.1	0.1	Orhon et al., 1993
Confectionary			
Plant I	0.8	0.1	Orhon et al., 1995
Plant II	1.0	0.1	Orhon et al., 1995
Integrated meat processing	1.5	0.13	Gorgun et al., 1995
Corn wet mill	3.0	0.5	Eremektar et al., 2002
Pharmaceutical	2.2	0.05	Ubay Cokgor et al., 2008a
Domestic sewage			
ASM2	3.0	0.03	Henze et al., 1987
Plant 1	3.0	0.05	Ubay Cokgor, 1997
Plant 2	3.0	0.1	Orhon et al., 1999b
Plant 3	2.0	0.4	Orhon et al., 2002b

Hydrolysis generates an equivalent amount of readily biodegradable substrate with no energy/oxygen utilization.

$$\frac{dS_S}{dt} = -\frac{dX_S}{dt} \qquad (3.26)$$

The rate expression for hydrolysis involves, similar to microbial growth, two major kinetics constants, *the maximum specific hydrolysis rate*, k_h and *the half saturation coefficient for hydrolysis*, K_X. The switch function for dissolved oxygen defining the limitations of aerobic conditions is also included in the expression. The rate coefficients k_h and K_X, defining hydrolysis of the slowly biodegradable COD fraction are much more important than their counterparts in microbial growth, as this mechanism solely dictates the effluent quality and the

Table 3.7. Kinetics of dual hydrolysis for different industrial wastewaters and domestic sewage

Wastewater type	Rapid hydrolysis		Slow hydrolysis		References
	k_{hS} day^{-1}	K_{XS} g COD/g COD	k_{hX} day^{-1}	K_{XX} g COD/g COD	
Textile					
Organized industrial district predominantly textile	2.5	0.4	0.1	0.5	Ubay Cokgor, 1997
Cotton knit fabric	3.0	0.05	1.0	0.5	Germirli Babuna et al., 1998
Cotton polyester mixture	3.6	0.05	1.0	0.2	Germirli Babuna et al., 1998
Denim processing	0.8	0.05	0.5	0.15	Orhon et al., 1998a
Acrylic carpet finishing	1.4	0.02	0.36	0.02	Yildiz et al., 2007
Polyamide carpet finishing	3.5	0.04	0.72	0.04	Yildiz et al., 2008
Tannery					
Plain settling effluent	1.1	0.2	0.3	0.2	Orhon et al., 1998b
Plain settling effluent	1.0	0.23	0.18	0.08	Murat et al., 2002
Poultry processing	3.3	0.07	2.3	0.6	Eremektar et al., 1999
Slaughterhouse	1.8	0.008	0.8	0.4	del Pozo et al., 2003
Domestic sewage					
Plant 1	3.1	0.2	1.2	0.5	Ubay Cokgor, 1997
Plant 2	3.8	0.2	1.9	0.18	Ubay Cokgor et al., 1998

Organic carbon removal

composition of the activated sludge in the reactor. As the values of k_h and K_X listed in Table 3.6 indicate, hydrolysis appears to be a much slower process for the majority of industrial wastewaters compared to domestic sewage.

For most industrial wastewaters, the magnitude of the overall slowly biodegradable fraction is so high that this fraction presumable incorporates a range of organics with markedly different rates of hydrolysis. In such cases, a *dual hydrolysis model* may be adopted, separately accounting for the hydrolysis of soluble (S_H) and particulate (X_S) slowly biodegradable compounds (Orhon *et al.*, 1998a).

$$\frac{dS_H}{dt} = -k_{hS} \frac{S_H/X_H}{K_{XS} + S_H/X_H} X_H \qquad (3.27)$$

$$\frac{dX_S}{dt} = -k_{hX} \frac{X_S/X_H}{K_{XX} + X_S/X_H} X_H \qquad (3.28)$$

Experimental results obtained for the kinetics of dual hydrolysis are outlined in Table 3.7. The data in this table appear to be very illustrative of the very marked difference between the kinetic coefficients related to two different slowly biodegradable COD fractions. It is interesting to note that the kinetic data derived for S_H are practically the same as the ones associated with a single hydrolysis model, with a clear indication that the experimental procedure tends to define S_H rather than the total slowly biodegradable COD (Henze *et al.*, 1987). Evidently, this is likely to involve a severe overestimation for the utilization of X_S. Furthermore, the hydrolysis rate for the particulate/slowly hydrolysable COD appears to be very close to what commonly defines endogenous respiration, especially for some industrial wastewaters. This observation challenges the validity of the values generally adopted for endogenous decay rate and underlines the possibility of significant interference of the slowly hydrolysable COD in the aerobic stabilization tests used for the experimental assessment of the endogenous decay rate.

3.5.3 Endogenous respiration

Endogenous metabolism can be defined as the sum of all biochemical activities of microorganisms in the absence of utilizable extra cellular compounds likely to serve as sources of energy and biosynthesis. Microorganisms deprived of substrate become exclusively dependent upon internal resources for maintaining their lives. At the stage of deprivation they find themselves with a complement of enzymes for access to endogenous carbon sources. Accordingly, the endogenous respiration rate of activated sludge can be defined in operational terms as the oxygen consumption rate in the absence of substrate from external sources (Spanjers *et al.*, 1998). This process is an integral part of activated sludge modelling as it accounts for a significant part of the total oxygen consumption in the system. It can be conveniently used to explain the decrease of the net sludge production and the reduced activity of the suspended organic matter with increasing sludge age.

In reality, the loss of suspended organic matter may be attributed, aside from endogenous respiration to a number of mechanisms like cellular decay, maintenance energy requirements, grazing by higher microorganisms like protozoa, toxic substances or adverse environmental conditions (pH or temperature) resulting in cell lysis. The dominant mechanism much depends on the operating conditions of the process (van Loosdrecht and Henze, 1999). Therefore, the endogenous respiration process associated with cellular decay should only be regarded as a convenient overall description of the activity of the mixed culture in activated sludge systems.

The concept of energy of maintenance is also advocated for explaining a lower observed sludge production: Some steps in endogenous metabolism may be essential to the vitality of microorganisms. Others may occur simply because substrate and enzyme are present and in contact with each other. In either case, metabolism of exogenous energy sources replaces the energy sources consumed by endogenous reactions. Maintenance requirements are mathematically expressed in different ways. Although different in principle, the concepts of maintenance and endogenous respiration may be interpreted as compatible processes for practical purposes, as the respective process rates are mathematically equivalent and cannot be easily distinguished from one another experimentally (Van Loosdrecht and Henze, 1999).

The overall decrease in biomass is traditionally defined by means of a first-order rate expression with respect to biomass concentration:

$$\frac{dX}{dt} = -k_d X \qquad (3.29)$$

where, k_d is the endogenous decay coefficient.

In this expression, biomass is defined in terms of the volatile suspended solids (VSS) and consequently, accurate description of the decay process is restricted with the limitations of this parameter. In fact, VSS measures, aside from active biomass, the particulate inert organic matter that accumulates in the reactor, non biodegradable microbial residues, and particulate organic substrate remaining after biodegradation within the mixed liquor or stored inside the cells. Therefore it cannot differentiate a number of mechanisms responsible for the fate of particulate components.

In new models involving detailed differentiation of soluble and particulate components, the endogenous decay process is introduced together with the concept of viability (Orhon and Artan, 1994). In this process, the cell decay only accounts for the fate of active biomass fraction of the activated sludge and becomes directly related to endogenous respiration. This way, the endogenous decay coefficient, b_H is defined as a function of the active heterotrophic biomass concentration, X_H in the reactor in a similar first-order rate expression (Marais and Ekama, 1976; Warner *et al.*, 1986):

$$\frac{dX_H}{dt} = -b_H X_H \qquad (3.30)$$

Table 3.8. outlines b_H values recently suggested based on experimental assessment or as default values for mathematical modelling.

Table 3.8. Reported values for the endogenous decay coefficient for different wastewaters

Wastewater type	b_H day^{-1}	References
Textile		
Organized industrial district predominantly textile	0.18	Ubay Cokgor, 1997
Cotton knit fabric	0.18	Germirli Babuna *et al.*, 1998
Cotton polyester mixture	0.14	Germirli Babuna *et al.*, 1998
Denim processing	0.14	Orhon *et al.*, 2001
Polyester finishing	0.12	Germirli Babuna *et al.*, 1998
Polyamide carpet finishing	0.10	Yildiz *et al.*, 2008
Tannery		
Plain settling effluent	0.12	Orhon *et al.*, 1999a
Plain settling effluent	0.12	Murat *et al.*, 2002
Dairy	0.14	Orhon *et al.*, 1993
Confectionary		
Plant I	0.10	Orhon *et al.*, 1995
Plant II	0.13	Orhon *et al.*, 1995
Meat processing	0.10	Gorgun *et al.*, 1995
Corn wet mill	0.11	Eremektar *et al.*, 2002
Domestic sewage		
Plant 1	0.18	Orhon *et al.*, 1998c
Plant 2	0.20	Orhon *et al.*, 2002b
Plant 3	0.09	Avcioglu *et al.*, 1998

3.5.4 Generation of residual microbial products

Models structured upon viability and active biomass inherently include generation of residual microbial products. These products are both particulate and soluble in nature and they are differentiated by virtue of their resistance to biodegradation under the operating conditions of activated sludge systems.

3.5.4.1 Particulate residual microbial products

Cellular decay is generally associated with generation of inert endogenous mass or particulate residual organic products: A fraction f_{EX} of the active biomass does not undergo any further reaction and accumulates in the activated sludge as particulate residual metabolic products, X_P. Accordingly, the rate of X_P generation can be defined as a first-order expression with respect to active biomass:

$$\frac{dX_P}{dt} = f_{EX} b_H X_H \qquad (3.31)$$

where, f_{EX} is the fraction of particulate inert COD generated in biomass decay.

In a system where endogenous decay prevails, the rate of overall biomass reduction has to be expressed as the sum of the endogenous decay rate of active biomass and the generation rate of the particulate inert metabolic products:

$$\frac{dX}{dt} = \frac{dX_H}{dt} + \frac{dX_P}{dt} \qquad (3.32)$$

Since,

$$\frac{dX_P}{dt} = -f_{EX} \frac{dX_H}{dt} \qquad (3.33)$$

The relationship between the two decay coefficients k_d and b_H can be defined with the following expression:

$$k_d = (1 - f_{EX}) \frac{X_H}{X} b_H \qquad (3.34)$$

This expression confirms the earlier observations that k_d is not a constant, but varies as a function of the active fraction of the biomass, leading to smaller k_d values for longer sludge ages.

The significance of particulate inert products for activated sludge modelling was recognized for the last 40 years. Symons and McKinney (1958) were the first researchers to note that the total oxidation of activated sludge was not possible and a small amount of non-oxidizable material remained. In a comprehensive review, McKinney (1962), defined an endogenous decay coefficient, b_H of 0.17/day at 20°C and concluded that a constant inert VSS fraction would be generated during endogenous metabolism at a rate that would correspond to an f_{EX} value of 0.21. Traditional modelling practically ignored inert VSS generation for a long period. As an integral part of scientific attempts to provide a new basis for activated sludge modelling, Marais and Ekama (1976), Dold *et al.* (1980) and Dold and Marais (1986) recognized inert particulate COD as a significant model parameter and suggested that the endogenous residue generated for unit active biomass destroyed should be taken as 0.20. Accordingly, Henze *et al.* (1987) proposed a default value of $f_{EX} = 0.2$ for activated sludge models incorporating endogenous decay.

In a recent study where a new experimental procedure was proposed for the assessment of the particulate inert COD in wastewaters, Orhon *et al.* (1999c) claimed that a constant value for the f_{EX} coefficient was not justified. In the study, the magnitude of X_P generation was quite specific and exhibited a significant difference for various organic carbon sources: It appeared to increase as the wastewater composition changed from simpler to more complex organic compounds. The experimentally calculated f_{EX} value was 0.4 for tannery wastewater, substantially higher than the other extreme of 0.11 for a synthetic mixture of readily biodegradable compounds. A constant f_{EX} value of 0.20 could only be approached with domestic sewage ($f_{EX} = 0.23$) requiring, however, further experimental support.

3.5.4.2 Soluble residual microbial products

Soluble residual (inert) microbial products, S_P is one of the most important parameters in the performance evaluation of activated sludge systems, as it controls, together with the inert COD of influent origin, the magnitude of effluent soluble COD. At high sludge age levels normally adopted for the treatment of industrial wastewaters, a significant part the soluble COD in the effluent is not the remaining fraction of the influent substrate which is depleted, but it is presumably organic matter released through microbial activities (Grady and Williams, 1975). A significant portion of this soluble organic matter is refractory or at least very slowly biodegradable for the usual range of activated sludge operation (Chudoba, 1985).

The kinetic description for the release of soluble residual products was first provided by Eckhoff and Jenkins (1967). They postulated a growth-associated mechanism where S_P could be expressed as a function of the influent biodegradable COD concentration, C_{S1}. Later, a similar decay associated process was conveniently defined on the basis of the hypothesis that a fraction f_{ES} of the endogenous biomass would not be oxidized, but converted into soluble inert products (Orhon and Artan, 1994):

Table 3.9. Experimental results on the assessment of initial soluble inert COD and generation of soluble inert microbial products (Orhon and Ubay Cokgor, 1997)

Wastewater type	C_{S1} (mg/l)	S_{I1} (mg/l)	Y_{SP}	f_{ES}	References
Municipal wastes					
Domestic	150	8	0.096	0.14	Orhon et al., 1994
Domestic	164	13	0.064	0.10	Orhon et al., 1994
Domestic	250	15	0.062	0.09	Lesouef et al., 1992
Municipal	190	15	0.086	0.13	Orhon et al., 1994
(Domestic-tanneries)					
Industrial wastes					
Tannery					
Plant 1	1500	323	0.040	0.06	Kabdasli et al., 1994
Plant 2	1075	262	0.040	0.06	Kabdasli et al., 1994
Plant 3	1870	464	0.045	0.07	Kabdasli et al., 1994
Textile					
Woven fabric	1176	90	0.044	0.07	Orhon et al., 1997
Knit fabric	800	88	0.040	0.06	Orhon et al., 1997
Knit fabric	535	117	0.088	0.13	Orhon et al., 1997
Cotton and synthetics dying and finishing	1000	190	0.083	0.12	Orhon et al., 1997
Dairy					
Integrated	480	—	0.068	0.10	Orhon et al., 1993
Yogurt and buttermilk	1190	—	0.062	0.09	Orhon et al., 1993
Pulp and paper	3340	137	0.057	0.09	Germirli et al., 1991
Meat processing	1990	110	0.057	0.09	Germirli et al., 1991
Antibiotics	9330	2520	0.100	0.15	Germirli et al., 1991
Strong wastes					
Cheese whey					
Anaerobic influent	60000	—	0.027	0.04	Germirli et al., 1993
Aerobic influent	1020	256	0.054	0.08	Germirli et al., 1993
Citric acid plant					
Anaerobic influent	29300	1870	0.054	0.08	Germirli et al., 1993
Aerobic influent	2025	804	0.210	0.31	Germirli et al., 1993
Anaerobic influent	28100	1600	0.060	0.09	Ubay Cokgor, 1997
Aerobic influent	4055	1900	0.078	0.12	Ubay Cokgor, 1997

$$\frac{dS_P}{dt} = f_{ES} b_H X_H \qquad (3.35)$$

Since,

$$S_P = Y_{SP} C_{S1} \qquad (3.36)$$

where, Y_{SP} denotes the ratio between soluble residual products and the initial biodegradable COD. The fraction of the endogenous biomass converted into soluble inert products may then be defined as a function of Y_{SP} and Y_H (Orhon et al., 1999c):

$$f_{ES} = \frac{Y_{SP}}{Y_H} \qquad (3.37)$$

The magnitude of initial inert COD together with f_{ES} and Y_{SP} coefficients, characterizing generation of soluble inert metabolic products experimentally assessed for a wide range of industrial wastewaters, are outlined in Table 3.9.

3.6 MULTI-COMPONENT MODELLING OF ORGANIC CARBON REMOVAL

The new conceptual approach of wastewater characterization relies on the existence and identification of a number of substrate fractions with different biodegradation characteristics. This approach also differentiates active biomass from other particulate biomass components. The revolutionary changes in the interpretation of substrate and biomass, supported by experimental findings, also affected the basic structure of mechanistic models for organic carbon removal. Compared with traditional models (Pearson, 1968; Jenkins and Garrison, 1968; Lawrence and McCarty, 1980) having only two components (*substrate/biomass*) and two processes (*growth/decay*), the number of model components is increased to account for all significant substrate and biomass fractions. Also, new processes such as *hydrolysis* are included for a better description of the fate of different model components. The expanded model structure also required a more elaborate description of stoichiometric relationships between these components.

A mechanistic model should now include a minimum of eight components for the modelling of organic carbon removal. They can be conveniently differentiated in two groups: substrate components and biomass components. The readily biodegradable COD, S_S, the slowly biodegradable COD, X_S, the soluble inert COD, S_I and the particulate inert COD, X_I define the biodegradation characteristics of the wastewater. The rapidly hydrolysable COD, S_H is often considered as an additional model component for most industrial wastewaters. The major biomass components obviously include active heterotrophic biomass, X_H, together with soluble and particulate inert microbial products, S_P and X_P. The soluble inert microbial products, S_P are sometimes conceived together with S_I, as a fictitious influent COD fraction. This approximation often creates analytical problems in the experimental assessment of other COD fractions and especially, X_I. Therefore, S_P should be included as a separate model component in the modelling of industrial wastewaters. Dissolved oxygen concentration, S_O is now the indispensable parameter for the model, mainly for respirometric evaluations and also for calculating the overall oxygen demands of the biological reactor. An overall nitrogen parameter, C_N and a phosphorus parameter, C_P may be included in the model, mainly to assess the potential of nutrient removal through incorporation into biomass during organic carbon utilization.

3.6.1 The endogenous decay model

As elaborated in the previous sections, this approach assumes that active biomass undergoes respiration generating energy to satisfy maintenance requirements. A part of the biomass is oxidized to CO_2 and H_2O and serves as fuel using dissolved oxygen as electron acceptor. The rate of endogenous respiration can be determined directly by respirometric measurements (Avcioglu *et al.*, 1998). This modelling approach was first proposed by Orhon and Artan (1994) as a modified version of Activated Sludge Model No.1, ASM1 (Henze *et al.*, 1987). Since then, it served in numerous evaluations carried out for domestic sewage and a wide spectrum of industrial wastewaters. The model incorporates three processes: *Heterotrophic growth, hydrolysis* and *endogenous respiration*. These processes determine, as schematically outlined in Figure 3.3, the fate of model components in the activated sludge reactor: The readily biodegradable COD, S_S is directly utilized by the heterotrophic biomass, X_H for growth. This process consumes dissolved oxygen, S_O as electron acceptor. The slowly biodegradable COD, X_S is converted to S_S through hydrolysis, a process which does not require energy input and dissolved oxygen consumption. Decay of the generated heterotrophs, X_H occurs as a simultaneous process as growth, where biomass is oxidized also using S_O as electron acceptor. Soluble and particulate

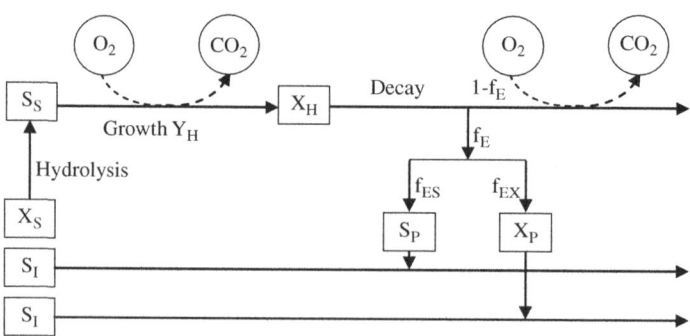

Figure 3.3. Mechanistic description of the endogenous decay model

Table 3.10. Matrix representation of the endogenous decay model

Component → Process ↓	1 S_I	2 X_I	3 S_S	4 S_H	5 X_S	6 X_H	7 X_P	8 S_P	9 S_O	10 C_N	Process rate $ML^{-3}T^{-1}$
1 Growth			$-\dfrac{1}{Y_H}$			1			$-\dfrac{1-Y_H}{Y_H}$	$-i_{XN}$	$\mu_H \dfrac{S_S}{(K_S+S_S)} X_H$
2 Decay						-1	f_{EX}	f_{ES}	$-(1-f_{EX}-f_{ES})$	$-i_{XN}(1-f_{EX})$	$b_H X_H$
3 Rapid hydrolysis			1	-1							$k_{hS} \dfrac{S_H/X_H}{K_{XS}+S_H/X_H} X_H$
4 Slow hydrolysis			1		-1						$k_{hX} \dfrac{X_S/X_H}{K_{XX}+X_S/X_H} X_H$
Parameter, ML^{-3}	COD	COD	COD	COD	COD	Cell COD	COD	COD	O_2	NH_3-N	

endogenous residues, S_P and X_P are produced as part of endogenous respiration. Influent soluble inert COD S_{I1} bypasses the reactor. Influent particulate COD, X_{I1} is entrapped and accumulated in the biomass without undergoing any biochemical reactions.

Table 3.10 gives the matrix representation of the process kinetics and stoichiometry the endogenous decay model. The rapidly hydrolysable COD, S_H is included in the matrix as a model parameter mainly because the utilization of the major portion of biodegradable COD in industrial wastewaters is properly explained by means of double hydrolysis of S_H and X_S fractions. Overall nitrogen parameter, C_N is also included in the matrix. This parameter may be useful for checking the amount of nitrogen that would be required for the removal of the organic carbon content of the wastewater. The minimum nutritional requirements are important since most industrial wastewaters are deficient of nutrients. It should also be noted that influent inert COD fractions, S_{I1} and X_{I1} do not include any stoichiometric coefficients, since by definition they do not take part in any biochemical reactions.

3.6.2 Death regeneration concept – ASM1

Activated sludge model no.1 (ASM1) has a pioneering role on multi-component activated sludge modelling. This approach originally proposed by Dold *et al.* (1980) and later adopted by Henze *et al.* (1987; 1995) in ASM1 and the following mechanistic models, essentially introduced COD fractionation for substrate and biomass components. ASM1 is basically structured on the assumption that active biomass is converted to a combination of particulate residual products and slowly biodegradable substrate, through *death* and *lysis* as schematically plotted in Figure 3.4. No loss of COD is involved and no direct electron acceptor is utilized. The slowly biodegradable substrate released is hydrolysed, generating an equivalent amount of readily biodegradable COD. Under aerobic conditions, new cells are produced at the expense of this COD fraction together with oxygen uptake. Under anoxic conditions, nitrate replaces dissolved oxygen.

The model particularly provides a mechanistic explanation for cell decay in an anaerobic system with no electron acceptors such as dissolved oxygen or nitrate nitrogen, where decay takes place without any loss of COD and external electron acceptors. In aerobic systems however, it may easily be noted that both endogenous decay and death regeneration models lead to the same quantitative results by selecting appropriate values for kinetic and stoichiometric coefficients (Orhon and Artan, 1994). The model does not involve the generation of soluble inert microbial products, which constitutes a serious problem for the accurate

Organic carbon removal

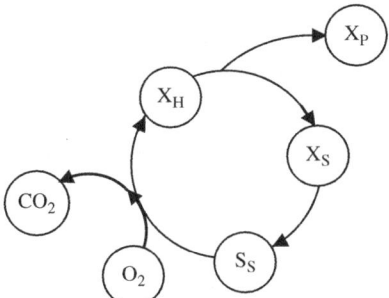

Figure 3.4. Death regeneration process in ASM1 (Orhon and Artan, 1994)

interpretation of the composition of the effluent COD for most industrial wastewaters. Table 3.11 gives the part of ASM1 structure associated with organic carbon removal in the same matrix format.

3.6.3 Substrate storage concept – ASM3

Models generally presume that microbial growth is the only biochemical mechanism for growth; substrate is directly converted into biomass which is subsequently used for endogenous respiration. It is now strongly argued that this mechanistic approach may not always be accurate, especially in the case where substrate is not continuously available in the reactor either temporally or spatially (van Loosdrecht *et al.*, 1997). In fact, some biological treatment systems like the *sequencing batch* reactor, (SBR), operate with an intermittent wastewater feeding which creates a *feast phase* followed by a *famine phase* after depletion of all available external substrate and these phases follow one another during the cyclic SBR operation. A similar set of conditions may also be sustained in a continuous flow biological reactor where the biodegradable COD in the wastewater is rapidly utilized and consumed in the initial section of the reactor so that the remaining volume of the reactor basically functions for endogenous respiration. Experimental findings indicate that under these operating conditions biomass tends to accumulate internal storage polymers which are subsequently utilized for growth when external substrate is no longer available (van den Eijnde *et al.*, 1984). The generation of internal storage polymers is now well documented (van Loosdrecht *et al.*, 1997; Majone *et al.*, 1999; Beun *et al.*, 2000). A new modelling approach, recently introduced as *activated sludge model no.3*, (ASM3) assumes that storage is the sole initial biochemical mechanism for the utilization readily biodegradable COD (Gujer *et al.*, 2000). ASM3 also uses the concept of COD fractionation for substrate

Table 3.11. Matrix representation of ASM1 for organic carbon removal

Component →	1	2	3	4	5	6	7	8	9	Process rate
Process ↓	S_I	X_I	S_S	X_S	X_H	X_P	S_O	S_{NH}	S_{ALK}	$ML^{-3}T^{-1}$
1 Growth			$-\dfrac{1}{Y_H}$		1		$-\dfrac{(1-Y_H)}{Y_H}$	$-i_{XN}$	$-\dfrac{i_{XN}}{14}$	$\mu_H\left(\dfrac{S_S}{K_S+S_S}\right)X_H$
2 Decay				$1-f_{EX}$	-1	f_{EX}				$b_H X_H$
3 Hydrolysis			1	-1						$k_h\left(\dfrac{X_S/X_H}{K_X+(X_S/X_H)}\right)X_H$
Parameter, ML^{-3}	COD	COD	COD	COD	Cell COD	COD	O_2	NH_3-N	Alkalinity – Molar units	

and biomass with two major differences compared with previous models. First, readily biodegradable substrate, S_S is conceived as the entire biodegradable COD in the soluble range determined by filtration through membrane filters with a pore size of 0.45 μm. This way, it covers both S_S and S_H (S_S+S_H) defined in ASM1 and endogenous decay models. Second, ASM3 assumes, as previously mentioned, that all biodegradable COD is initially converted into internal storage products and growth occurs only at the expense of stored polymers. Accordingly, ASM3 includes an additional *storage process*. Storage is generally observed as a faster process compared with growth and it is characterized in terms of a kinetic coefficient, k_{STO} defining the maximum storage rate. The concentration of storage products, X_{STO} is inevitably incorporated into the model structure as a particulate model component, an additional biomass fraction. The conversion of S_S into X_{STO} is defined by means of the storage yield coefficient, Y_{STO}. ASM3 also accounts for the generation of soluble inert microbial products, S_P. Matrix representation of ASM3 structure is given in Table 3.12.

In Table 3.12, the basic ASM3 structure is slightly modified in a way to associate microbial products generation with endogenous decay, for providing a unified basis with the other modelling alternatives.

3.6.4 Modelling of complex industrial wastewaters – case studies

3.6.4.1 Alternative models

The common feature of growth models (endogenous decay, ASM1) and storage models (ASM3) is that they both represent extreme cases, where external biodegradable substrate is either entirely consumed for growth or for storage. It should be remembered that mechanistic models can only give a simplified image of the real picture but sometimes the approximation provided may not be acceptable. In this case, a specific model sets a more reliable basis for process evaluation and performance estimation, provided that it is justified by experimental data. A model structure including simultaneous growth and storage has often proved to be a better tool for the evaluation of activated sludge behaviour (Krishna and van Loosdrecht, 1999; Karahan- Gul *et al.*, 2003).

Dizdaroglu-Risvanoglu *et al.*, (2007) reported an experimental study where the oxygen uptake rate (OUR) profiles obtained for filtered tannery effluent at two different F/M ratios of 0.07 and 0.2 g COD/g VSS.day were calibrated using ASM1, ASM3 and three different models based on simultaneous growth and storage. It was observed that ASM1 was not suitable for an accurate mechanistic

Table 3.12. Matrix representation of ASM3 modified for the generation of microbial products

Component	1	2	3	4	5	6	7	8	Process rate
Process	S_I	X_I	S_S	X_H	X_P	S_P	X_{STO}	S_O	$ML^{-3}T^{-1}$
Storage of readily biodegradable substrate			-1				Y_{STO}	$-(1-Y_{STO})$	$k_{STO}\dfrac{S_S}{K_S+S_S}X_H$
Growth on storage product				1			$-\dfrac{1}{Y_H}$	$-\dfrac{1-Y_H}{Y_H}$	$\mu_H\dfrac{X_{STO}/X_H}{K_{STO}+X_{STO}/X_H}X_H$
Endogenous respiration				-1	f_{EX}	f_{ES}		$-(1-f_E)$	$b_H X_H$
Respiration of storage product							-1	-1	$b_{STO}X_{STO}$
Parameter, ML^{-3}	COD	COD	COD	Cell COD	COD	COD	COD	O_2	

description of the biodegradation of the wastewater studied, for two main reasons: (i) it did not give a correct prediction of different COD and biomass fractions in the wastewater. The level of the readily biodegradable COD fraction associated with ASM1 calibration was much lower than the 820 mg/l of VFA concentration measured in the tannery sample; (ii) it did not provide a consistent set of process coefficients for the two feeding rates.

Similarly, the results suggested that storage as defined in ASM3 did not correctly reflect the actual storage mechanism which occurred during the calibration experiment. The inherent assumption in ASM3 that all readily biodegradable COD is converted in storage products which then becomes the only substrate for growth also resulted in different kinetics at different feeding rates in the experiment. ASM3 also defined a higher amount of S_{S1} than the measured level of VFA in the influent and this discrepancy was reflected in the simulation results. Figure 3.5 illustrates the calibration results of the OUR profiles obtained at an F/M ratio of 0.2 g COD/g VSS day with ASM1 and ASM3. The simultaneous growth and storage models, although providing a better fit for the observed OUR profiles, also suffered from the S_{S1} definition adopted from ASM3. They showed that the relative impact of growth and storage changes as the feast and famine pattern is affected by the selected feeding regime.

3.6.4.2 Expanded COD fractionation

The approach commonly adopted today is to differentiate two or three COD fractions with different biodegradation rates. The readily biodegradable COD, S_{S1} is usually associated with a fast utilization rate by means of direct microbial growth, whereas the soluble and particulate slowly biodegradable fractions, S_{H1} and X_{S1} generally represent organic compounds that need to undergo hydrolysis, acting as the slower rate controlling mechanism for their removal. These fractions are incorporated into mechanistic models as major process components. Respirometric experiments impart a specific shape to the oxygen uptake rate (OUR) profiles depending on the magnitude and rate of utilization of these COD fractions. Calibration of the adopted models reveals the relevant information on COD fractions and process kinetics.

Depending on the complexity of organics in the wastewater analyzed, the OUR curve may deviate from its typical pattern. The observed OUR perturbations are commonly explained with the existence of additional COD fractions and biochemical processes with different kinetics. The OUR profile displayed in Figure 3.6 obtained for a chemical industry effluent represents an unusual example for a complex substrate mixture (Ubay Cokgor et al., 2008b). It is quite different from a typical OUR curve generated with commonly accepted

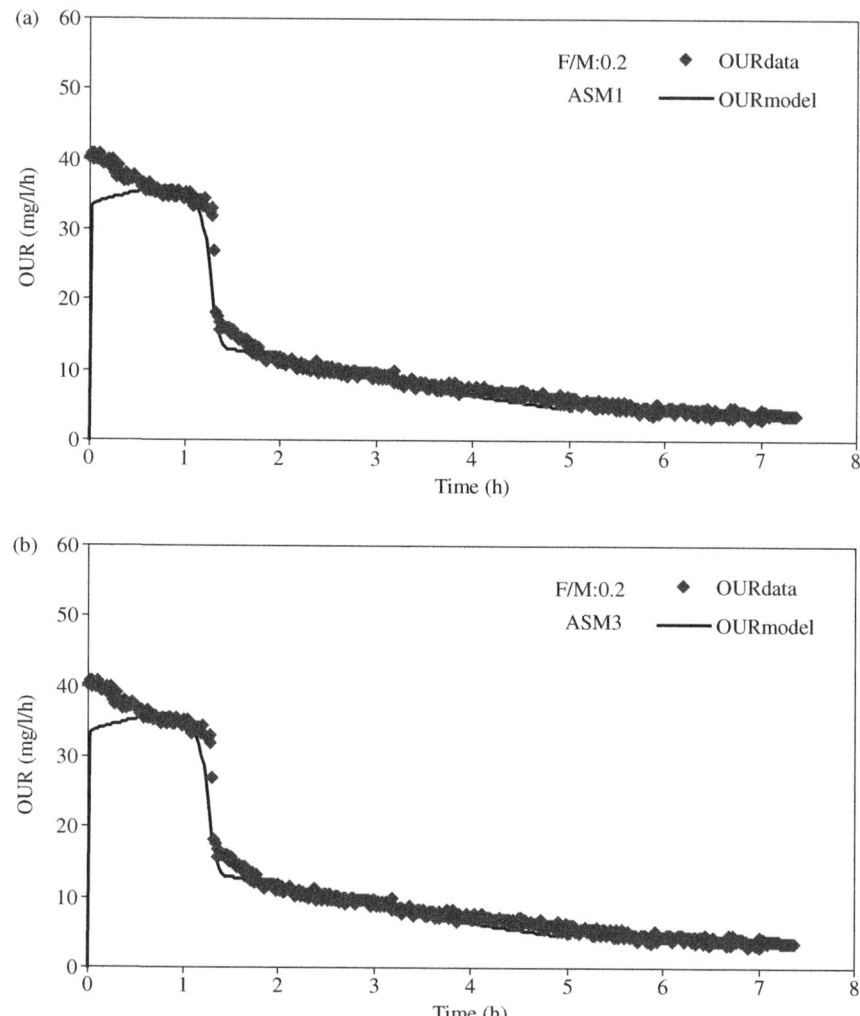

Figure 3.5. Calibration of the OUR profile obtained at F/M = 0.2 g COD/g VSS.day with (a) ASM1 and (b) ASM3

COD fractions, indicating the existence of at least four different biodegradable groups of organics.

Therefore, the model used to explain and simulate the experimental data included four COD fractions, namely three readily biodegradable fractions, S_{S1},

Organic carbon removal

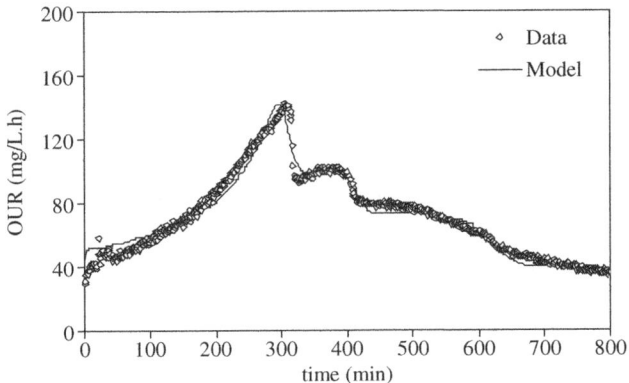

Figure 3.6. The experimental OUR profile of the chemical industry effluent (Ubay Cokgor *et al.,* 2008b)

S_{S2}, S_{S3} and a slowly biodegradable fraction, S_{H1}. The process rates of all three S_S components were defined in terms of a commonly adopted Monod-type of biodegradation kinetics for direct microbial growth. The sequential substrate utilization observed on the OUR curve was simulated by switch functions. The final phase of the OUR profile, much lower than the preceding plateaus and gradually declining towards the endogenous decay level indicated a typical hydrolysis process and it was defined by means of soluble hydrolysable substrate, S_H also adopted as a model component. The structure of the corresponding model is outlined in the usual matrix format in Table 3.13.

3.7 OPTIMUM SYSTEM DESIGN BASED ON MODELLING – A CASE STUDY

Conventional analysis of wastewater characteristics does not often generate essential information for an appropriate treatment design. As previously underlined, the evaluation of the wastewater has to cover beyond traditional parameters, COD fractionation together with experimental assessment of the biodegradation kinetics of different COD fractions. A mechanistic model, structured on the basis of kinetic data derived from the specific experiments, may be an important asset for practical application. The study reported by Yildiz *et al.,* (2008) represents a typical case indicating the merit of process modelling for optimal system design.

Table 3.13. Selected mechanistic model for the chemical industry effluent (Ubay Cokgor et al., 2008b)

Processes	S_{O2}	S_{S1}	S_{S2}	S_{S3}	S_H	X_H	Rate
Growth on S_{S1}	$-\dfrac{1-Y_H}{Y_H}$	$-\dfrac{1}{Y_H}$				1	$\mu_{H1} \dfrac{S_{S1}}{(K_S + \dfrac{S_{S1}^2}{K_I}) + S_{S1}} X_H$
Growth on S_{S2}	$-\dfrac{1-Y_H}{Y_H}$		$-\dfrac{1}{Y_H}$			1	$\mu_{H2} \dfrac{S_{S2}}{K_{S2} + S_{S2}} \dfrac{K_S}{K_S + S_S} X_H$
Growth on S_{S3}	$-\dfrac{1-Y_H}{Y_H}$			$-\dfrac{1}{Y_H}$		1	$\mu_{H3} \dfrac{S_{S3}}{K_{S3} + S_{S3}} \dfrac{K_S}{K_S + S_S} X_H$
Hydrolysis of S_H		1			−1		$k_h \dfrac{S_H/X_H}{K_X + S_H/X_H} X_H$
Biomass decay	$1-f_E$					−1	$b_H X_H$

Table 3.14. COD fractionation in polyamide carpet finishing wastewaters (Yildiz et al., 2008)

Process	Total C_{T1}	Soluble S_{T1}	Particulate X_{T1}	Soluble inert S_{I1}	S_{S1}	Biodegradable COD S_{H1}	X_{S1}
Pre-washing	2750	2500	250	110	0	173	2217
Dyeing & Softening	1030	820	210	140	100	116	464
Composite	1890	1660	230	125	12	40	1485

Table 3.15. Process stoichiometry and kinetics for polyamide carpet finishing wastewaters (Yildiz et al., 2008).

Process	Y_H cell COD/ COD	$\hat{\mu}_H$ day^{-1}	K_S mg/L	k_{h1}	k_{h2} day^{-1}	b_H	K_X COD/ cell COD
Pre-Washing		6	1.0	3.12	0.73	0.1	0.025
Dyeing & Softening	0.66	6	1.0	4.60	0.33		0.100
Composite		6	1.0	3.50	0.72		0.040

The study involved an investigation on polyamide carpet finishing wastewaters. The manufacturing process generated 67 m^3 per ton of polyamide fiber processed. The amount of wastewater was equally divided between the *pre-washing* step and the following *dyeing* and *softening* process. The composite effluent had a total COD of around 1900 mg/l, mostly soluble in nature and

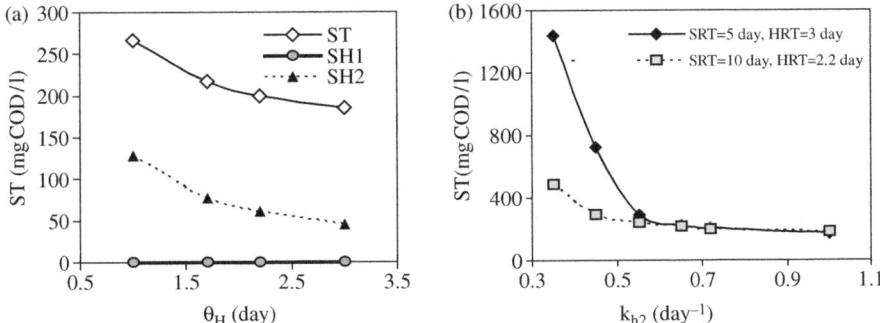

Figure 3.7. Model simulation of effluent soluble COD for a sludge age of 10 days (Yildiz et al., 2008)

mainly induced by pre-washing. Model calibration of the OUR profiles obtained from individual wastewater streams and the composite effluent indicated that (i) soluble slowly hydrolysable COD was the major fraction constituting around 97% of the total biodegradable COD and 78% of the total COD content, and (ii) it controlled the COD removal efficiency. COD fractionation and biodegradation kinetics of different COD fractions are given in Tables 3.14 and 3.15.

Model interpretation of the experimental data showed that an unusually long hydraulic retention time was a prerequisite for the hydrolysis and further utilization of the slowly hydrolysable COD fraction. Based on model simulation, a total soluble COD level below 200 mg/l in the effluent could only be obtained in an activated sludge system operated at a sludge age of 10 days and a hydraulic retention time of 10 days, as illustrated in Figure 3.7.

3.8. REFERENCES

Avcioglu, E., Orhon, D. and Sozen, S. (1998) A new method for the assessment of heterotrophic respiration rate under aerobic and anoxic conditions. *Water Sci. Technol.* **38**(8–9), 95–103.

Beun, J.J., Paletta, F., van Loosdrecht, M.C.M. and Heijnen, J.J. (2000) Stoichiometry and kinetics of poly-β-hydroxybutyrate metabolism in aerobic, slow growing, activated sludge cultures. *Biotechnol. Bioeng.* **67**(4), 379–389.

Chudoba, J. (1985) Quantitative estimation in COD units of refractory organic compounds produced by activated sludge microorganisms. *Water Res.* **19**(1), 37–43.

del Pozo, R., Tas, D.O., Dulkadiroglu, H., Orhon, D. and Diez, V. (2003) Biodegradability of slaughterhouse wastewater with high blood content under anaerobic and aerobic conditions. *J. Chem. Technol. Biot.* **78**(4), 384–391.

Dizdaroglu-Risvanoglu, G., Karahan, O., Ubay Cokgor, E., Orhon, D. and van Loosdrecht, M.C.M. (2007) Substrate storage concepts in modelling activated sludge systems for tannery wastewaters. *J. Environ. Sci. Heal. A.* **42**(14), 2159–2166.

Dold, P.L., Ekama, G.A. and Marais, G.v.R. (1980) A general model for the activated sludge process. *Prog. Wat. Tech.* **12**(6), 47.

Dold P.L. and Marais G.v.R. (1986) Evaluation of the general activated sludge model proposed by the IAWPRC Task Group. *Water Sci. Technol.* **18**(6), 63–89.

Eckhoff, D. and Jenkins, D. (1967) *Activated Sludge System Kinetics of the Steady and Transient States.* SERL Report No. 67-2. University of California, Berkeley.

Eremektar, G., Ubay Cokgor, E., Ovez, S., Germirli Babuna, F. and Orhon, D. (1999). Biological treatability of poultry processing plant effluent: a case study. *Water Sci. Technol.* **40**, 323–329.

Eremektar, G., Karahan-Gul, O., Germirli-Babuna, F., Ovez, S., Uner, H. and Orhon, D. (2002) Biological treatability of a corn wet mill effluent. *Water Sci. Technol.* **45**(12), 339–346.

Germirli, F., Orhon, D. and Artan, D. (1991) Assessment of the initial inert soluble COD in industrial wastewaters. *Water Sci. Technol.* **23**(4–6), 1077–1086.

Germirli, F., Orhon, D., Artan, N., Ubay, E. and Gorgun, E. (1993) Effect of two-stage treatment on the biological treatability of strong industrial wastewaters. *Water Sci. Technol.* **28**(2) 145–154.

Germirli Babuna, F., Orhon, D., Ubay-Cokgor, E., Insel, G. and Yapraklı, B. (1998) Modelling of activated sludge for textile wastewaters. *Water Sci. Technol.* **38**(4–5), 9–17.

Germirli Babuna, F., Soyhan, B., Eremektar, G. and Orhon, D. (1999) Evaluation of treatability for two textile mill effluents. *Water Sci. Technol.* **40**(1), 145–152.

Gorgun, E., Ubay Cokgor, E., Orhon, D., Germirli, F. and Artan, N. (1995) Modelling biological treatability for meat processing effluent. *Water Sci. Technol.* **32**(12), 43–52.

Grady, C.P.L. Jr. and Williams, D.E. (1975) Effects of influent substrate concentrations on the kinetics of natural microbial pollutions in continuous culture, *Water Res.* **9**, 171–180.

Grady, C.P.L., Daigger, G.T. and Lim, H.C. (1999) *Biological Wastewater Treatment.* 2nd edition, Marcal Dekker Inc., New York.

Gujer, W., Henze, M., Mino, T. and van Loosdrecht, M. (2000) *Activated Sludge Model No.3.* IWA Scientific and Technical Report No.9, (eds. Activated Sludge Models ASM1, ASM2, ASM2D and ASM3, Henze, M., Gujer, W., Mino, T., van Loosdrecht, M.), London, IWA.

Henze, M., Grady, C.P.L. Jr., Gujer, W., Marais, G.v.R. and Matsuo, T. (1987) *Activated Sludge Model No.1.* IAWPRC Scientific and Technical Report No.1, IAWPRC, London.

Henze, M., Gujer, W., Mino, T., Matsuo, T., Wentzel, M.C. and Marais, G.v.R. (1995) *Activated Sludge Model No.2.* IAWPRC Scientific and Technical Report No.2, IAWQ, London.

Jenkins, D. and Garrison, W.E. (1968) Control of activated sludge by mean cell residence time. *J. Wat. Pollut. Control Fed.* **40**, 1905.

Kabdasli, I., Tunay, O. and Orhon, D. (1994) The treatability of chromium tannery wastes. *Water Sci. Technol.* **28**(2), 97–106.

Karahan-Gul, O., van Loosdrecht, M.C.M. and Orhon, D. (2003) Modification of activated sludge model No. 3 considering direct growth on primary substrate. *Water Sci. Technol.* **47**(11), 219–225.

Krishna, C. and van Loosdrecht, M.C.M. (1999) Substrate flux into storage and growth in relation to activated sludge modelling. *Water Res.* **33**(14), 3149–3161.

Lawrence, A.W. and McCarty, P.L. (1980) Unified Basis Unified basis for biological treatment design and operation. *J. SED. Proc. ASCE* **96**(SA3), 757.

Lesouef, A., Payrandeau, A., Rogalla, F. and Kleiber, B. (1992) Optimizing nitrogen removal reactor configurations by on-site calibration of the IAWPRC Activated Sludge Model. *Water Sci. Technol.* **25**(6), 105–120.

Levenspiel, O. (1972) *Chemical Reaction Engineering.* 2nd Edition, John Wiley and Sons. New York, N.Y.

Majone, M., Dircks, K. and Beun, J.J. (1999) Aerobic storage under dynamic conditions in activated sludge processes. The state of the art. *Water Sci. Technol.* **39**(1), 61–73.

Marais G.v.R. and Ekama G.A. (1976) The activated sludge process. Part I: steady state behaviour. *Water S.A.* **2**(4), 163–199.

McKinney, R.E. (1962) Mathematics of complete-mixing activated sludge. *Proc. Am. Soc. Civil Eng.* **88**(SA3), 87–113.

Murat, S., Ates Genceli, E., Tasli, R., Artan, N. and Orhon, D. (2002) Sequencing batch reactor treatment of tannery wastewater for carbon and nitrogen removal. *Water Sci. Technol.* **46** (9), 219–227.

Orhon, D., Gorgun, E., Germirli, F. and Artan, N. (1993) Biological treatability of dairy wastewaters. *Water Res.* **27**(4) 625–33.

Orhon, D. and Artan, N. (1994) *Modelling of Activated Sludge Systems.* Technomic Publishing Co., Lancaster.

Orhon, D., Sozen, S. and Ubay, E. (1994) Assessment of nitrification - denitrification potential of Istanbul domestic wastewaters. *Water Sci. Technol.* **30**(6), 21–30.

Orhon, D., Yildiz, G., Ubay Cokgor, E. and Sozen, S. (1995) Respirometric evaluation of the biodegradability of confectionary wastewaters. *Water Sci. Technol.* **32**(12), 11–19.

Orhon, D. and Ubay Cokgor, E. (1997) COD fractionation in wastewater characterization – the state of the art. *J.Chem. Technol. Biot.* **68**, 283–293.

Orhon, D., Cokgor, E.U. and Sozen, S. (1998a) Dual hydrolysis model of the slowly biodegradable substrate in activated sludge systems. *Biotech. Technol.* **12**(10), 731–741.

Orhon, D., Sozen, S., Cokgor, E.U. and Ates Genceli, E. (1998b) The effect of chemical settling on the kinetics and design of activated sludge for tannery wastewaters. *Water Sci. Technol.* **38**(4–5), 355–362.

Orhon D., Hanhan O., Gorgun, E. and Sozen, S. (1998c) A unified basis for the design of nitrogen removal activated sludge process – The Braunschweig exercise. *Water Sci. Technol.* **38**(1), 227–236.

Orhon, D., Ubay Cokgor, E. and Sozen, S. (1999a) Experimental basis for the hydrolysis of slowly biodegradable substrate in different wastewaters. *Water Sci. Technol.* **39**(1), 87–95.

Orhon, D., Tasli, R. and Sozen, S. (1999b) Experimental basis of activated sludge treatment for industrial wastewaters – the state of the art. *Water Sci. Technol.* **40**(1), 1–11.

Orhon, D., Karahan, O. and Sozen, S. (1999c) The Effect of microbial products on the experimental assessment of the particulate inert COD in wastewaters. *Water Res.* **33**(14), 3191–3203.

Orhon, D., Germirli Babuna, F. and Insel, G. (2001) Characterization and modelling of denim processing wastewaters for activated sludge. *J. Chem. Technol. Biot.* **76**(6), 919–931.

Orhon, D., Okutman, D. and Insel, G. (2002a) Characterization and biodegradation of settleable organic matter for domestic wastewater. *Water S.A.* **28**(3), 299–305.

Orhon, D., Kabdasli, I., Ubay Cokgor, E., Tunay, O. and Artan, N. (2002b) Experimental assessment of optimum operation strategy for large industrial wastewater treatment plants – A case study. *Environ. Eng. Sci.* **19**(1), 47–58.

Pearson, E.A. (1968) *Kinetics of Biological Treatment, in Advances in Water Quality Improvement.* Vol. 2, E. F. Gloyna and W. W. Eckenfelder, Jr., eds., University of Texas Press, Austin, TX.

Petersen, E.E. (1975) *Chemical Reaction Analysis.* Prentice-Hall, Englewood Cliffs, NJ.

Spanjers, H., Vanrolleghem, P.A., Olsson, G. and Dold, P.L. (1998) *Respirometry in Control of the Activated Sludge Process: Principles.* IAWQ Scientific and Technical Report No. 7. London, UK.

Symons, J.M. and McKinney, R.E. (1958) The biochemistry of nitrogen in the synthesis of activated sludge. *Sewage Indust. Wastes* **30**(7), 874–890.

Ubay Cokgor, E. (1997). Respirometric evaluation of process kinetics and stoichiometry for aerobic systems. Ph.D. Thesis, Istanbul Technical University.

Ubay Cokgor, E., Sozen, S., Orhon, D. and Henze, M. (1998) Respirometric analysis of activated sludge behaviour – I. Assessment of the readily biodegradable substrate. *Water Res.* **32**(2), 461–475.

Ubay Cokgor, E., Arslan-Alaton, I., Erdinc, E., Insel, G. and Orhon, D. (2007) Effect of photochemical pretreatment on COD fractionation of a non-ionic textile surfactant. *Water Sci Technol.* **55**(10), 155–163.

Ubay Cokgor, E., Karahan, O. and Orhon, D. (2008a) The effect of mixing pharmaceutical and tannery wastewaters on the biodegradation characteristics of the effluents. *J. Hazard. Mater.* **156**(1–3), 292–299.

Ubay Cokgor, E., Insel, G., Aydin, E. and Orhon, D. (2008b) Respirometric evaluation of a mixture of organic chemicals with different biodegradation kinetics. *J Hazard. Mater.* (Accepted for publication, doi:10.1016/j.jhazmat.2008.03.051).

van den Eijnde, E., Vriens, L., Wynants, M. and Verachtert, H. (1984) Transient behaviour and time aspects of intermittently and continuously fed bacterial cultures with regard to filamentous bulking of activated sludge. *Appl. Microbiol. Biot.* **19**, 44–52.

van Loosdrecht, M.C.M., Pot M.A. and Heijnen, J.J. (1997) Importance of bacterial storage polymers in bioprocesses. *Water Sci. Technol.* **35**(1), 41–47.

van Loosdrecht, M.C.M. and Henze, M. (1999) Maintenance, endogenous respiration, lysis, decay and predation. *Water Sci. Technol.* **39**(1), 107–117.

Warner, A.P.C., Ekama, G.A. and Marais, G.v.R. (1986) The Activated Sludge Process Part IV – Application of the General Kinetic Model to Anoxic-Aerobic Digestion of Waste Activated Sludge. *Water Res.* **20**(8), 943.

Yildiz, G., Insel, G., Cokgor, E.U. and Orhon, D. (2007) Respirometric assessment of biodegradation for acrylic fibre-based carpet finishing wastewaters. *Water Sci. Technol.* **55**(10), 99–106.

Yildiz, G., Insel, G., Ubay Cokgor, E. and Orhon, D. (2008) Biodegradation kinetics of the soluble slowly biodegradable substrate in polyamide carpet finishing wastewater. *J. Chem. Technol. Biot.* **83**(1), 34–40.

4
Modelling of nutrient removal

4.1 INTRODUCTION

Nutrient removal from wastewaters is of vital importance for the protection of sensitive water bodies from eutrophication. In recent years, a greater concern for water quality has been the major issue for the implementation of stringent effluent regulations. The increasingly stricter nitrogen and phosphorus limits on wastewater discharges have stimulated studies on improving the activated sludge process and developing single sludge configurations for biological nutrient removal.

A single sludge activated sludge process for nitrogen and phosphorus removal is one of the most complex microbiological systems ever engineered for a specific purpose. It should control and sustain *autotrophs* for nitrification, together with *heterotrophs* for organic carbon removal under aerobic conditions. It should provide for the same heterotrophs anoxic conditions suitable for the removal of the oxidized nitrogen compounds (nitrite, nitrate) by means of anoxic respiration.

© 2009 IWA Publishing. *Industrial Wastewater Treatment by Activated Sludge*, by Derin Orhon, Fatos Germirli Babuna and Ozlem Karahan. ISBN: 9781789065282. Published by IWA Publishing, London, UK.

It should also favour selective growth of a specific group of heterotrophs, commonly called *phosphate accumulating microorganisms*, (PAOs) responsible for enhanced biological phosphorus removal (EBPR) through an appropriate sequence of anaerobic/aerobic phases. Therefore, biological nitrogen and phosphorus removal should be based upon an accurate interpretation of the behaviour of at least three different groups of microorganisms under a sequence of anoxic, anaerobic and aerobic conditions. The microbial basis of these processes is evidently the same for industrial wastewaters and domestic sewage.

This chapter outlines the fundamental mechanisms of nitrification, biological nitrogen removal and enhanced biological phosphorus removal. These mechanisms establish the basis for the mechanistic models developed for evaluating the fate of nitrogen and phosphorus in activated sludge systems.

4.2 BIOLOGICAL NITROGEN REMOVAL

Wastewaters usually contain non-oxidized forms of nitrogen, either organic nitrogen or ammonia. Therefore, biological nitrogen removal proceeds as a sequence of two different processes. The first process is *nitrification* which oxidizes ammonia to nitrite, (NO_2^-) and nitrate, (NO_3^-). *Denitrification* is the following process in the sequence, reducing nitrate and returning it to the atmosphere as molecular nitrogen (N_2). The main biochemical reactions that occur in activated sludge systems in relation to nitrogen removal may be summarized as:

(i) Microbial breakdown of complex organic nitrogen and release of ammonia, S_{NH},
(ii) Incorporation of nitrogen into cells through biosynthesis and its removal by excess sludge, and
(iii) Microbial mediated oxidation and reduction of nitrogen forms – nitrification and denitrification.

Biological conversion of organic nitrogen to ammonia occurs by means of hydrolysis of the particulate organic nitrogen fraction into simpler soluble compounds which then undergo *ammonification* reactions resulting in ammonia release (Dold and Marais, 1986; Henze *et al.*, 1987; Orhon and Artan, 1994). Nitrification is usually the rate limiting step in the sequence and for practical purposes, the influent total nitrogen content, C_{TKN1} may be approximated as ammonia nitrogen, S_{NH} for system design.

It should be noted that nitrification does not remove nitrogen but it acts as the decisive step in controlling the level of oxidized nitrogen, the only nitrogen fraction converted to N_2 by denitrification. Process stoichiometry for nitrogen removal should account for the amount of nitrogen incorporated into active biomass through biosynthesis. Characteristics of industrial wastewaters are quite different from domestic sewage in terms of their nitrogen content and the resulting COD/N ratio. Whilst only 20–30% N removal can be attributed to microbial growth in domestic sewage, most industrial wastewaters may even be deficient in nitrogen for effective COD removal.

4.2.1 Modelling of nitrification

4.2.1.1 Major nitrogen fractions

Major nitrogen fractions can be determined by direct measurements. This is a significant asset for evaluating nitrogen removal systems. The total *Kjeldahl* nitrogen (TKN) analysis yields the total unoxidized nitrogen content. The ammonia nitrogen analysis gives both the free ammonia and the ammonium ion concentrations. Thus, the organic nitrogen concentration may be computed as the difference between the TKN and the ammonia nitrogen. The oxidized nitrogen (*nitrite* and *nitrate*) compounds are not covered by the TKN test and they require specific analyses for their assessment.

Wastewater characterization for different nitrogen fractions is usually made using the same basic concept for organic compounds, in a way to emphasize the biodegradation and interaction with major biochemical reactions. In this framework, characterization primarily identifies biodegradable and non biodegradable nitrogen fractions. In this approach, differentiation of soluble and particulate components is conveniently adopted using the same size threshold – 0.45 μm filtration size – as in COD fractionation. The resulting sub-division for unoxidized nitrogen content of wastewaters is shown in Figure 4.1. Ammonia nitrogen, S_{NH1} is the most important component of this subdivision as it is readily available for biosynthesis or for oxidation to nitrite or nitrate through nitrification. *Particulate biodegradable organic nitrogen*, X_{ND} has to be hydrolyzed for conversion into *soluble biodegradable organic nitrogen*, S_{ND} by means of heterotrophic activity. Further breakdown of S_{ND} into S_{NH} is accomplished with the *ammonification* process which occurs similarly as part of the heterotrophic action.

The non biodegradable nitrogen components are usually defined in terms of their soluble and particulate COD counterparts, S_{I1} and X_{I1}:

$$S_{NI1} = i_{NSI}S_{I1} \tag{4.1}$$

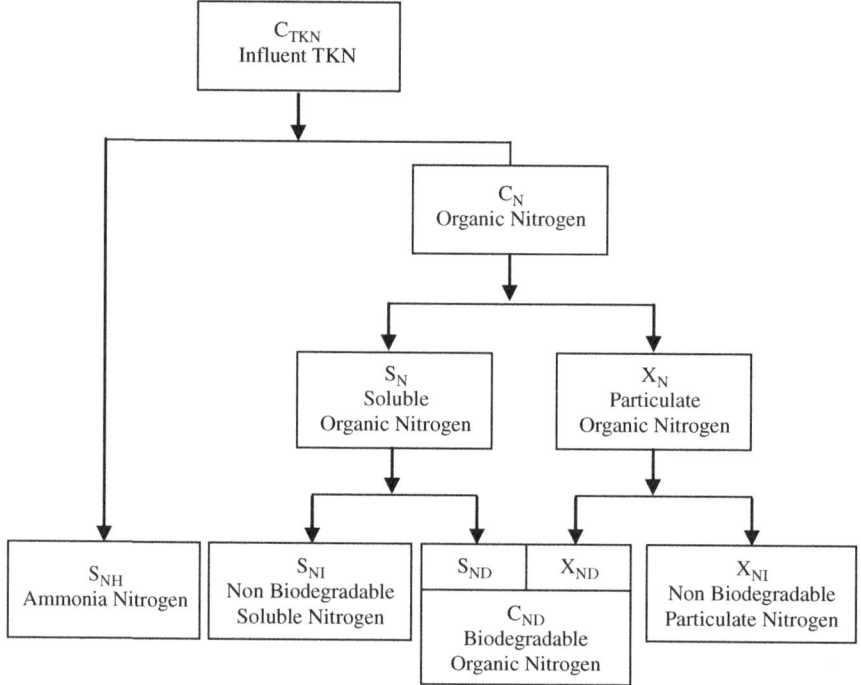

Figure 4.1. Major unoxidized nitrogen fractions in wastewaters

$$X_{NI1} = i_{NXI} X_{I1} \tag{4.2}$$

where,

i_{NSI} = nitrogen fraction of the influent soluble inert COD, S_{I1}
i_{NXI} = nitrogen fraction of the influent particulate inert COD, X_{I1}

The total biodegradable organic nitrogen of the wastewater, C_{ND1} can be calculated by subtracting the non biodegradable fractions and the ammonia from the influent TKN, C_{TKN1}:

$$C_{ND1} = S_{ND1} + X_{ND1} = C_{TKN1} - S_{NH1} - S_{NI1} - X_{NI1} \tag{4.3}$$

Similarly, the total ammonia concentration potentially available for biological reactions can be calculated as:

$$S_{NH1} + C_{ND1} = C_{TKN1} - S_{NI1} - X_{NI1} \tag{4.4}$$

Nutrient removal

These nitrogen fractions serve as process components in the modelling of nitrification and denitrification processes taking place in the activated sludge systems operated for nitrogen removal.

4.2.1.2 Stoichiometry of nitrification

Nitrification is an essential part of the metabolic activities of a group of *chemoautotrophic* bacteria depending on inorganic nitrogen compounds for energy. It proceeds as a sequence of two distinct aerobic processes: In the first step energy is derived from the oxidation of ammonia to nitrite (NO_2^-) and in the second step, oxidation of the generated nitrite to nitrate (NO_3^-) serves as the energy source:

$$NH_4^+ + 3/2O_2 \rightarrow NO_2^- + 2H^+ + H_2O \quad (4.5)$$

$$NO_2^- + 1/2O_2 \rightarrow NO_3^- \quad (4.6)$$

In nitrification, cellular carbon for biosynthesis is solely derived mainly from CO_2 and also from other inorganic compounds such as HCO_3^- and $CO_3^=$. Generally species of the genus *Nitrosomonas* is associated with the first step and *Nitrobacter* with the second step. Studies reveal that several other groups of chemoautotrophs are also capable of performing nitrification.

In this context, the stoichiometry of nitrification basically defines a process where ammonia serves as the electron donor and oxygen as the electron acceptor: The corresponding half reaction for complete oxidation of ammonia to nitrate by two different groups of bacteria may be written as:

$$NH_4^+ + 3H_2O \rightarrow NO_3^- + 10H^+ + 8e^- \quad (4.7)$$

This reaction consumes 1/8 mole or 1.75 g of nitrogen for each electron released (1.75 g N/e⁻.e). A portion of the electrons released by the oxidation of ammonia is transferred to O_2 in the energy reaction:

$$1/8NH_4^+ + 3/8H_2O \rightarrow 1/8NO_3^- + 10/8H^+ + e^-$$

$$1/4O_2 + H^+ + e^- \rightarrow 1/2H_2O$$

$$\overline{1/8NH_4^+ + 1/4O_2 \rightarrow 1/8NO_3^- + 2/8H^+ + 1/8H_2O} \quad (4.8)$$

The above overall energy reaction shows that for each gram of nitrogen utilized as electron donor, 4.57 g of O_2 is required, if biosynthesis is not taken into account. As will be indicated in the following parts, the value of 4.57 g O_2/g N is an important parameter for the stoichiometry of nitrification. A part of the electrons released is used as the reducing power for converting CO_2 to the oxidation level of cellular organic carbon. Accordingly, the biosynthesis reaction may be formulated as follows using the simplifying assumption of $C_5H_7NO_2$ for biomass composition:

$$1/8NH_4^+ + 3/8H_2O \rightarrow 1/8NO_3^- + 10/8H^+ + e^-$$

$$1/4CO_2 + 1/20NH_3 + H^+ + e^- \rightarrow 1/20C_5H_7O_2N + 8/20H_2O$$

$$\overline{1/8NH_4^+ + 1/4CO_2 + 1/20NH_3}$$

$$\rightarrow 1/20C_5H_7O_2N + 1/8NO_3^- + 2/8H^+ + 1/40H_2O \qquad (4.9)$$

The overall growth stoichiometry may be used to define a yield coefficient, Y_A, as 0.24 g cell COD/g N, a value which is commonly accepted as the autotrophic yield characterizing nitrification (Henze et al., 1987).

Another important feature of nitrification is the proton release as indicated by equation (4.9). Accordingly, the process consumes alkalinity equivalent to the amount of protons released. There is also an additional demand of alkalinity corresponding to the amount of ammonia used as the nitrogen source in biosynthesis and converted into organic nitrogen in the cell.

Using the balance in equation (4.9) and the yield coefficient, Y_A, the oxygen requirement associated with nitrification may be expressed as:

$$(4.57 - Y_A) g\ O_2/g\ N \qquad (4.10)$$

or,

$$\left(\frac{4.57 - Y_A}{Y_A}\right) g\ O_2/g\ cell\ COD \qquad (4.11)$$

The above expression is used as the stoichiometric coefficient relating the dissolved oxygen parameter, S_O to the growth of nitrifiers in the process kinetics.

System stoichiometry also indicates that nitrification consumes one mole of HCO_3^- per mole of ammonia nitrogen incorporated into autotrophic biomass and two moles of HCO_3^- per mole of ammonia nitrogen oxidized. In other words 7.14 g $CaCO_3$ equivalent alkalinity will be consumed per one gram of oxidized nitrogen. The corresponding stoichiometric coefficient may be calculated as follows (Orhon and Artan, 1994):

$$\left(-\frac{i_{XN}}{14} - \frac{1}{7Y_A}\right) moles\ HCO_3^-/g\ cellCOD \qquad (4.12)$$

This coefficient, defines alkalinity consumption in terms of autotrophic growth.

4.2.1.3 Growth of autotrophs

A Monod type of rate expression is usually adopted for expressing the overall growth kinetics of nitrification (Stratton and McCarty, 1967; Henze et al., 1987). This is used as an empirical rate limiting relationship for the sequence of two different microbial activities, between the initial energy yielding substrate and autotrophic growth:

$$\mu_A = \hat{\mu}_A \frac{S_{NH}}{K_{NH} + S_{NH}} \qquad (4.13)$$

where, μ_A = specific growth rate of autotrophic biomass
$\hat{\mu}_A$ = maximum specific growth rate of active autotrophic biomass
S_{NH} = ammonia nitrogen concentration in the reactor
K_{NH} = ammonia half saturation concentration.

Similar to heterotrophic growth, the yield coefficient defines the linear relationship between the nitrification rate and the autotrophic growth rate:

$$\frac{dS_{NH}}{dt} = -\frac{1}{Y_A}\frac{dX_A}{dt} = -\frac{\hat{\mu}_A}{Y_A}\frac{S_{NH}}{K_{NH} + S_{NH}}X_A \qquad (4.14)$$

where, X_A = autotrophic biomass concentration.

The specific growth rate of autotrophs, $\hat{\mu}_A$ is the vital parameter for nitrification mainly for two reasons:

(i) It has a much lower value compared to its counterpart for heterotrophic growth, $\hat{\mu}_H$.
(ii) It is highly susceptible to adverse effects of environmental parameters such as, dissolved oxygen concentration, temperature, pH and the presence of inhibitors.

The effect of dissolved oxygen on the growth kinetics of autotrophs is defined on the basis of a double saturation function relating the rate of autotrophic growth to S_{NH} and S_O:

$$\mu_A = \hat{\mu}_A \left(\frac{S_{NH}}{K_{NH} + S_{NH}}\right)\left(\frac{S_O}{S_O + K_{OA}}\right) \qquad (4.15)$$

where, K_{OA} = the half saturation constant for dissolved oxygen.

The second term in this expression becomes rate limiting at low S_O concentrations and serves as a switching function for autotrophic growth under aerobic conditions. The magnitude of K_{OA} to be adopted for design varies from as low as 0.002 mg/l (Dold and Marais, 1986) to 1.25 mg/l (Sarioglu *et al.*, 2008) with the wastewater and the operating conditions of the selected activated sludge configuration. In mechanistic activated sludge models, a K_{OA} level of 0.4–0.5 mg/l is recommended as a default value for model simulation studies (Henze *et al.*, 1987; 1995).

Nitrification kinetics is significantly affected by temperature. This effect is commonly observed on the maximum specific growth rate, $\hat{\mu}_A$. The temperature dependency of $\hat{\mu}_A$ has an important design implication, because it indicates the minimum sludge age at which nitrification may be sustained. Experimental observations suggest that the temperature effect on $\hat{\mu}_A$ may best be expressed by an Arrhenius-type of an equation in the range of 7–30°C:

$$\hat{\mu}_{AT} = \hat{\mu}_{A20} \theta^{T-20} \tag{4.16}$$

Different values in the range of 1.08–1.12 are reported in the literature for the temperature coefficient, θ (Bohnke, 1989; Ekama and Marais, 1984). Experimental results indicate that nitrification is also sensitive to pH and define an optimum pH range of 7.5–8.5 for the growth of nitrifiers.

4.2.1.4 Decay of autotrophs

The decay of nitrifiers is commonly defined by the following conventional first-order rate expression:

$$\frac{dX_A}{dt} = -b_A X_A \tag{4.17}$$

The endogenous decay rate, b_A is defined as 0.15 /day at 20°C and 0.05 /day at 15°C, somewhat lower than its counterpart b_H for heterotrophs under aerobic conditions (Gujer *et al.*, 2000).

4.2.1.5 Conversion of organic nitrogen to ammonia

Generally two different components are considered for the generation of ammonia from organic nitrogen, particulate organic nitrogen fraction, X_{ND}, and its soluble counterpart, S_{ND}. As previously mentioned, generation of ammonia is assumed to take place in a sequence of two consecutive reactions. First, X_{ND} is hydrolyzed to S_{ND}, which in turn is converted into ammonia by means of an *ammonification* reaction. The rate expression defined for the slowly

Nutrient removal

biodegradable COD is also assumed to apply to the hydrolysis of X_{ND}:

$$\frac{dX_{ND}}{dt} = -k_H \frac{X_{ND}/X_H}{K_X + (X_{ND}/X_H)} X_H \quad (4.18)$$

The following rate expression is used to approximate the conversion of the soluble organic nitrogen fraction, S_{ND} into ammonia:

$$\frac{dS_{ND}}{dt} = -k_a S_{ND} X_H \quad (4.19)$$

where, k_a = ammonification rate.

It should be noted that the proposed mechanisms for the conversion of organic nitrogen to ammonia are not fully substantiated by experimental support. Ekama and Marais (1984) suggest a simpler, single-stage rate expression for the breakdown of the total organic nitrogen, C_{ND} for the release of ammonia:

$$\frac{dC_{ND}}{dt} = -k_r C_{ND} X_H \quad (4.20)$$

where, k_r = overall conversion rate of total organic nitrogen, and

$$\frac{dS_{NH}}{dt} = -\frac{dC_{ND}}{dt} \quad (4.21)$$

4.2.1.6 Process kinetics

The process kinetics for nitrification is presented in a matrix format in Table 4.1, in a way similar to organic carbon removal. The matrix also includes applicable stoichiometric coefficients establishing the necessary relationships between model components. Major nitrogen fractions in the wastewater are defined as the first five model components. By definition, the soluble and particulate inert nitrogen fractions, S_{NI1} and X_{NI1} do not participate to any biochemical reactions; S_{NI1} bypasses the reactor and X_{NI1} becomes entrapped in the biomass and leaves the reactor with the excess sludge wastage. Organic nitrogen fractions, X_{ND} and S_{ND} are converted into ammonia through the sequence of hydrolysis and ammonification reactions. Ammonia nitrogen, S_{NH} represents the main substrate for autotrophic biomass and includes both ionized and un-ionized forms. S_{NO} is the main product of nitrification and represents both nitrite and nitrate nitrogen. X_A is the active autotrophic biomass responsible for nitrification. X_P defines the inert residue resulting from endogenous respiration of autotrophs. Dissolved oxygen, S_O is one of the vital model components of nitrification kinetics, as the

Table 4.1. Process kinetics and stoichiometry for nitrification

Component	1	2	3	4	5	6	7	8	Process rate $ML^{-3}T^{-1}$
i	X_{ND}	S_{ND}	S_{NH}	X_A	X_P	S_O	S_{NO}	S_{ALK}	
j Process									
1 Growth of autotrophs			$-i_{XN} - \dfrac{1}{Y_A}$	1		$-\dfrac{4.57 - Y_A}{Y_A}$	$\dfrac{1}{Y_A}$	$-\dfrac{i_{XN}}{14} - \dfrac{2}{14 Y_A}$	$\hat{\mu}_A \dfrac{S_{NH}}{K_{NH} + S_{NH}} X_A$
2 Decay of autotrophs				-1	f_{EX}	$-(1 - f_{EX})$			$b_A X_A$
3 Hydrolysis of particulate organic nitrogen	-1	1							$k_H \dfrac{X_{ND}/X_H}{K_X + (X_{ND}/X_H)} X_H$
4 Ammonification of soluble organic nitrogen		-1	1					$\dfrac{1}{14}$	$k_a S_{ND} X_H$
Observed conversion rates, $M/L^3 T$					$r_i = \sum\limits_j v_{ij} \rho_j$				

indispensable ingredient for autotrophic metabolism and an important design parameter. The final component, alkalinity is included in the model for evaluating the likely impact of pH variations on the process kinetics. As shown in the table, the following four processes:

  (i) growth of autotrophs
  (ii) decay of autotrophs
  (iii) hydrolysis of particulate organic nitrogen, and
  (iv) ammonification of soluble organic nitrogen

provide the mathematical description of process kinetics and stoichiometry for nitrification.

4.2.2 Modelling of denitrification

Denitrification is essentially a biochemical process converting oxidized inorganic nitrogen into gaseous nitrogen in the absence of dissolved oxygen. The nitrogen gas, N_2 is generally observed as the major end product of denitrification. Denitrification proceeds when oxidized nitrogen compounds, nitrate and/or nitrite are available in the biological reactor. As domestic sewage and most industrial wastewaters do not generally contain nitrogen in the oxidized form, biological nitrogen removal requires a process configuration which couples denitrification with nitrification.

Denitrification relies on the reduction of oxidized nitrogen which serves as the final electron acceptor in the electron transport chain, in the absence of dissolved oxygen. This process is called *anoxic respiration* and requires a separate *anoxic volume*. Anoxic conditions may also be sustained within the microbial flocs under favorable conditions. The biochemical mechanism involved is also called *nitrate dissimilation* where nitrate or nitrite replaces dissolved oxygen in aerobic respiration.

Nitrate dissimilation is coupled with the oxidation of biodegradable organic compounds serving as energy and carbon sources for heterotrophic growth. It is believed that the same facultative heterotrophic biomass sustained under aerobic conditions is capable of switching from oxygen to nitrate as the terminal electron acceptor when anoxic conditions prevail. In fact, the electron transfer process is almost identical for aerobic respiration and nitrate dissimilation, with the exception of a single enzyme, *nitrate reductase*, generated in the absence of oxygen. Oxygen, when introduced in the reactor, represses the synthesis of nitrate reductase and switches the metabolic activity to aerobic respiration. In a simplified way, denitrification involves two steps. In the first step, nitrate is reduced to nitrite with two electrons transferred from the oxidation of organic compounds. In the second step, nitrite is further reduced to any of the end products below:

$$NO_3^- \rightarrow NO_2^- \rightarrow NO \rightarrow N_2O \rightarrow N_2 \qquad (4.22)$$

4.2.2.1 Stoichiometry of denitrification

The stoichiometry of denitrification relies on the same principles as organic carbon removal under aerobic conditions, except for the fact that it utilizes a different electron acceptor. However, the two processes have totally opposite objectives when evaluated in terms of the corresponding treatment involved. In fact in aerobic systems, the main objective is total removal of biodegradable organic carbon. Accordingly, system design primarily envisages the required amount of electron acceptor – dissolved oxygen – to be supplied for biodegradable COD removal. In denitrification systems taking place under anoxic conditions however, the main objective is to remove the electron acceptor – nitrate – and appropriate design makes sure that a stoichiometrically sufficient amount of biodegradable COD, serving as both electron donor and carbon source, is present for the reduction of available nitrate in the wastewater. In this context, three different sources of organic carbon may be envisaged as electron donors in the denitrification process:

(i) External carbon – methanol, etc., – added during the anoxic phase of the process;
(ii) Internal carbon – biodegradable COD present in the wastewater
(iii) Endogenous organic carbon generated and consumed as part of the endogenous decay process.

For all carbon sources, the stoichiometry of denitrification remains the same, a chemotrophic process where biodegradable COD serves both as electron donor and the carbon source and nitrate, NO_3^- acts as the final electron acceptor with the following half reaction:

$$\frac{1}{5}NO_3^- + \frac{6}{5}H^+ + e^- \rightarrow \frac{1}{10}N_2 + \frac{3}{5}H_2O \qquad (4.23)$$

Utilization of nitrate as the electron acceptor sets a different energetic balance affecting the stoichiometry of denitrification. Under anoxic conditions, the overall energy reaction generates lower energy as compared to aerobic processes. Consequently, the anoxic yield coefficient, Y_{HD} remains lower than its counterpart Y_H characterizing aerobic heterotrophic growth. A value of 0.50 g cell COD/g COD has been proposed for Y_{HD} on the basis of energetics of related metabolic processes and interpretation of experimental results (Orhon et al., 1996). The differentiation of Y_{HD} from Y_H is now introduced into mechanistic activated sludge models. A similar Y_{HD} value of 0.54 g cell COD/g COD has been suggested as a default value in ASM3 (Gujer, et al., 2000).

The basic half reaction given in equation (4.23) indicates that the transfer of each electron reduces 2.8 g of nitrate nitrogen (2.8 g NO_3^-–N/e$^-$.e). Calculated on a O_2 or COD equivalent of substrate, 8 g of COD will be consumed to liberate one e$^-$ (8 g COD/e$^-$.e). Thus, $(1 - Y_{HD})$ 2.8 g NO_3^-–N will be reduced by 8 g COD. Consequently the stoichiometric relationship between COD consumed and nitrate reduced may be formulated as:

$$(1 - Y_{HD})\frac{2.8}{8} = \frac{(1 - Y_{HD})}{2.86} g\, NO_3^- - N/g\, COD \qquad (4.24)$$

Expressed in terms of active heterotrophic biomass produced, the above equation becomes:

$$\frac{(1 - Y_{HD})}{2.86 Y_{HD}} g\, NO_3^- - N/g\, cell\, COD \qquad (4.25)$$

This expression is used as the stoichiometric coefficient relating growth of nitrifiers to the concentration of oxidized nitrogen, S_{NO}, in the process kinetics of denitrification. In real systems, the above expression should be corrected by considering that nitrate will also serve as the electron acceptor for endogenous respiration, so that the overall stoichiometry between the amount of nitrate reduced and COD removed becomes:

$$\frac{(1 - Y_{NHD})}{2.86} g\, NO_3^- - N/g\, COD \qquad (4.26)$$

where, $\qquad YNHD = \dfrac{Y_{HD}}{1 + b_{HD}\theta_{XD}} \qquad (4.27)$

and b_{HD} = rate coefficient for anoxic endogenous decay

θ_{XD} = anoxic sludge age.

The value of 2.86 g COD/g N is quite significant in the stoichiometry of denitrification. As the biodegradable COD equivalent of nitrate reduced under anoxic conditions, it is one of the most meaningful parameters used in process design. It gives a theoretical nitrate requirement of 0.35 g NO_3^-–N for each g of COD utilized for the energy requirements of nitrifiers. However, a portion of the biodegradable COD will be used as carbon source for microbial growth so that the amount of nitrate reduced will be always lower than 0.35 g NO_3^-–N/g COD with a correction factor $(1 - Y_{NHD})$. With the assumption that ammonia is available as the nitrogen source for biological growth the overall N/COD ratio may be expressed as:

$$N/COD = [0.35(1 - Y_{NHD}) + i_{XN} Y_{NHD}] \; g \; NO_3^- - N/g \; COD \qquad (4.28)$$

where, i_{XN} = 0.086 gN/g cell COD.

A more accurate calculation of the applicable COD/N ratio should take into account nitrite, NO_2^-–N and dissolved oxygen. If nitrite is also generated in the preceding nitrification phase, it will be reduced as indicated in the following half-reaction:

$$\frac{1}{3} NO_2^- + \frac{4}{3} H^+ + e^- \rightarrow \frac{1}{6} N_2 + \frac{2}{3} H_2O \qquad (4.29)$$

According to this reaction:

$$1 e^- .e \; NO_2^- - N = \frac{14}{3} = 4.67 \; g \; NO_2^- - N$$

and $$(1 - Y_{NHD}) \frac{4.67}{8} = \frac{(1 - Y_{NHD})}{1.71} g \; NO_2^- - N/g \; COD \qquad (4.30)$$

Furthermore, an additional COD removal is likely to occur through preferential utilization of the dissolved oxygen that may be introduced to the reactor by means of sludge recycle and internal recycle from the aerobic volume. Then, the total biodegradable COD requirement for the reduction of nitrate and nitrite together with de-oxygenation may be computed as follows:

$$\Delta C_{S1} = \frac{2.86}{1 - Y_{NHD}} S_{NO_3} + \frac{1.71}{1 - Y_{NHD}} S_{NO_2} + \frac{1}{1 - Y_{NH}} S_{O1} \qquad (4.31)$$

where, ΔC_{S1} = required biodegradable COD consumption
S_{NO3} = initial nitrate-N concentration
S_{NO2} = initial nitrite-N concentration
S_{O1} = initial dissolved oxygen concentration.

Assuming that the effect of nitrite and dissolved oxygen may be neglected, expression (4.31) may be simplified as:

$$\frac{\Delta C_{S1}}{\Delta S_{NO_3^-}} \cong \frac{2.86}{1 - Y_{NHD}} \qquad (4.32)$$

It should be noted that the anoxic volume fraction and substrate removal kinetics under anoxic conditions should be suitable for the utilization of ΔC_{S1} to ensure total depletion of the available nitrate. The balance between biodegradable COD and available nitrate is better defined in terms of the *denitrification potential*, N_{DP}, which will be introduced as a significant design parameter in the following chapters.

The stoichiometry of the overall energy reaction for denitrification indicates that each mole of nitrate nitrogen used as electron acceptor consumes one mole of proton, leading to the generation of one mole of HCO_3^- alkalinity (Orhon and Artan, 1994). On the other hand, denitrification removes from solution one mole of HCO_3^- alkalinity for each mole of ammonia nitrogen incorporated into biomass. The combined effect of the two mechanisms may be expressed by the following relationship:

$$\frac{1}{14}\left(\frac{1-Y_{HD}}{2.86Y_{HD}} - i_{XN}\right) moles\ HCO_3^-/g\ cell\ COD \qquad (4.33)$$

4.2.2.2 Growth of denitrifiers

Activated sludge process designed for simultaneous carbon and nitrogen removal is operated as a single sludge system. In this process the same heterotrophic biomass is subjected to a sequence of aerobic/anoxic phases. After the aerobic phase, heterotrophic biomass recycled to the anoxic volume readily adapts its metabolic functions to consume nitrate as the electron acceptor instead of dissolved oxygen. From a process kinetics standpoint, the major feature of this cyclic sequential operation is the observation that the maximum specific growth rate of heterotrophs is reduced under anoxic conditions. Consequently, for single sludge systems, current mechanistic models are structured with the simplifying assumption that for the heterotrophic growth the same rate expression applies with the addition of an empirical correction factor $\eta_g < 1.0$ acting as an overall rate reducing factor for anoxic conditions (Batchelor, 1982; Henze et al., 1987; Orhon and Artan, 1994). Growth rates observed under aerobic and anoxic conditions are reasonably close to one another, as the suggested value for the correction factor, η_g varies in the range of 0.7–1.0 (Dold and Marais, 1986; Orhon et al., 1996).

In this context, the specific growth rate of heterotrophs under anoxic conditions may be approximated by the same Monod-type of function that relates the specific anoxic growth rate, μ_{HD} to the readily biodegradable COD, S_S:

$$\mu_{HD} = \eta_g \hat{\mu}_H \frac{S_S}{K_S + S_S} \qquad (4.34)$$

and,
$$\frac{dX_H}{dt} = \mu_{HD} X_H$$

Equation (4.34) is also expressed as a double Monod function reflecting the simultaneous effect of the readily biodegradable substrate, S_S and nitrate, S_{NO} on the specific growth rate (Orhon and Artan, 1994):

$$\mu_{HD} = \eta_g \hat{\mu}_H \left(\frac{S_S}{K_S + S_S}\right)\left(\frac{S_{NO}}{K_{NO} + S_{NO}}\right) \quad (4.35)$$

where, K_{NO} = the half saturation constant for nitrate nitrogen.

This expression implies that nitrate nitrogen is likely to exert a similar impact as dissolved oxygen on aerobic growth. However, experimental values reported for K_{NO} are in the low range of 0.1 – 0.5 mg NO_3^-–N/l (Henze et al., 1987; Grady et al., 1999). Considering the low K_{NO} values applicable in practice, the term related to the effect of S_{NO} on growth kinetics may be safely neglected and the denitrification rate may be assumed zero order with respect to S_{NO}, when it safely exceeds the level indicated by K_{NO}.

Then, the removal rate of the readily biodegradable COD may be related to the process rate for anoxic growth by the anoxic yield coefficient, Y_{HD}:

$$\frac{dS_S}{dt} = -\frac{\mu_{HD}}{Y_{HD}} X_H = -\frac{1}{Y_{HD}} \eta_g \hat{\mu}_H \frac{S_S}{K_S + S_S} X_H \quad (4.36)$$

A similar expression may be derived for nitrate removal rate:

$$\frac{dS_{NO}}{dt} = -\frac{1 - Y_{HD}}{2.86 Y_{HD}} \mu_{HD} X_H \quad (4.37)$$

or,
$$\frac{dS_{NO}}{dt} = -\frac{\mu_{HD}}{Y_D} X_H \quad (4.38)$$

This expression may also be used to define the *denitrification yield*, Y_D which gives the amount of cell COD produced per unit amount of NO_3^-–N/l consumed:

$$Y_D = -\frac{2.86 Y_{HD}}{1 - Y_{HD}} \quad (4.39)$$

4.2.2.3 Hydrolysis of the slowly biodegradable COD

Similar to domestic sewage, the readily biodegradable COD fraction, S_S accounts only for a minor part of the total COD content of most industrial wastewaters. Therefore the bulk of the organic carbon required in the anoxic phase is supplied through hydrolysis of the slowly biodegradable COD. Then, the hydrolysis clearly becomes the rate limiting step for the anoxic growth and the efficiency of nitrate removal. Using the same simplifying assumption for anoxic growth, the hydrolysis rate may be expressed as:

$$\frac{dX_S}{dt} = -\eta_h k_h \frac{X_S/X_H}{K_X + (X_S/X_H)} X_H \qquad (4.40)$$

where, η_h = the correction factor for the rate of hydrolysis under anoxic conditions.

In the case where the rapidly hydrolysable COD, S_H, and the slowly hydrolysable COD, X_S, must be differentiated for appropriate wastewater characterization, expression (4.37) may be expanded to cover the corresponding dual hydrolysis mechanism:

$$\frac{dS_H}{dt} = -\eta_h k_{hS} \frac{S_H/X_H}{K_{XS} + (S_H/X_H)} X_H \qquad (4.41)$$

$$\frac{dX_S}{dt} = -\eta_h k_{hX} \frac{X_S/X_H}{K_{XX} + (X_S/X_H)} X_H \qquad (4.42)$$

There is certainly no reliable information regarding the magnitude of the correction factor, η_h for industrial wastewaters. Even for domestic sewage, while a default η_h value of 0.40 is suggested for modelling (Henze et al., 1987) detailed experimental studies yielded a much higher average value of 0.90 (Sozen et al., 1998). It is recommended that the applicable value for η_h be assessed on a case by case basis from specific biodegradation experiments, if needed.

4.2.2.4 Decay of denitrifiers

Experimental observations also indicate that endogenous decay proceeds at a reduced rate under anoxic conditions. Adopting a similar approach, a simplified rate expression including a correction factor, η_E, is generally used for the anoxic endogenous respiration of heterotrophs:

$$\frac{dX_H}{dt} = -\eta_E b_H X_H \qquad (4.43)$$

where, η_E = the correction factor for the rate of endogenous decay under anoxic conditions.

Experimental studies suggest values in the range of 0.55 to 0.80 for this coefficient (Henze, *et al.*, 1987; Kristensen *et al.*, 1992; Avcioglu *et al.*, 1998).

4.2.2.5 Process kinetics

With the simplifying assumption that the process kinetics for heterotrophic behavior under aerobic conditions is also applicable with the introduction of correction factors for rate adjustment, the denitrification kinetics may be summarized as shown in Table 4.2, which indicates relevant rate expressions and the stoichiometric relationships between major processes and model components. The table includes two hydrolysable COD fractions, S_H and X_S, as dual hydrolysis provides a better characterization for the biodegradation of most industrial wastes. The process kinetics in the table defines *the nitrogen utilization rate,* NUR, as a counterpart of the oxygen utilization rate, OUR of heterotrophs under aerobic conditions. NUR is widely used as a significant parameter for modeling and especially for the experimental evaluation of biodegradation under anoxic conditions (Orhon and Artan, 1994; Sozen *et al.*, 2002). It is expressed in the same way as OUR as a function of anoxic microbial growth and decay.

$$NUR = \frac{dS_{NO}}{dt} = -\frac{1-Y_{HD}}{2.86 Y_{HD}} \eta_g \mu_H X_H - \frac{(1-f_E)}{2.86} \eta_E b_H X_H \qquad (4.44)$$

4.3. ENHANCED BIOLOGICAL PHOSPHORUS REMOVAL

Biological phosphorus removal from wastewater can take place in two different ways: Stoichiometric coupling to microbial growth and enhanced polyphosphate (poly-P) storage in the biomass. In both cases, the physical way to achieve phosphorus removal is through withdrawal and discharge of excess biomass. Polyphosphate storage has long been identified as the key mechanism responsible for the *enhanced biological phosphorus removal* (EBPR) process (Levin and Shapiro, 1965). The EBPR process can only be sustained if an *anaerobic phase* is incorporated in the activated sludge flow scheme. The system has to be operated in a sequence of anaerobic/aerobic phase and the influent is first collected into the anaerobic volume (Barnard, 1975).

The anaerobic/aerobic flow configuration favors the predominance of a certain group of microorganisms, commonly called *phosphorus accumulating organisms*, (PAOs). During the anaerobic period, PAOs have the ability of storing volatile fatty acids (VFAs) like acetate, S_A, which is mainly produced by regular heterotrophs using fermentable substrate, S_F during this period. Storage of VFAs as poly-hydroxyalkanoates, (PHA), proceeds using the internal polyphosphate (poly-P) pool as the energy source resulting in a release of orthophosphate (P_i).

Table 4.2. Process kinetics and stoichiometry for denitrification

Component → / Process ↓	1 S_S	2 S_H	3 X_S	4 X_{HD}	5 X_P	6 S_{NH}	7 S_{NO}	8 S_{ALK}	Process rate
Growth	$-\dfrac{1}{Y_{HD}}$			1		$-i_{XN}$	$-\dfrac{1-Y_{HD}}{2.86 Y_{HD}}$	$\dfrac{1}{14}\left[\dfrac{1-Y_{HD}}{2.86 Y_{HD}} - i_{XN}\right]$	$\eta_g \hat{\mu}_H \dfrac{S_S}{K_S + S_S} X_H$
Rapid hydrolysis	1	-1							$\eta_h k_{hS} \dfrac{S_H/X_H}{K_{XS} + (S_H/X_H)} X_H$
Slow hydrolysis	1		-1						$\eta_h k_{hX} \dfrac{X_S/X_H}{K_{XX} + (X_S/X_H)} X_H$
Endogenous decay				-1	f_{EX}	$i_{XN}(1-f_{EX})$	$-\dfrac{(1-f_{EX})}{2.86}$	$\dfrac{i_{XN}(1-f_{EX})}{14}$	$\eta_E b_{HD} X_H$
	COD	COD	COD	MLVSS	MLVSS	NH_3–N	NO_3^-–N	Molar concentration	

In the subsequent aerobic period, PAOs grow on the stored PHA and take up P_i for the replenishment of poly-P reserves, thus increasing the P content of the sludge. As regular heterotrophs cannot take up substrates like acetate due to the lack of an electron acceptor, the anaerobic conditions favor the enrichment of PAOs within the activated sludge.

Historically PAOs are associated with *Acinetobacter* spp (Fuchs and Chen, 1975). However, the experimental evidence reported so far indicate that *Acinetobacter* is not solely or primarily responsible for EBPR (Mino *et al.*, 1998).

Since the recognition of EBPR in activated sludge systems, various biochemical models have been proposed for explaining the process. Since PHA is a reduced polymer, reducing power is required for its synthesis. Two different modeling approaches have been suggested with different biochemical explanations for the source of the reducing power: The Mino model (Mino and Matsuo, 1984) and the Comeau-Wentzel model (Comeau *et al.*, 1986; Wentzel *et al.*, 1986). In the light of accumulated information a mechanistic model, commonly known as *ASM2*, defined the basic stoichiometry and kinetics of EBPR (Henze *et al.*, 1995). This was a simple model primarily intended as a scientific instrument for the prediction performance treatment plants operated for EBPR. It did not include a number of important model components and processes because of simplicity and lack of reliable supporting information. The model was later improved as *ASM2d* (Henze *et al.*, 1995) mainly to include denitrification by PAOs.

In this context, modeling for EBPR involves a number of additional components and processes: First, the volatile fatty acid content of the wastewater, expressed in terms of acetate, S_A, has to be separately identified. This requires subdivision of the readily biodegradable substrate, S_S into *fermentable substrate*, S_F, and *fermented substrate*, S_A. Fermentation of S_F into S_A is done by means of regular heterotrophs under anaerobic conditions. Obviously phosphate, S_{PO4} is the major parameter of interest. The model has to include the PAO concentration, X_{PAO} aside from active heterotrophic biomass, X_H. Also, X_{PHA} defines the concentration of *polyhydroxyalkanoates* (mainly *polyhydroxybutyrate*, PHB when acetate is utilized), the internal organic storage products formed at the expense of available S_A and X_{PP} is used to identify the concentration of stored poly-phosphate inside PAOs. Major biochemical processes associated with EBPR are summarized below:

(i) *Growth of PAOs:* The model assumes that PAOs grow only at the expense of internal organic storage products, X_{PHA}, under aerobic conditions. S_A is not considered as an alternative substrate because it is quite unlikely that it remains available in the aerobic phase.

(ii) *Storage of X_{PHA}:* It is assumed that X_{PHA} storage occurs as a direct conversion of S_A since required energy is supplied from the hydrolysis of X_{PP} and resulting release of S_{PO4}. The rate of X_{PHA} formation is expressed to depend on S_A and the X_{PP}/X_{PAO} ratio.

(iii) *Storage of X_{PP}:* X_{PP} is stored under aerobic conditions using the energy derived from the utilization of X_{PHA}. The rate of X_{PP} storage is observed to slow down as the X_{PP}/X_{PAO} ratio is increased. Accordingly, the rate expression for X_{PP} storage includes an inhibition term which is triggered when the X_{PP}/X_{PAO} ratio approaches the maximum allowable level, K_{MAX}.

(iv) *Lysis of PAOs and their storage products:* The model assumes a first order rate expression using the concept of death-regeneration, which produces slowly biodegradable substrate, X_S and residual particulate microbial products, X_P.

The stoichiometry and kinetics of EBPR are presented in Tables 4.3 and 4.4. The EBPR stoichiometry includes three major coefficients, namely: The yield coefficient for PAOs, Y_{PAO}; the coefficient indicating the PP requirement (S_{PO4} release) for PHA storage, Y_{PO4} (g P/g COD), and the coefficient indicating PHA requirement for PP storage, Y_{PHA} (g COD/g P). The same value is usually assumed for Y_{PAO} as Y_H characterizing heterotrophic growth. A Y_{PO4} of 0.40 g P/g COD and a Y_{PHA} of 0.20 g COD/ g P are suggested as default values in

Table 4.3. Stoichiometry of the EBPR process in ASM2 (Henze *et al.*, 1995)

Process	S_{O2}	S_A	S_{PO4}	X_P	X_S	X_{PAO}	X_{PP}	X_{PHA}
Storage of PHA on S_A		−1	Y_{PO4}				$-Y_{PO4}$	1
Aerobic storage of X_{PP}	$-Y_{PHA}$		−1				1	$-Y_{PHA}$
Anoxic storage of X_{PP}			−1				1	$-Y_{PHA}$
Aerobic growth	$1-\dfrac{1}{Y_H}$		$-i_{PBM}$			1		$-\dfrac{1}{Y_H}$
Anoxic growth			$-i_{PBM}$			1		$-\dfrac{1}{Y_H}$
Lysis of X_{PAO}			$v_{15,PO4}$	f_{XE}	$1-f_{XE}$	−1		
Lysis of X_{PP}			1				−1	
Lysis of X_{PHA}	1							−1

$v_{15,PO4} = i_{PBM} - (1-f_{XI})i_{PXS} - f_{XI}i_{PXI}$

Table 4.4. Process kinetics for EBPR in ASM2 (Henze et al., 1995)

Process	Rate expressions
Storage of PHA on S_A	$q_{PHA} \dfrac{S_A}{K_A + S_A} \dfrac{X_{PP}/X_{PAO}}{K_{PP} + X_{PP}/X_{PAO}} X_{PAO}$
Aerobic storage of X_{PP}	$q_{PP} \dfrac{S_{PO4}}{K_{PS} + S_{PO4}} \dfrac{X_{PHA}/X_{PAO}}{K_{PHA} + X_{PHA}/X_{PAO}} \dfrac{K_{MAX} - X_{PP}/X_{PAO}}{K_{IPP} K_{MAX} - X_{PP}/X_{PAO}} X_{PAO}$
Anoxic storage of X_{PP}	$\eta_{NO3} q_{PP} \dfrac{K_{O2}}{K_{O2} + S_{O2}} \dfrac{S_{PO4}}{K_{PS} + S_{PO4}} \dfrac{X_{PHA}/X_{PAO}}{K_{PHA} + X_{PHA}/X_{PAO}} \dfrac{K_{MAX} - X_{PP}/X_{PAO}}{K_{IPP} K_{MAX} - X_{PP}/X_{PAO}} X_{PAO}$
Aerobic growth	$\mu_{PAO} \dfrac{S_{O2}}{K_{O2} + S_{O2}} \dfrac{X_{PHA}/X_{PAO}}{K_{PHA} + X_{PHA}/X_{PAO}} X_{PAO}$
Anoxic growth	$\eta_{NO3} \mu_{PAO} \dfrac{K_{O2}}{K_{O2} + S_{O2}} \dfrac{X_{PHA}/X_{PAO}}{K_{PHA} + X_{PHA}/X_{PAO}} X_{PAO}$
Lysis of X_{PAO}	$b_{PAO} X_{PAO}$
Lysis of X_{PP}	$b_{PP} X_{PP}$
Lysis of X_{PHA}	$b_{PHA} X_{PHA}$

ASM2 (Henze et al., 1995). The process kinetics given in Table 4.4 also includes different switch functions for dissolved oxygen, ammonia, alkalinity, etc; in ASM2 but the validity of the selected saturation coefficients (K_{O2}, K_{NH}, K_{ALK}, etc.) in these functions are not fully substantiated.

REFERENCES

Avcioglu, E., Orhon, D. and Sozen, S. (1998) A new method for the assessment of heterotrophic endogenous respiration rate under aerobic and anoxic conditions. *Water Sci. Technol.* **38**(8-9), 95 – 103.

Barnard, J.L. (1975) Biological Nutrient Removal without the Addition of Chemicals. *Water Res.* **9**, 485.

Batchelor, B. (1982) Kinetic Analysis of Alternative Configurations for Single Sludge nitrification/Denitrification. *J. Wat. Pollut. Control Fed.* **54**, 1493.

Bohnke, B. (1989) Design of Nitrogen Removal Stage in a Sewage Treatment Plant. *Korrespondenz Abwasser*, **36**(13), 65.

Comeau, Y., Hall, K.J., Hancock, R.E.W and Oldham, W.K. (1986) Biological Model for Enhanced Biological Phosphorus Removal. *Water Res.* **20**, 1511.

Dold P.L. and Marais G.v.R. (1986) Evaluation of the general activated sludge model proposed by the IAWPRC Task Group. *Water Sci. Technol.* **18**(6), 63–89.

Ekama, G.A. and Marais, G.v.R. (1984) Nature of Municipal Wastewaters. In *Theory, Design and Operation of Nutrient Removal Activated Sludge Process*, Chapter 2, Water Research Commission, Pretoria, South Africa.

Fuchs, G.V. and Chen, M. (1975) Microbiological Bases of Phosphate Removal in the Activated Sludge Process for the Treatment of Wastewater. *Microb. Ecol.* **2**, 119.

Grady, C.P.L., Daigger, G.T. and Lim, H.C. (1999) *Biological Wastewater Treatment.* 2nd edition, Marcal Dekker Inc., New York.

Gujer, W., Henze, M., Mino, T. and van Loosdrecht, M. (2000) *Activated Sludge Model No.3.* IWA Scientific and Technical Report No.9, (eds. Activated Sludge Models ASM1, ASM2, ASM2D and ASM3, Henze, M., Gujer, W., Mino, T., van Loosdrecht, M.), London, IWA.

Henze, M., Grady, C.P.L. Jr., Gujer, W., Marais, G.v.R. and Matsuo, T. (1987) *Activated Sludge Model No.1.* IAWPRC Scientific and Technical Report No.1, London: IAWPRC.

Henze, M., Gujer, W., Mino, T., Matsuo, T., Wentzel, M.C. and Marais, G.v.R. (1995) *Activated Sludge Model No.2.* IAWPRC Scientific and Technical Report No.2, London, IAWQ.

Kristensen, G.H., Jorgensen, P.E. and Henze, M. (1992) Characterization of functional microorganism groups and substrate in activated sludge and wastewater by AUR, NUR and OUR. *Water Sci. Technol.* **25**(6), 43–57.

Levin, G.V. and Shapiro, J. (1965) Metabolic Uptake of Phosphorus by Wastewater Organisms. *J. Wat. Pollut. Control Fed.*, **37**, 800.

Mino, T. and Matsuo, T. (1984) Basic mechanism of biological phosphorus removal process. *Jap. J. Wat. Pollut. Res.* **10**, 605–609.

Mino, T., van Loosdrecht, M.C.M. and Heijnen, J.J. (1998) Microbiology and biochemistry of the enhanced biological phosphate removal process. *Water Res.* **32**(11), 3193–3207.

Orhon, D. and Artan, N. (1994) *Modelling of Activated Sludge Systems.* Technomic Publishing Co., Lancaster.

Orhon, D., Sozen, S. and Artan, N. (1996) The effect of heterotrophic yield on the assessment of the correction factor for anoxic growth. *Water Sci. Technol.* **34**(5-6), 67–74.

Sarioglu, M., Insel, G., Artan, N. and Orhon, D. (2008) Modeling nitrogen removal performance of a membrane bioreactor under dissolved oxygen dynamics. *Environ. Eng. Sci.* (accepted for publication).

Sozen, S., Ubay-Cokgor, E., Orhon, D. and Henze, M. (1998) Respirometric analysis of activated sludge behaviour – II. Heterotrophic growth under aerobic and anoxic conditions. *Water Res.* **32**(2), 476–488.

Sozen, S., Artan, N., Orhon, D. and Avcioglu, E. (2002) Assessment of the denitrification potential for biological nutrient removal using OUR/NUR measurements. *Water Sci. Technol.* **46**(9), 237–246.

Stratton, T.E. and McCarty, P.L. (1967) Prediction of Nitrification Effects on the Dissolved Oxygen Balance of Streams. *Environ. Sci. Technol.* **1**(5), 405.

Wentzel, M.C., Lotter, L.H., Lowenthal, R.E. and Marais, G.v.R. (1986) Metabolic Behaviour of *Acinetobacter* spp. In *Enhanced Biological Phosphorus Removal – A Biochemical Model. Water SA*, **12**(4), 209.

5

Experimental assessment of biodegradation

5.1 EXPERIMENTAL BASIS OF BIODEGRADATION

Activated sludge modeling has been a useful tool in the design and operation of wastewater treatment plants. Modeling efforts have been used to a great extent for predicting system performance and evaluation of different treatment alternatives. In this context modeling should be incorporated with accurate and reliable information which provides the characteristics of wastewaters to be treated and which defines the biochemical mechanisms correctly.

The use of modeling for the design and operation of activated sludge systems will only be possible upon the accurate assessment of (i) wastewater fractions that are used as model components, (ii) biomass fractions present in the activated sludge system, and (iii) the kinetic and stoichiometric coefficients that are used to determine the characteristics of the biochemical conversions. Experimental

© 2009 IWA Publishing. *Industrial Wastewater Treatment by Activated Sludge*, by Derin Orhon, Fatos Germirli Babuna and Ozlem Karahan. ISBN: 9781789065282. Published by IWA Publishing, London, UK.

assessment of the model components and coefficients has been the challenging area of research since the contemporary understanding of modeling has been introduced by IWA task Group with ASM1 (Henze et al., 1987).

Conventional wastewater characterization where the organic substrate is expressed in terms of TOC, COD or BOD_5 will not be appropriate for modeling purposes since TOC presents the organic carbon content without providing the oxidation state; although expressing oxidation state, COD presents both biodegradable and non biodegradable organic carbon and BOD_5 presents only a small fraction of the biodegradable organic matter in most of the cases, especially when complex wastewaters like industrial effluents are concerned. It is crucial that the biodegradable and non biodegradable fractions of organic matter are determined experimentally for predicting activated sludge system responses. The biochemical conversions occurring in activated sludge processes should be described for the biodegradable fraction of organic matter and thus the COD fractionation in terms of biodegradability is one of the main concerns for modeling.

5.2 PRINCIPLES OF RESPIROMETRY

5.2.1 Concept of oxygen uptake rate (OUR)

The basic understanding and interpretation of complex biochemical reactions taking place in the activated sludge process has been strengthened with the recognition of dissolved oxygen as a significant model component and its incorporation into mechanistic models. Dissolved oxygen (DO) and nitrate are the final electron acceptors for the biological processes involved in carbon and nutrient removal in activated sludge systems. DO and nitrate are not only significant for accurate modeling and design but are also practically used as system control parameters in activated sludge wastewater treatment plants. Dissolved oxygen is particularly important for setting the electron balance between biodegradable COD, biomass and electron acceptor for aerobic processes. Nitrate is also crucial to establish this balance for anoxic processes occurring in the biological systems. Recent activated sludge models like *Activated Sludge Model No.1*, *ASM1*, (Henze et al., 1987), *ASM2* (Henze et al., 1995) and *ASM3* (Gujer et al., 2000) incorporate dissolved oxygen and nitrate as model parameters as final electron acceptors.

The *oxygen utilization rate* (OUR) has been defined as the rate of oxygen utilization in the biochemical processes defined in the activated sludge models as given in equation (5.1).

$$OUR = -\frac{dS_O}{dt} \qquad (5.1)$$

In other words, OUR, is the change observed in the dissolved oxygen concentration (S_O) in time due to biochemical transformations. Thus, OUR is an overall process rate reflecting the cumulative impact of all oxygen/energy consuming reactions.

In addition to that, OUR and *nitrate utilization rate* (NUR) have been proved to be very effective tools for the experimental determination of biodegradable COD fractions as well as other kinetic and stoichiometric model coefficients. The first pioneering study was conducted by Ekama *et al.* (1986) for the assessment of biodegradable COD fractions and model parameters. Related principles and procedures have been incorporated first into ASM1 (Henze *et al.*, 1987), and later into the models that followed ASM1. Then, a wide spectrum of different applications have been developed to improve the experimental assessment of wastewater characterization and process kinetics (Sollfrank and Gujer, 1991; Kappeler and Gujer, 1992; Spanjers and Vanrolleghem, 1995; Ubay Cokgor *et al.*, 1998; Sozen *et al.*, 1998; Avcioglu *et al.*, 1998; Insel *et al.*, 2003; Karahan *et al.*, 2002a).

The emerging instrumentation techniques have encouraged the researchers to use the respirometric techniques to a great extent. It is now possible to conduct online OUR measurements in a defined experimental system or in the wastewater treatment plants for process control purposes. The respirometric measurements are conducted using highly developed dissolved oxygen electrodes. Mainly two techniques are used to determine OUR. The first technique is mostly used for research purposes with lab-scale online respirometers (Figure 5.1). The respirometric analysis system is composed of an aerated reactor vessel where

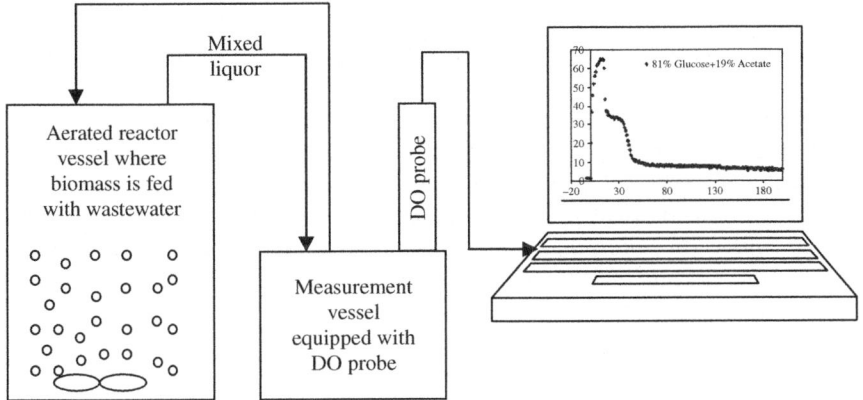

Figure 5.1. Schematic representation of a lab-scale respirometer set-up

the activated sludge system is simulated with biomass fed with wastewater. The aerated reactor vessel is sampled continuously for OUR measurement. The samples are transferred to a closed vessel equipped with a dissolved oxygen sensor, where the depletion of dissolved oxygen is monitored for a defined period of time, i.e. one minute. The decrease in the dissolved oxygen concentration in the given time period is then reported as the OUR value at the time of sampling. Contemporary respirometers are mostly equipped with computers which have data processors that read the dissolved oxygen data from the measurement vessel and calculate OUR value at the end of each measurement interval. The computer also generates the graphical representation of the OUR profile during the experimental run and provides means of extracting numerical data after the completion of the test.

The second technique involves continuous dissolved oxygen measurement in the aerated vessel, whether a lab-scale reactor or the aeration chamber of a wastewater treatment plant. Provided that the oxygen mass transfer coefficient, $k_L a$, and the dissolved oxygen saturation concentration, S_O^{sat}, is known and/or determined under the test conditions, OUR value can be estimated using the dissolved oxygen data by the expression (5.2) given below:

$$\left[\frac{dS_O}{dt}\right]_{System} = -OUR + k_L a(S_O^{sat} - S_O) \qquad (5.2)$$

OUR will be then equal to the mass transfer rate of oxygen since the change in the oxygen concentration of the system open to atmosphere will be zero.

Example 5.1 Calculation of OUR in an aerobic respirometric test

A batch aerobic experiment has been conducted to obtain the OUR response of an activated sludge system fed with an industrial effluent. Calculate the OUR values and draw the OUR profile:

(a) if OUR values are obtained from dissolved oxygen (DO) measurements of mixed liquor samples taken at different times. DO data obtained in the measurements conducted during 2 minute batch tests are given in Table 5.1.

(b) if OUR values are obtained by DO measurements in the reactor at different times. The saturation concentration of oxygen in the reactor and the $k_L a$ value are given below. The DO data measured in the reactor is presented in Table 5.2.

$$S_O^{Sat} = 8.17 \, mg \, O_2/l$$
$$k_L a = 25 \, h^{-1}$$

Table 5.1. Dissolved oxygen data obtained from mixed liquor samples taken at different times during 2 minutes

Sampling time (min)	DO values (mg/l)	
	Initial	Final
−10	7.85	7.52
−2	7.85	7.52
0	6.37	5.30
10	6.25	5.08
20	6.29	5.16
30	7.17	6.24
40	7.41	6.68
60	7.57	7.00
90	7.73	7.23
120	7.81	7.34
150	7.81	7.34

Table 5.2. Dissolved oxygen data obtained in the reactor at different times

Sampling time (min)	DO values (mg/l)
−10	7.85
−2	7.85
0	6.37
10	6.25
20	6.29
30	7.17
40	7.41
60	7.57
90	7.73
120	7.81
150	7.81

(a) Given the DO data in Table 5.1 the OUR values can be estimated as given by the following expression:

$$OUR_t = \frac{(DO_{initial} - DO_{final})}{\Delta t}$$

Thus, the OUR at 10 minutes before feed addition can be calculated as:

$$OUR_{-10} = \frac{(7.85 - 7.52)}{2} \times 60 = 10 \, mg/l.h$$

Industrial Wastewater Treatment by Activated Sludge

The OUR values obtained all through the experiment are listed below:

Sampling time (min)	OUR (mg/l.h)
−10	10
−2	10
0	32
10	35
20	34
30	28
40	22
60	17
90	15
120	14
150	14

The OUR profile of the test can be drawn as shown in Figure 5.2(a).

(b) Given the DO data in Table 5.2 the OUR values at time t can be estimated as dictated by equation (5.2) as follows:

$$OUR_t = k_L a (S_O^{sat} - S_{Ot})$$

Thus, the OUR at 10 minutes after feed addition can be calculated as below:

$$OUR_{10} = 25 \times (8.17 - 6.25) = 48 \ mg/l.h$$

The OUR values calculated all through the experiment are listed as below:

Sampling time (min)	OUR (mg/l.h)
−10	8
−2	8
0	45
10	48
20	47
30	25
40	19
50	17
60	15
90	11
120	9
150	9

The OUR profile of the test can be drawn as shown in Figure 5.2 (b).

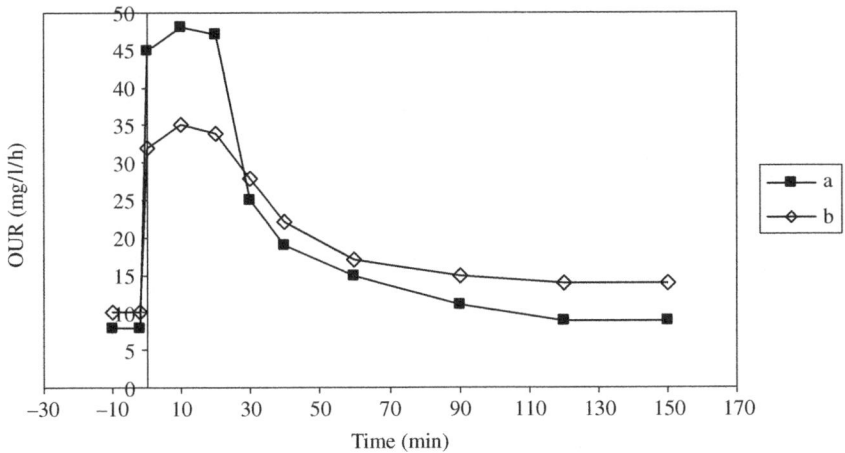

Figure 5.2. OUR curve obtained (a) from the respirometric experiment with mixed liquor samples taken at different times during 2 minutes, and (b) from DO measurements in the reactor at different times

5.2.2 Interpretation of respirometric data

The OUR profile reflects the true electron acceptor utilization response of the microbial community under the operating conditions of the experimental reactor. This response is interpreted in terms of a model that assumedly describes the kinetic behaviour of the community. A typical OUR profile obtained in an aerated batch reactor fed with readily biodegradable substrate can be interpreted for the dominant kinetic coefficients at different fractions of the curve according to ASM1 as given in Figure 5.3.

Parts of the OUR curve reflects different levels of sensitivity to the sequence of biological process in the batch reactor, which is a valuable asset for kinetic evaluation such as:

(i) the initial OUR level before substrate addition is determined by the endogenous decay activity, $b_H X_H$,
(ii) the starting level of the first plateau is set by the growth activity, $\hat{\mu}_H X_H$,
(iii) the ascending slope of the first plateau, if any, is due to biomass growth,

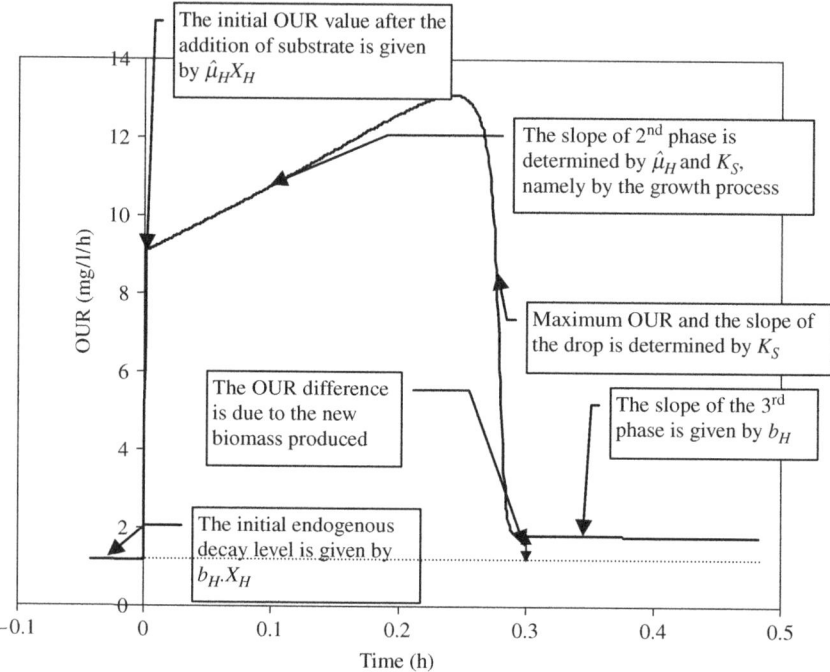

Figure 5.3. Interpretation of the OUR profile for readily biodegradable substrate according to ASM1

(iv) the half saturation constant, K_S is most effective on the slope of the drop after the first plateau, and
(v) last phase reflects again endogenous respiration. The difference between the initial and final OUR levels are due to new biomass produced.

When slowly hydrolysable matter is present in the wastewater sample, the shape of the OUR curve after the first plateau is mainly governed by the hydrolysis rate, namely, k_h and K_X.

The same OUR curve should be interpreted according to ASM3 with a completely different point of view, since the biochemical mechanisms participating in oxygen utilisation in the case of ASM3 are storage, growth, endogenous decay and respiration of storage products. Assuming that the concentration of the storage products is much less than that of heterotrophic biomass, the first phase of the OUR curve will be dictated by the aerobic heterotrophic decay rate, b_H, similar to ASM1. The initial OUR level, however, is associated with $k_{STO}X_H$, provided that the stoichiometry of storage is well defined. The maximum heterotrophic growth rate, $\hat{\mu}_H$, can be estimated from the slope of the storage (second) phase of the OUR curve, as shown in Figure 5.4(a), since the slope is determined by the growth process, namely by $\hat{\mu}_H$ and the half saturation coefficient for storage products, K_{STO}, when the growth yield, Y_H, is known. Although the substrate half saturation coefficient, K_S, also has

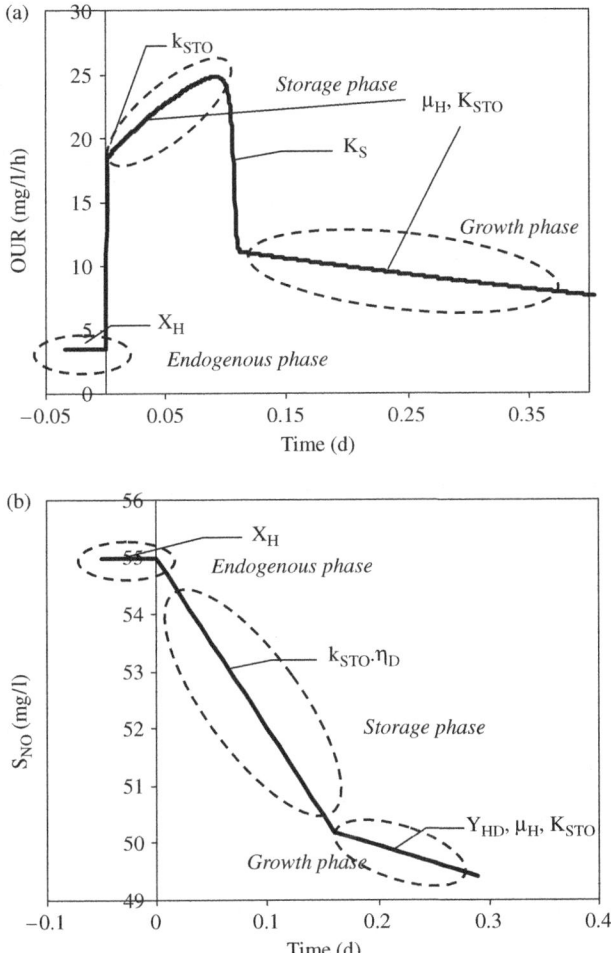

Figure 5.4. Interpretation of (a) OUR and (b) NUR profiles according to ASM3 (Avcioglu et al., 2003)

considerable effect on this slope and the maximum attainable value of OUR, K_S can be directly and almost independently estimated by the sharpness and inclination of the drop in the OUR. The slope of the third phase, namely the growth phase, is given by the $\hat{\mu}_H$ and K_{STO} couple, thus, these coefficients can be estimated via model simulations by using both the storage (second) and growth (third) phases of the OUR curve.

In the same manner, a nitrate utilisation rate (NUR) profile can be evaluated according to ASM3 as shown in Figure 5.4(b). The rate for the first phase is associated with endogenous respiration and the respiration of storage products, which can be neglected. Similar to the aerobic case, the anoxic heterotrophic yield coefficient, Y_{HD}, $\hat{\mu}_H$ and K_{STO} can be estimated from the slope of the growth phase using model simulation. Substrate affinity coefficient, K_S can be adjusted by the inclination and the smoothness of the transition from the second phase to the third phase. The coefficient η_D, which describes the reduction in the reaction rates under anoxic conditions, can be determined by comparative evaluation of experimental data obtained with aerobic and anoxic batch tests as presented by Avcioglu et al. (2003).

5.3 ASSESSMENT OF BIODEGRADATION CHARACTERISTICS

The assessment of biodegradation characteristics is an integral part of activated sludge modelling. Respirometry is extensively used in the experimental assessment of the biodegradable COD fractions and the non biodegradable fractions are mostly identified by long term batch tests.

The respirometric approach involves conducting a batch test with activated sludge fed with the wastewater sample under concern. OUR profile obtained is interpreted to estimate the amount of biodegradable COD present in the wastewater sample. The area under the OUR profile, above the endogenous decay level gives the amount of oxygen consumed for the amount of substrate available, which in other words will be equal to the amount of total biodegradable COD present in the wastewater sample. Considering that the heterotrophic yield coefficient, Y_H, is defined as the amount of biomass produced by the consumption of a certain amount of biodegradable COD, C_S, the total amount of oxygen consumed for the utilized amount of biodegradable substrate is given as in equation (5.3):

$$\Delta O_2 = C_S(1 - Y_H) \tag{5.3}$$

5.3.1 Assessment of inert fractions

The assessment of the inert fractions present in industrial wastewaters is very important because the inert fractions indicate indirectly the biodegradable COD fraction. Several methods have been proposed for the estimation of inert COD fractions in wastewaters. The assessment of the initial particulate inert organics, X_{I1} has gained the attention of many researchers. In this context, one of the first research efforts was the work of Ekama *et al.* (1986), which involved the analysis of a laboratory-scale completely mixed activated sludge system with a sludge age of more than 5 days, where X_{I1} was calculated by comparing the measured MLVSS concentration with the computed value on the basis of process kinetics. A similar approach was recommended by Henze *et al.* (1987). The approach was based on the comparison of the observed and calculated sludge production values. In 1992 an empirical method was proposed by Pedersen and Sinkjaer (1992). The method suggested that X_{I1} would approximately be equal to the difference between the particulate portions of the COD and the ultimate BOD in the effluent of a low-loaded activated sludge plant. Obviously, this method only provided a rough approximation, as it used BOD and COD together and did not account for the particulate residual COD generated during microbial activity. A more elaborate method was described by Kappeler and Gujer (1992) which involved using an experimental aerobic batch reactor and an evaluation procedure defined in the method was based upon model simulation and curve fitting. However this method ignored the generation of soluble residual microbial products, directly reflecting on the magnitude of the computed X_{I1} value.

Recent studies on the assessment of inert COD fractions were composed considering the generation of both particulate and soluble residual product. The majority of the methods proposed for the calculation of the initial particulate inert COD fraction in wastewaters rely on simulations and experimental verifications using newly developed multi-component activated sludge models (Henze *et al.*, 1987; Orhon and Artan, 1994). Upon the recognition of the importance of residual microbial products for the assessment of inert fractions, Orhon *et al.* (1994) proposed another procedure involving monitoring of the particulate COD in a batch reactor, both accounting for soluble and particulate inert products.

The methods involving multi-component modeling require that two kinetic constants, namely the heterotrophic yield coefficient, Y_H and the endogenous decay constant, b_H, be correctly determined by independent experimental methods. These methods generally adopted a default value of 0.2 (0.08 for the death-regeneration model) for the inert fraction of biomass produced during endogenous respiration, f_{EX}. The particulate inert fraction f_{EX} plays an important role in the evaluation of the experimental results during the assessment of inert COD fractions, since f_{EX} differentiates the initially present particulate inert COD from particulate inert microbial products generated during the experiments. The significance of particulate inert products for activated sludge modeling was recognized in the late 1950's when Symons and McKinney (1958) noted that the total oxidation of activated sludge was not possible. They were the first researchers to announce that a small amount of non-oxidizable polysaccharide material remained through the biochemical conversions. Kountz and Fourney (1959) also observed that 58% of substrate reappeared as new cells and 23% by weight of the new activated sludge produced was non-oxidizable with their experimental studies conducted with skimmed milk as substrate. Washington and Symons (1962) have presented that the volatile solids accumulation was in the range of 11.5±14.5% of the ultimate BOD removed on the basis of a long term carbon balance study with sodium acetate. These researchers have estimated that 20±25% of the weight of the activated sludge produced was accumulated as volatile solids by taking a yield value of 0.58 g VSS/g ultimate BOD as suggested by Kountz and Fourney (1959).

In parallel with these experimental studies, McKinney (1962) stated that 1/3 of the ultimate oxygen demand of the organic matter metabolized was oxidized and the remaining 2/3 was converted into biomass ($Y_H = 0.67$ g cell COD/g COD), defined an endogenous decay coefficient, b_H of 0.17 d^{-1} at 20°C, which remain as well established and accepted values for heterotrophic yield and endogenous decay coefficients for the modeling of activated sludge systems treating domestic wastewaters. McKinney (1962) concluded that a constant inert VSS fraction would be generated at a rate of 0.036 d^{-1} at 20°C during endogenous metabolism, yielding an f_{EX} value of 0.21. Later, Washington and Hettling (1965) have experimentally found that the volatile sludge accumulation was observed to be 10.3% of the organic matter removed, presenting additional experimental support for particulate inert products generation. However, in the study of Washington and Hettling (1965), the f_{EX} value has been estimated as 0.15, lower than the results of McKinney (1962) when both removed organic matter and accumulated volatile solids are calculated as oxygen equivalent, with an assumed Y_H value of 0.67 g cell COD/g COD.

With the establishment of new activated sludge modeling concepts Marais and Ekama (1976), Dold *et al.* (1980) and Dold and Marais (1986) have all reported in their studies that inert VSS generation was observed and agreed on the fact that the endogenous residue generated for unit active biomass destroyed should be taken as 0.20. Similarly, with the introduction of ASM1 (Henze *et al.*, 1987) the default value of f_{EX} was proposed as 0.20 for endogenous decay-type activated sludge models, when domestic wastewaters are concerned. Thus, it is now established that the assessment of the inert COD fractions present in the effluents is strictly bound to the accurate determination of the amount of the soluble and particulate microbial products generated. Microbial products are defined as a fraction of the decaying biomass that has been produced through the conversion of the biodegradable COD present in the wastewaters. The inert metabolic products generated during endogenous decay are determined with the fractions f_{ES} and f_{EX} for soluble and particulate products, respectively.

Recognition of the importance of soluble and particulate metabolic product generation in multi-component modeling of activated sludge, leads to the need for reliable experimental methods based on COD rather than VSS in order to build the electron equivalence balance for modeling. Germirli *et al.* (1991) have proposed a new method involving the setup of 2 batch reactors fed with filtered wastewater and glucose solution. Orhon *et al.* (1994) have introduced three different numerical evaluation methods for a similar experimental setup, with the addition of another reactor fed with raw wastewater. The proposed estimation methods were structured for either using all 3 reactors or only 2 of them. Orhon *et al.* (1999a), further developed the numerical evaluation of the experimental results and applied the method to domestic and tannery wastewaters. Orhon *et al.* (1999a) have found that the f_{EX} for tannery effluents was surprisingly high at an average value of 0.40 g COD/g COD, while that of the domestic sewage was estimated as 0.20 g COD/g COD in parallel with the generally accepted value found in the literature.

5.3.1.1 *Estimation of Inert Fractions by Batch Experiments*

The procedure involves running three aerated batch reactors, the first one fed with the raw wastewater sample, the second one with the soluble (filtered) wastewater sample and the last one with a glucose solution approximately having the same COD content as the soluble wastewater reactor, as previously suggested by Germirli *et al.* (1991) and Orhon *et al.* (1994). The batch reactors are seeded with minimal amount of biomass (around 40–50 mg/l VSS) that has previously been acclimated in an aerated fill and draw reactor fed with wastewater sample and glucose, such that 50% of the COD is provided by wastewater, 50% of the COD by glucose. The necessary amounts of nutrient buffer solutions are added to each reactor for nutritional requirements and in order to attain enough buffer capacity. The reactors are monitored for total and filtered COD and the pH of the reactors are controlled. The experiment is continued until all biodegradable COD is depleted and the biomass is mineralized. The experiment should be ended when subsequent samples have the same unchanging COD values.

Assessment of S_{I1} and f_{ES}: The approach proposed by Germirli et al. (1991) relies on the assumption that the reactors fed with soluble (filtered) wastewater sample and with glucose may be characterized with the same f_{ES} value. This assumption is biochemically sound as glucose is the central compound of almost all metabolic cycles of heterotrophic microorganisms, where residual metabolic products are generated in succeeding steps. It is also in compliance with the basic structure of the current activated sludge models, where all complex organic compounds are assumed to undergo hydrolysis and break down into simpler compounds like glucose, before they can be introduced into the microbial growth mechanism.

In the experiment, when all biodegradable substrates in the two reactors (filtered wastewater and glucose reactors) are depleted, the soluble COD concentration in the filtered wastewater reactor will be observed to drop from S_{T1} to $(S_T)_2$, where $(S_T)_2$ defines the sum of the initial soluble inert COD, S_{I1} and the COD reflecting soluble residual metabolic products, $(S_P)_2$ generated in the reactor. The soluble COD concentration in the glucose reactor will decrease from S_{G1} to $(S_G)_1$, where $(S_G)_1$ represents only the soluble residual COD, $(S_P)_G$. Then, the fraction of biodegradable COD converted into soluble inert metabolic products, Y_{SP} and f_{ES} may be estimated as given below:

$$Y_{SP} = f_{ES} Y_H = \frac{(S_P)_G}{S_{G1}} \tag{5.4}$$

$$f_{ES} = \frac{1}{Y_H} \frac{(S_P)_G}{S_{G1}} \tag{5.5}$$

The mass balance constructed for the final value of soluble COD of the filtered wastewater reactor can be interpreted as given below for the estimation of the soluble inert fraction.

$$(S_P)_2 = (S_T)_2 - S_{I1} = f_{ES} T_H (S_{T1} - S_{I1}) \tag{5.6}$$

S_{I1} may be calculated from the following expression:

$$S_{I1} = \frac{(S_T)_2 - f_{ES} Y_H S_{T1}}{(1 - f_{ES} Y_H)} \tag{5.7}$$

Assessment of f_{EX}: The experimental determination of f_{EX} is structured upon the concept that in an aerated batch reactor operated long enough for the completion of all biological activity, the level of the particulate metabolic products generated will be proportional to the amount of biodegradable substrate utilized. The reactor fed by the soluble wastewater sample is very convenient for this purpose, as $(X_P)_2$ may be expressed as follows, being the only particulate COD component at the end of the experiment:

$$(X_P)_2 = (C_T)_2 - (S_T)_2 = Y_{XP}(S_{T1} - S_{I1}) \tag{5.8}$$

where, Y_{XP} is the fraction of biodegradable COD converted into particulate inert metabolic products, and is defined as given below:

$$Y_{XP} = f_{EX} Y_H \tag{5.9}$$

f_{EX} may be calculated as follows, using the last two expressions:

$$f_{EX} = \frac{1}{Y_H} \frac{[(C_T)_2 - (S_T)_2]}{(S_{T1} - S_{I1})} \tag{5.10}$$

Assessment of X_{I1}: Results of the first reactor fed with the raw wastewater sample are used for the estimation of X_{I1}. The stoichiometric expressions that may be derived on the basis of experimental observations are listed below:

$$(X_T)_1 = (C_T)1 - (S_T)_1 = X_{I1} + (X_P)_1 \tag{5.11}$$

$$C_{S1} = C_{T1} - X_{I1} - S_{I1} \tag{5.12}$$

$$(X_P)_1 = f_{EX} Y_H (C_{T1} - X_{I1} - S_{I1}) \tag{5.13}$$

As S_{I1} and f_{EX} values associated with the wastewater tested are already ascertained, the corresponding value of X_{I1} may be computed as follows:

$$X_{I1} = \frac{(X_T)_1 - f_{EX} Y_H (C_{T1} - S_{I1})}{(1 - f_{EX} Y_H)} \tag{5.14}$$

The proposed procedure only requires the value of the heterotrophic yield coefficient, Y_H defined for the wastewater under investigation.

Example 5.2 Assessment of inert fractions using batch tests

Determine the inert COD fractions of a tannery wastewater sample according to the experimental results obtained by running three aerated batch reactors; the first one fed with raw wastewater sample, the second with soluble/filtered wastewater sample and the last one with a glucose solution approximately having the same COD content as the soluble wastewater reactor. Experimental results are presented in Table 5.3.

118   Industrial Wastewater Treatment by Activated Sludge

Table 5.3. Experimental results for a tannery wastewater sample

Time (day)	Raw wastewater reactor (1st Reactor)		Filtered wastewater reactor (2nd Reactor)		Glucose reactor (3rd Reactor)
	Total COD (mg/l)	Soluble COD (mg/l)	Total COD (mg/l)	Soluble COD (mg/l)	Soluble COD (mg/l)
0	1680	790	790	790	795
2	1352	419	606	419	93
6	1060	625	530	230	80
9	1030	300	526	320	90
14	891	229	502	206	70
20	845	205	365	205	50
23	815	256	370	186	23
27	792	256	361	117	37
29	808	280	396	210	33
33	796	225	387	114	23
35	700	225	405	205	25
37	700	225	405	205	25
41	700	225	360	205	25
55	700	225	360	205	25
69	700	225	360	205	25

(a) Assessment of S_{I1} and f_{ES}

COD results of the filtered wastewater reactor (2nd reactor) and the glucose reactor (3rd reactor), as illustrated in Figure 5.5, are required at this stage. The necessary experimental data are listed below:

S_{T1} = 790 mg/l
$(S_T)_2$ = 205 mg/l
S_{G1} = 795 mg/l
$(S_P)_G$ = 25 mg/l
Y_H = 0.64 g cell COD/gCOD

The value of f_{ES} will be obtained from the glucose reactor as below:

$$Y_{SP} = f_{ES} Y_H = \frac{(S_P)_G}{S_{G1}} = \frac{25}{795} = 0.032$$

$$f_{ES} = \frac{1}{Y_H} \frac{(S_P)_G}{S_{G1}} = \frac{1}{0.64} \frac{25}{795} = 0.05$$

Experimental assessment

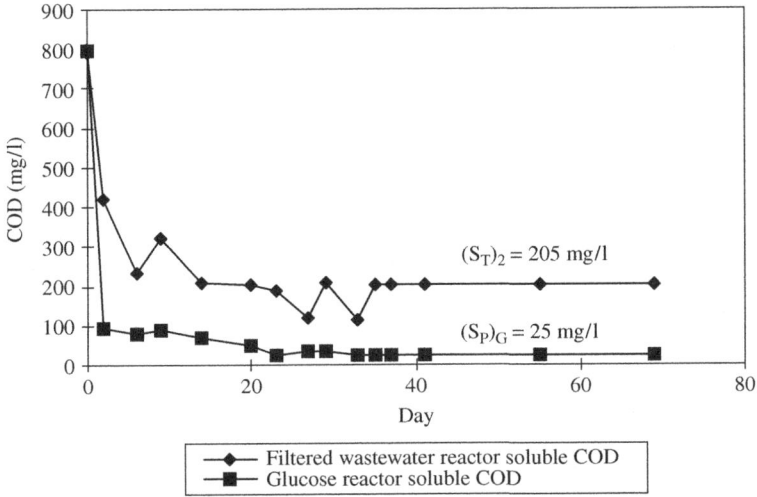

Figure 5.5. Soluble COD results obtained for filtered wastewater and glucose reactors

The value of S_{I1} can be estimated using the data obtained from the soluble wastewater reactor from equation (5.7);

$$S_{I1} = \frac{(S_T)_2 - f_{ES}Y_H S_{T1}}{(1-f_{ES}Y_H)} = \frac{205 - 0.05 \times 0.64 \times 790}{(1-0.05 \times 0.64)} = 185\, mg/l\, COD$$

(b) Assessment of f_{EX}

Results of the filtered wastewater reactor (2nd reactor), as illustrated in Figure 5.6, are required at this stage. The necessary experimental data are listed below:

$$S_{T1} = 790\, mg/l$$
$$(C_T)_2 = 360\, mg/l$$
$$(S_T)_2 = 205\, mg/l$$

The stoichiometric relationships for the evaluation of f_{EX} :

$$X_{P2} = (C_T)_2 - (S_T)_2 = 360 - 205 = 155\, mg/l\, COD$$
$$S_{S1} = S_{T1} - S_{I1} = 790 - 185 = 605\, mg/l\, COD$$

$$Y_{XP} = f_{EX}Y_H = \frac{(X_P)_2}{S_{S1}} = \frac{155}{605} = 0.256$$

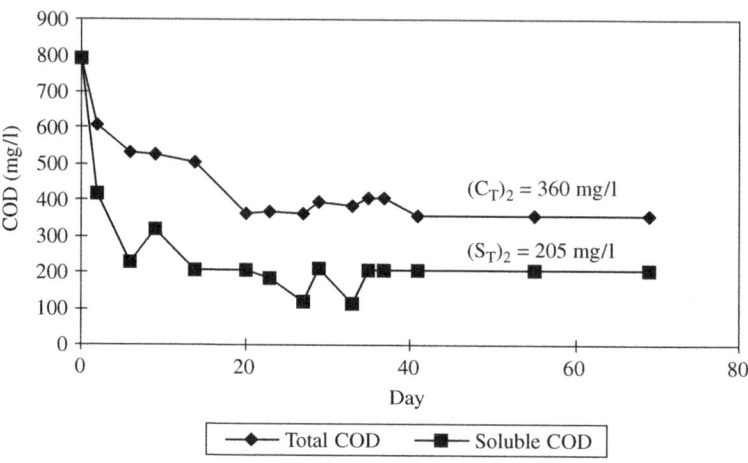

Figure 5.6. Total and soluble COD results obtained for filtered wastewater reactor

$$f_{EX} = \frac{1}{Y_H} \frac{[(C_T)_2 - (S_T)_2]}{(S_{T1} - S_{I1})} = \frac{1}{0.64} \frac{(360 - 205)}{(790 - 185)} = 0.40$$

(c) Assessment of X_{I1}

Results of the raw wastewater reactor (1st reactor), as illustrated in Figure 5.7, are required at this stage. The necessary experimental data are listed below:

$C_{T1} = 1680\,mg/l$
$S_{T1} = 790\,mg/l$
$(C_T)_1 = 700\,mg/l$
$(S_T)_1 = 225\,mg/l$

The stoichiometric relationships for the evaluation of X_{I1}:

$$(X_T)_1 = (C_T)_1 - (S_T)_1 = X_{I1} + (X_P)_1$$
$$= 700 - 225 = 475\,mg/l\,COD$$
$$C_{S1} = C_{T1} - X_{I1} - S_{I1}$$
$$(X_P)_1 = f_{EX} Y_H (C_{T1} - X_{I1} - S_{I1})$$

$$X_{I1} = \frac{(X_T)_1 - f_{EX} Y_H (C_{T1} - S_{I1})}{1 - f_{EX} Y_H}$$

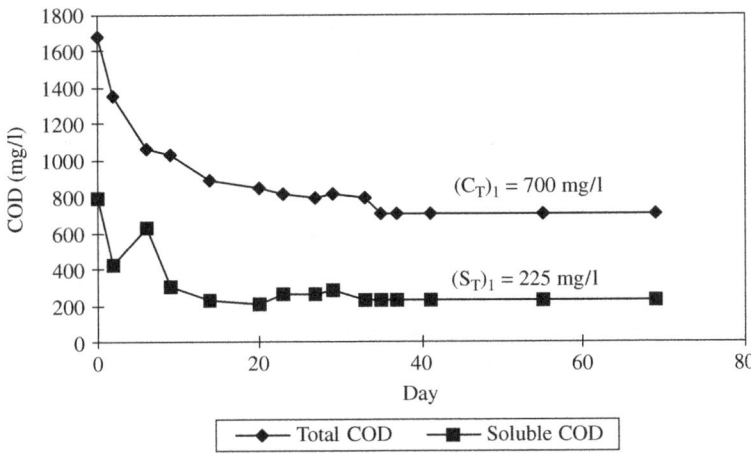

Figure 5.7. Total and soluble COD results obtained for raw wastewater reactor

$$= \frac{475 - 0.40 \times 0.64 \times (1680 - 185)}{(1 - 0.40 \times 0.64)} = 125 \, mg/l \, COD$$

5.3.1.2 Estimation of Inert Fractions by Respirometry

Orhon and Okutman (2003) have also defined a respirometric method to assess soluble and particulate inert COD components, S_{I1} and X_{I1}. The method involves a sequence of OUR measurements on raw and filtered (soluble) wastewater samples and a simple mass balance is constructed to estimate soluble and particulate inert fractions. The OUR profile generated in an aerated batch reactor initially fed with a pre-selected ratio of wastewater and biomass is used as a tool for the equivalence between the biodegradable COD fractions and the corresponding dissolved oxygen utilization (Henze et al., 1987; Ubay Cokgor et al., 1998). Since the area delineated by the OUR curve above the endogenous respiration level directly gives the amount of oxygen consumed at the expense of all available substrate, the biodegradable COD in the wastewater can be calculated from the fundamental mass balance expression given in equation (5.3) between substrate utilization, biomass growth and electron acceptor consumption:

$$C_S = C_{SR} \frac{V_R}{V_{WW}} = \frac{\Delta O_2}{1 - Y_H} \frac{V_T}{V_{WW}} \quad (5.15)$$

The total amount of biodegradable COD can be calculated provided that the heterotrophic yield coefficient (Y_H) is known. The calculation shows that the obtained value for the biodegradable COD fraction (C_{SR}) actually present in the wastewater sample used in the experiment (V_{WW}) has to be corrected using the total volume of the reactor vessel ($V_R = V_T$), since the wastewater sample is diluted with the activated sludge biomass and nutrient solutions added to the reactor in the respirometric test.

In this context, the OUR curves obtained from the reactors fed with the raw and filtered wastewater samples yield the total and soluble biodegradable COD fractions. The soluble and particulate inert COD fractions may then be calculated from mass balance using the initial total (C_{T1}) and soluble (S_{T1}) COD concentrations in the wastewater as given below:

$$S_{I1} = S_{T1} - C_{S,\text{Soluble}} \qquad (5.16)$$

$$X_{I1} = C_{T1} - C_{S,\text{Total}} - X_{H1} - S_{I1} \qquad (5.17)$$

where, X_{H1} is the COD equivalent of the heterotrophic biomass present in the wastewater sample.

The concentration of biomass present in the wastewater sample can be estimated by conducting an extra OUR measurement without additional biomass seeding and estimating the biomass concentration via modelling. In most industrial effluents however, X_{H1} can be neglected since no biomass is expected to be present in most of these wastewaters.

Example 5.3 Assessment of inert fractions using respirometry
Raw and filtered wastewater samples are subjected to respirometric tests, where the initial F/M ratios of the tests are adjusted to be 0.06 g COD/g VSS. The effluent is characterized with a total COD of 520 mg/l, a soluble portion of 130 mg/l after filtration. The volume of the filtered wastewater is 0.75 l and the total volume of the reactor is 2.77 l. The volume of the raw wastewater sample is selected as 0.3 l wastewater and the total reactor volume as 2.32 l, in order to maintain the same initial F/M ratio as the test conducted with the filtered effluent.

The amount of oxygen consumed in the test conducted with filtered wastewater is calculated (the area under the OUR curve) as 8.4 mg O_2/l and that of the test conducted with raw wastewater is found as 13.4 mg O_2/l from the areas under the OUR curve, above the endogenous level as shown in Figure 5.8.

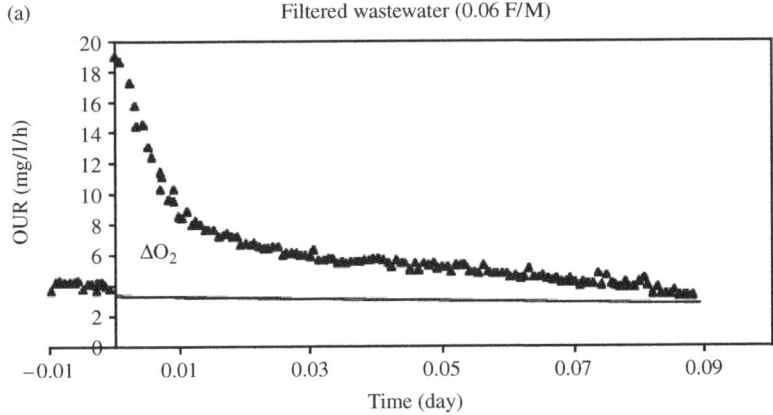

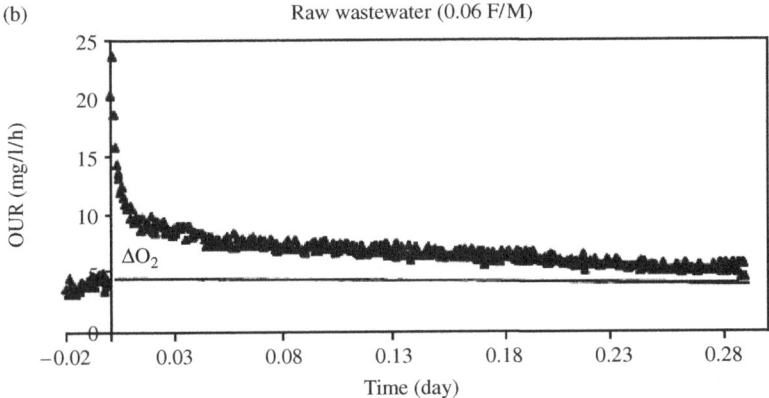

Figure 5.8. Respirometric test conducted with (a) filtered wastewater sample, and (b) raw wastewater sample (Orhon and Okutman, 2003)

Assuming that Y_H value was previously computed as 0.67 g cell COD/g COD for the same wastewater (Orhon et al., 2002; Orhon and Okutman, 2003) and the value of $X_{HI} = 95$ mg cell COD/l (Orhon et al., 2002), find;
  (a) Total and soluble biodegradable COD fractions, and
  (b) Soluble and particulate inert COD fractions.

(a) The amount of biodegradable COD present in the reactor fed with filtered wastewater, $C_{SR,filtered}$, can be estimated as follows:.

$$C_{SR,filtered} = \frac{\Delta O_2}{(1 - Y_H)} = \frac{8.4}{(1 - 0.67)} = 25.5 \, mg/l$$

The volume correction factor for the reactor fed with filtered wastewater is:

$$\frac{V_R}{V_{ww}} = \frac{2.77}{0.75} = 3.69$$

The biodegradable COD of filtered wastewater can be found as below:

$$C_{S,filtered} = C_{SR,filtered} \frac{V_R}{V_{ww}} = 25.5 \times 3.69 = 94\,mg/l$$

Similarly, the biodegradable fraction of COD for raw wastewater is:

$$C_{SR,raw} = \frac{\Delta O_2}{(1-Y_H)} = \frac{13.4}{(1-0.67)} = 40.6\,mg/l$$

The volume correction factor for raw wastewater reactor is:

$$\frac{V_R}{V_{ww}} = \frac{2.32}{0.3} = 7.73$$

The biodegradable COD of raw wastewater is then found as:

$$C_{S,raw} = C_{SR,raw} \frac{V_R}{V_{ww}} = 40.6 \times 7.73 = 314\,mg/l$$

(b) The respirometric experiments have revealed that soluble biodegradable COD ($C_{S,Soluble}$) was 94 mg/l and the total biodegradable COD fraction ($C_{S,Total}$) was 314 mg/l. The soluble inert COD fraction in the domestic sewage sample, S_{I1} can then be calculated as:

$$S_{I1} = 130 - 94 = 36\,mg/l$$

Particulate inert COD fraction, X_{I1} can be estimated as given below:

$$X_{I1} = 520 - 314 - 36 - 95 = 75\,mg/l$$

5.3.2 Assessment of biodegradable fractions

Readily biodegradable substrate S_S, is determined in accordance with the method suggested by Dold et al. (1986). The method involves running an aerated batch reactor, seeded with biomass, fed with wastewater. OUR profile is monitored in the reactor. The high OUR level in the beginning of the experiment after feed addition is dictated by the utilization of readily biodegradable COD present in the wastewater (S_{S1}) and the hydrolyzed portion of slowly biodegradable COD (X_{S1}). The OUR profile drops to a second plateau as initial

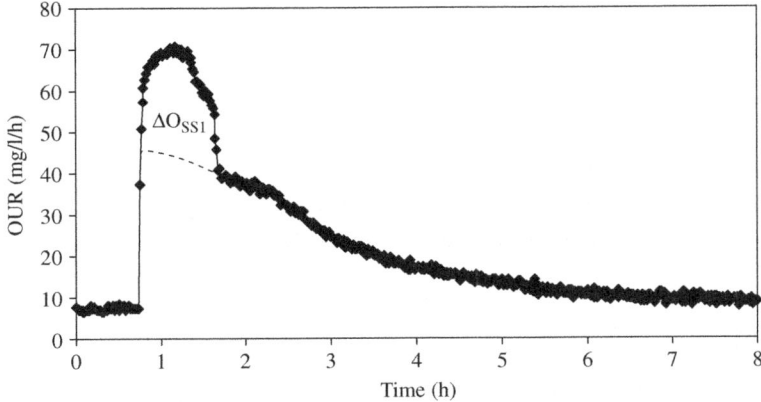

Figure 5.9. Estimation of the graphical area to calculate the amount of utilized for S_{S1}

readily biodegradable COD (S_{S1}) is depleted. The second plateau represents the OUR due to the readily biodegradable COD generated by the breakdown of slowly biodegradable COD (X_{S1}). The oxygen consumed for the utilization of S_{S1} is determined by the graphical interpretation of the OUR profile as the area of the region above the second plateau as shown in Figure 5.9.

The critical parameter for conducting the respirometric experiment is the initial F/M ratio. If F/M is too high, it may not be possible to depict the initial high OUR level, due to instrumental insufficiencies. If the F/M ratio is too low, the difference between the first and the second plateaus may not be detected easily. However, it has been noted that the initial F/M ratio should be selected as low as possible (around 0.1–0.2 g COD/g VSS) in order to get the best dynamic response in the respirometric assessment (Insel et al., 2003).

The initial F/M ratio is defined as:

$$\frac{F}{M} = \frac{C_{T1}}{X_T} \frac{V_{ww}}{V_{sludge}} \qquad (5.18)$$

where;
V_{ww} = volume of wastewater sample added (l)
V_{sludge} = volume of activated sludge seeded (l)
C_{T1} = total COD concentration of the wastewater sample (mg/l)
X_T = VSS concentration in the mixed liquor (mg/l)

The readily biodegradable fraction will be obtained using the oxygen utilized in the test as follows:

$$S_{S1} = \frac{\Delta O_{SS1}}{1 - Y_H} \frac{V_T}{V_{ww}} \qquad (5.19)$$

where;

ΔO_{SS1} = oxygen utilized for the degradation of S_{S1} (mg/l)
V_T = total volume of the reactor vessel (l)

5.4 EXPERIMENTAL ASSESSMENT OF MODEL COEFFICIENTS

Currently basic respirometric techniques are widely used also for the experimental assessment of individual stoichiometic and kinetic coefficients such as Y_H, $\hat{\mu}_H$ and b_H. Another significant model component, namely the active heterotrophic biomass, X_H, is also defined by similar techniques. The experimental assessment of hydrolysis coefficients k_h and K_X is not possible by direct experimental measurements but requires model calibration of the experimental OUR data, provided that necessary information on wastewater characterization and other kinetic and stoichiometric coefficients are previously obtained.

5.4.1 Assessment of yield coefficient

5.4.1.1 Heterotrophic growth yield (Y_H)

The respirometric method suggested by Dold et al. (1986) for the determination of readily biodegradable substrate S_{S1}, can also be used for the assessment of the heterotrophic yield coefficient, Y_H. However, accurate evaluation of the experimental OUR data for the determination of Y_H necessitates the identification of the relevant mechanism for substrate utilization and application of the appropriate mechanistic model. It is stated that substrate utilization may be due to direct microbial growth (Henze et al., 1987) or storage (Krishna and van Loosdrecht, 1999), depending on the operating conditions applied in the experimental setup. In the case where direct growth applies, the OUR profile will yield the amount of oxygen utilized for the direct growth of heterotrophic biomass and thus, the heterotrophic growth yield coefficient, Y_H (Sollfrank and Gujer 1991; Ubay Cokgor et al., 1998). This technique has been extensively used for domestic sewage and industrial wastewaters (Orhon et al., 1995; 1999b).

Heterotrophic yield coefficient Y_H, can be determined employing the method suggested by Ubay Cokgor et al., (1998) as given below:

$$Y_H = 1 - \frac{\Delta O_{SS} + \Delta O_{XS}}{\Delta COD} \qquad (5.20)$$

An example for the experimental determination of the heterotrophic yield coefficient is presented in Figure 5.10. The experiment has been conducted on tannery wastewaters with an F/M ratio employed as 0.2 g COD/g VSS. For the assessment of Y_H, oxygen utilization was monitored together with the soluble COD measurements in the reactor. The amount of oxygen utilized for growth on initial readily biodegradable substrate present in wastewater (ΔO_{SS}) and on the hydrolysis products of slowly biodegradable substrate (ΔO_{XS}) is determined by the area under the first plateau of OUR curve and above endogenous OUR level. Heterotrophic yield coefficient, Y_H has been calculated as 0.64 g cell COD/g COD.

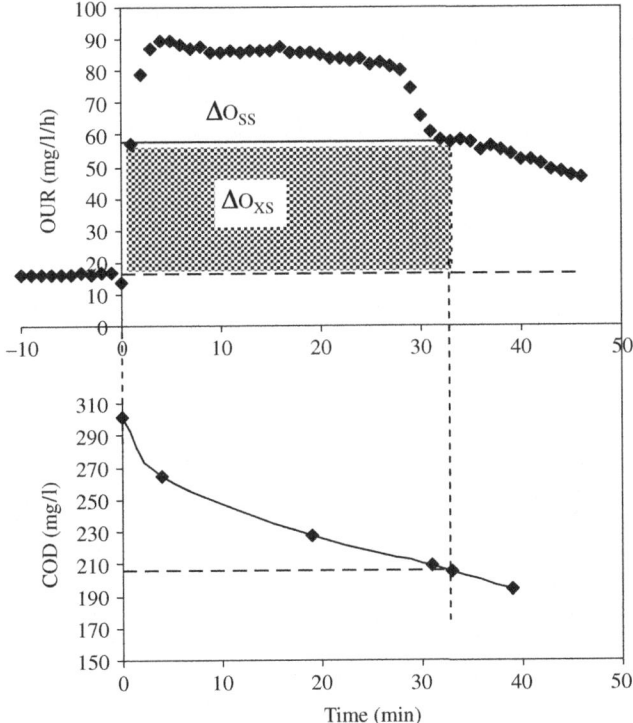

Figure 5.10. Assessment of Y_H using OUR and Soluble COD profiles of the repirometric experiment conducted with tannery wastewater (Tunay et al., 2004)

5.4.1.2 Substrate storage yield (Y_{STO})

A similar OUR profile may result from the conversion of biodegradable substrate to internal storage material and may be used for the computation of the storage yield, Y_{STO} (Karahan et al., 2002a). ASM3 (Gujer et al., 2000) is the first multi-component model incorporating substrate storage into activated sludge modelling. The basic stoichiometry of the substrate storage process can be expressed as follows:

$$\Delta O_{STO} = (1 - Y_{STO})\Delta S_{S1} \quad (5.21)$$

At the depletion of all initially available readily biodegradable COD, (S_{S1}), the above expression can be manipulated to give the storage yield, Y_{STO}:

$$Y_{STO} = \left(1 - \frac{\Delta O_{STO}}{S_{S1}}\right) \quad (5.22)$$

As shown in equation (5.22), it is possible to estimate the storage yield, Y_{STO}, as a function of the oxygen utilized for storage. The storage yield, Y_{STO}, can be computed if the oxygen used for the storage of a known amount of readily biodegradable COD (ΔO_{STO}) can be determined by means of respirometric measurements. However, it is not possible to determine the amount of oxygen used for storage individually from respirometric batch tests and modelling has been used to identify the fraction of consumed oxygen solely for substrate storage (Karahan et al., 2002b).

In order to estimate the amount of oxygen utilized for substrate storage the OUR profile can be fractionated, as shown in Figure 5.11(a), for each separate process incorporated in ASM3. It is also possible, through model simulation to subtract the OUR values of all three processes except that of storage from the overall OUR profile, to get the graphical magnitude of the area representing storage, as indicated in Figure 5.11(b).

As described in the mentioned figure, it is possible to determine the amount of ΔO_{STO} graphically without model simulation. This fact can be considered as a big advantage when evaluating experimental OUR data, since modelling would necessitate the determination of a number of kinetic and stoichiometric parameters. The area under OUR curve that will graphically represent the amount of oxygen utilized for storage can be depicted by drawing a line connecting the initial OUR level due to endogenous decay (the initial/average OUR_{EndDec} level), up to the inflection point of the overall OUR curve, as shown in Figure 5.12. The inflection point on the OUR profile reflects the depletion of the initially available readily biodegradable COD and therefore, the end point of storage of the substrate. This is also confirmed by the sudden drop observed in

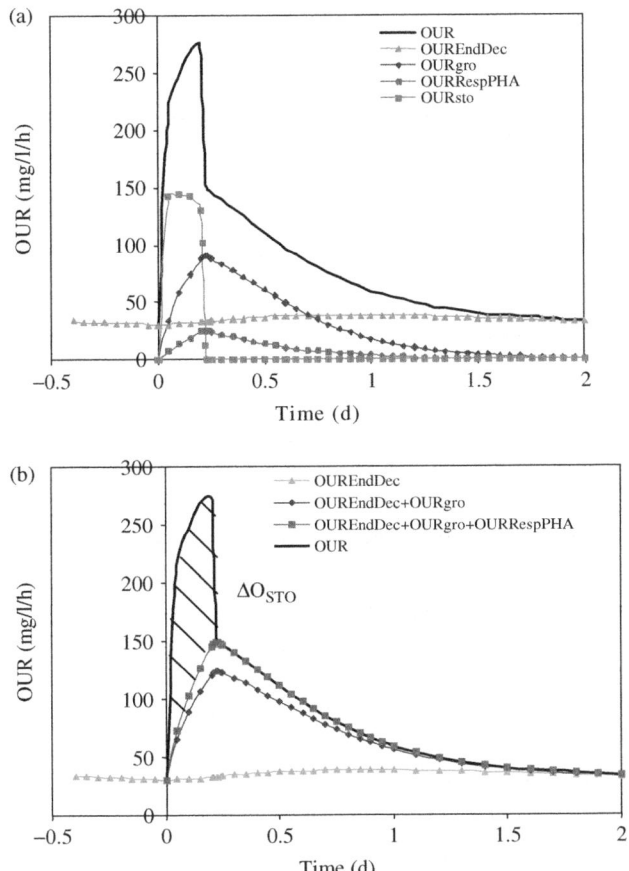

Figure 5.11. (a) Components of the OUR curve (b) Cumulative OUR curves for the identification of ΔO_{STO}

the OUR value which can be incorporated with the completion of external substrate utilization through storage. The area above the drawn line directly yields oxygen used for storage, ΔO_{STO}.

Karahan *et al.* (2002b) have calculated the Y_{STO} values for acetate and glucose as 0.78 and 0.90 g COD/g COD, respectively, similar to the values estimated using biochemical models (Beun *et al.*, 2000; Dircks *et al.*, 2000). The OUR profile obtained for acetate and the area determined for the storage of acetate is presented in Figure 5.13.

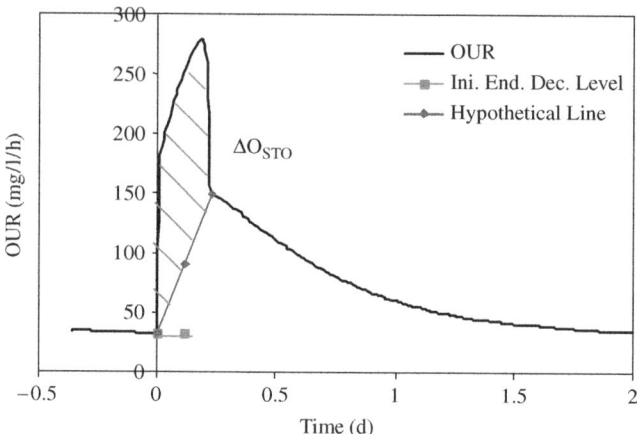

Figure 5.12. Graphical representation of the area for the determination of ΔO_{STO}

5.4.2 Assessment of endogenous decay coefficient (b_H)

The traditional modelling defined the endogenous decay coefficient as a lump sum term, k_d, from the linear relationship between the sludge age and the specific substrate removal rate (Pearson, 1966). This approach is, obviously only capable of determining an apparent decay coefficient of around $k_d = 0.05$ d^{-1}, lower than the true b_H value and the underestimation becomes more pronounced

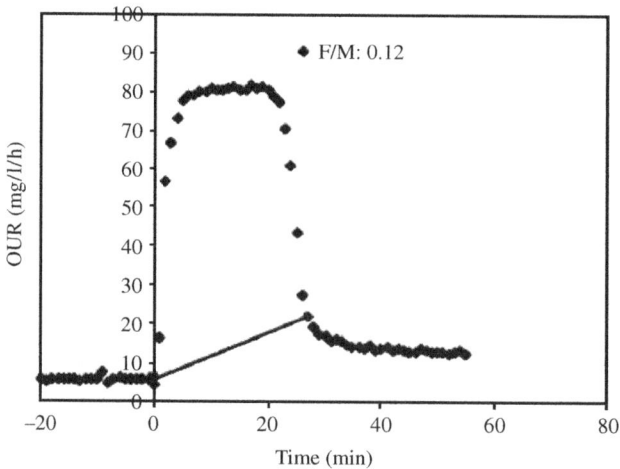

Figure 5.13. OUR profiles for the storage of acetate (Karahan et al., 2002b)

as the sludge age increases. Since k_d has been presented as a function of sludge age a wide range of k_d values have been reported in the literature.

In the current modelling approach, the active biomass and the inert organic particulate matter have been differentiated and the experimental assessment of b_H is conducted with the use of respirometry. In fact, the simply defined endogenous decay is actually composed of different mechanisms such as maintenance energy requirements, decay of cells, endogenous respiration, grazing by higher animals, toxic substances or adverse conditions that result in cell lysis (van Loosdrecht and Henze, 1999).

Ekama et al. (1986) have proposed a widely recognized and routinely used method for the determination of endogenous decay rate, b_H, which involves monitoring the changes in the OUR profile of a continuously aerated activated sludge sample (aerobic digestion) with no external substrate addition for several days, until 10% of initial OUR is reached. The slope of the plot of *ln OUR* versus time yields the value of b_H, as shown in the following set of equations. The active heterotrophic biomass concentration at time t in the batch test can be expressed in terms of the initial heterotrophic biomass concentration as:

$$X_H = X_{H0} e^{-b_H t} \qquad (5.23)$$

OUR obtained from the aerobic digester is given as:

$$OUR_t = (1 - f_E) b_H X_H \qquad (5.24)$$

where, $f_E = f_{ES} + f_{EX}$ \qquad (5.25)

Substituting the value of X_H given in equation (5.23) into the OUR expression presented in equation (5.24) and rearranging the equation, *ln OUR* will be driven as a function of b_H as shown below:

$$\ln OUR_t = \ln[(1 - f_E) b_H X_{H0}] - b_H t \qquad (5.26)$$

Thus, the first part of the above equation gives the initial *ln OUR* and the slope of the plot of *ln OUR* versus time is estimated as endogenous decay rate, b_H. The experimental OUR data is used for the determination of b_H, as shown in Figure 5.14, where the endogenous decay rate, b_H, is estimated as 0.21 d^{-1} for tannery effluents.

Avcioglu et al. (1998) have defined another respirometric method which is based on measuring changes in the growth response, in other words the change in the OUR value dictated by $\mu_H X_H$. This method enables monitoring the variation of X_H with time by always sustaining a constant μ_H level during the course of the experiment. Avcioglu et al. (1998) have obtained a significantly lower b_H value of around 0.1 d^{-1} at 20°C for domestic sewage.

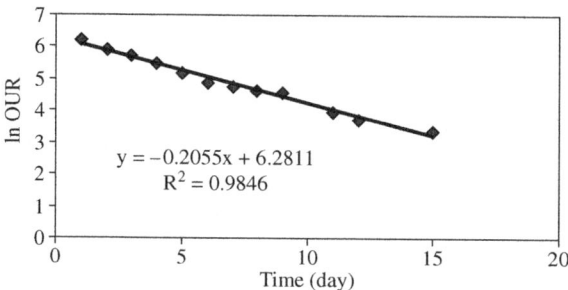

Figure 5.14. Experiments for the determination of b_H tannery wastewater (Tunay et al., 2004)

The procedure developed by Avcioglu et al. (1998) allows the estimation of b_H independent from X_H concentration. The procedure is based on batch OUR experiments conducted with endogenous biomass sampled from a mother aerated reactor at different time intervals. The method is based on the well established approach that the initial OUR in the batch reactor fed with sufficient amount of substrate for maximum growth conditions is given by the equation below:

$$OUR_1(t) = \left[\frac{1-Y_H}{Y_H}\hat{\mu}_H + (1-f_E)b_H\right]X_H \qquad (5.27)$$

Since the heterotrophic biomass can be defined as given in equation (5.23), the natural logarithm of $OUR_1(t)$ can be used to determine the endogenous decay coefficient, b_H, from the slope of the $ln\ OUR_1$ values plotted with respect to time, as dictated by the equation below:

$$\ln OUR_1(t) = \ln\left[\frac{1-Y_H}{Y_H}\hat{\mu}_H + (1-f_E)b_H\right]X_{H0} - b_H t \qquad (5.28)$$

Thus, the endogenous decay parameter (b_H) can be estimated based on the decrease in initial OUR levels with time (T = 0–7 days) as seen in Figure 5.15. The estimated b_H values for domestic sewage under aerobic conditions were found in the range of 0.08–0.11 day^{-1} (Avcioglu et al., 1998).

The method proposed by Avcioglu et al. (1998) provides a more accurate estimation of endogenous decay coefficient since it also reflects the activity loss of heterotrophic biomass.

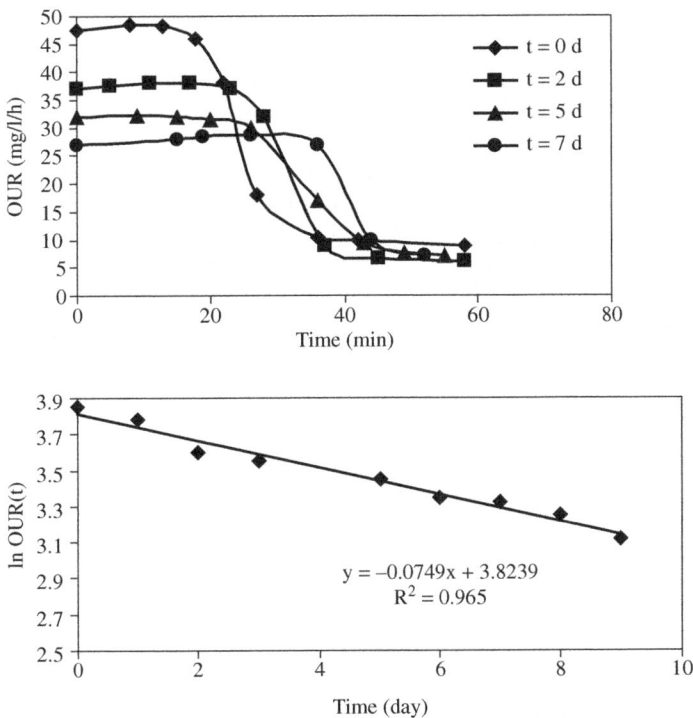

Figure 5.15. Endogenous decay rate estimation from batch OUR profiles (Avcioglu et al., 1998)

5.4.3 Assessment of maximum heterotrophic growth coefficient $(\hat{\mu}_H)$

The traditional methods using linearized Monod expression where the experiments are conducted by monitoring total COD and VSS data posses inherent difficulties which lead to consistently low $\hat{\mu}_H$ values. The use of respirometric techniques has been proposed in accordance with the structure of the new models, based upon the interpretation of the maximum OUR observed during a period with no substrate limitation. The procedure suggested by Ekama et al. (1986) using an aerated batch reactor started at a suitably low initial F/M ratio that would enable the detection of the initial OUR plateau. The procedure estimated the growth rate based on the assumption that the OUR level observed at the plateau was proportional to $\hat{\mu}_H X_H$ and X_H was determined by a tedious procedure. The procedure was subject to errors caused by the possible interferences imposed on the OUR value due to the neglected endogenous decay process at low F/M ratios recommended for the test. A similar test has been suggested by Kappelar and Gujer (1992) with a high F/M ratio, where an ascending OUR profile is observed. The OUR data obtained was used to estimate $\hat{\mu}_H$, with the assumption that maximum growth will be sustained as long as the readily biodegradable substrate concentration remains sufficiently high in the reactor. Thus, the evaluation procedure does not necessitate the determination of the initial level of active biomass, X_H, present in the reactor.

The heterotrophic growth rate, $\hat{\mu}_H$ can be obtained using heterotrophic biomass concentration at any time t as given by the expression presented in equation (5.23). Thus, the initial OUR at time t may be expressed as:

$$OUR(t) = \left[\frac{1-Y_H}{Y_H}\hat{\mu}_H + (1-f_E)b_H\right]X_H$$

$$= \left[\frac{1-Y_H}{Y_H}\hat{\mu}_H + (1-f_E)b_H\right]X_{H0}e^{(\hat{\mu}_H-b_H)t} \quad (5.29)$$

If initial OUR at time t is divided by initial OUR at time t_0, then this will lead to the following expression:

$$\frac{OUR(t)}{OUR(t_0)} = e^{(\hat{\mu}_H-b_H)t} \quad (5.30)$$

Based on the above expression the plot of $ln\ (OUR_t/OUR_0)$ will be used to estimate $\hat{\mu}_H$ as given below:

$$\frac{OUR_t}{OUR_0} = \hat{\mu}_H - b_H \quad (5.31)$$

An example for the experimental determination of maximum growth rate $\hat{\mu}_H$ in accordance with the method suggested by Kappaler and Gujer (1992) is given in Figure 5.16. For the assessment of $\hat{\mu}_H$, a batch respirometric test was conducted where the F/M ratio was selected as 4 g COD/g VSS for tannery wastewaters. The slope of the $ln(OUR_t/OUR_0)$ corresponds to $(\hat{\mu}_H - b_H)$ and $\hat{\mu}_H$ is determined with the assumption that b_H is 5% of the $\hat{\mu}_H$ value. Maximum growth rate values were determined as 2.59 d^{-1} (Tunay et al., 2004).

5.4.4 Assessment of k_h and K_X

Since the slowly biodegradable COD fraction (X_{S1}) is present at high amounts in both domestic and industrial effluents, hydrolysis of the slowly biodegradable substrate, X_{S1} is one of the most important steps of activated sludge modelling.

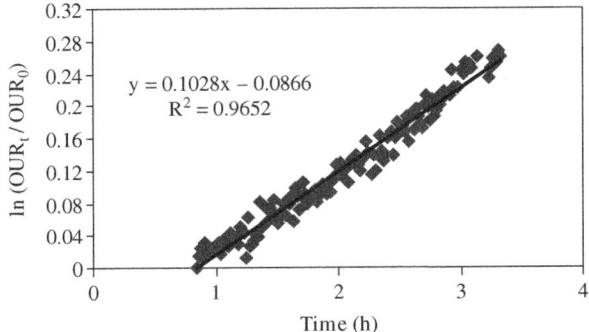

Figure 5.16. Assessment of $\hat{\mu}_H$ from batch respirometric test for tannery wastewaters (Tunay *et al.*, 2004)

The hydrolysis mechanism is defined by means of surface-limited reaction kinetics, with k_h and K_X as the rate coefficients. It is not possible to determine these coefficients by direct experimental procedures but their assessment can be achieved by model calibration of an OUR profile for the most appropriate values of these coefficients. The model is simulated with known values of COD fractions and kinetic coefficients, either defined by direct experiments or adopted by experience prior to curve fitting with the selected OUR profile. Orhon *et al.* (1999c) argued that it might be difficult and sometimes misleading to characterize the entire slowly biodegradable COD fraction by a single hydrolysis rate and suggested a dual hydrolysis mechanism for rapidly hydrolysable COD, S_{H1} and slowly hydrolysable COD, X_{S1} as different model components, as shown in Figure 5.17.

5.5 ASSESSMENT OF INHIBITION AND TOXICITY

Industrial wastewaters are greatly complicated by the presence of biologically resistant substances, such as heavy metals, persistent organic compounds, etc. These substances are commonly called inhibitors. Inhibitors interfere and impair the performance of biological treatment systems. The presence of inhibitors cause reduction removal of organic carbon, impairment of solid separation and modification of sludge compacting properties for the operation of activated sludge systems (Wong *et al.*, 1997). As commonly observed inhibitors, heavy metals such as nickel, chromium, etc may be present in some industrial wastewaters. The mechanisms by which heavy metals affect biological processes

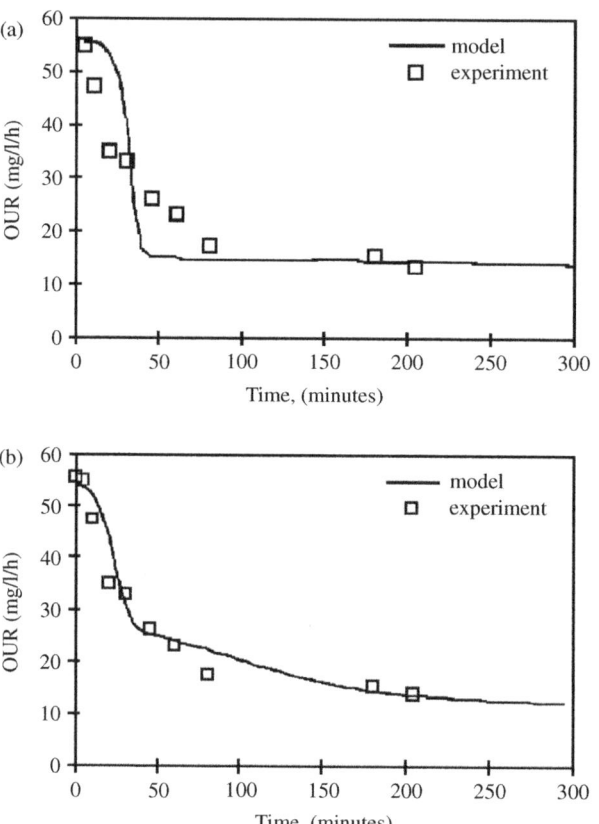

Figure 5.17. Calibration of the OUR data for domestic sewage using (a) a single hydrolysis model (b) a dual hydrolysis model (Orhon et al., 1999c)

are not fully defined but relatively low concentrations of various heavy metals may stimulate activated sludge while increased doses lead to partial or total impairment of system performance (Bagby and Sherrard 1981; Sujarittanota and Sherrard, 1981; Yetis et al., 1999).

Inhibition effects imposed on the activated sludge systems should be determined with simple experimental procedures; however these procedures should be sensitive and relevant to the activated sludge population. The applied experimental procedure should yield a numerical index that would identify the inhibitory effect with an acceptable accuracy and that would be used to set a threshold value for the prevention of inhibition in activated sludge systems.

There are a number of different assays to screen presence and effects of toxicants and inhibitors in wastewater, such as chemical and microscopic analyses; measurement of the inhibition of growth and viability of bacterial cells; respirometric procedures; and bioluminescent analyses (Kelly *et al.*, 2004; Ricco *et al.*, 2004). However, these assays generate markedly different indices and therefore the application of appropriate threshold values for the tested inhibitors cannot be determined trivially. These different index values may also result in stringent and overprotective limits. Respirometric approaches have also been used in the assessment of toxicity and inhibition, since it has been observed that the OUR response of activated sludge systems decreases when the wastewater contains toxicants or inhibitors (Albek *et al.*, 1997; Kelly *et al.*, 1997).

A well adapted respirometric procedure for the determination of inhibitory effects imposed on activated sludge is the *ISO 8192 Test* (ISO 8192, 2007). The method describes the experimental procedure for obtaining the EC_{50} value which defines the effective concentration of the inhibitor inducing 50% reduction in the oxygen uptake rate.

The ISO 8192 procedure is based on the measurement of OUR under defined conditions in the presence of a defined biodegradable substrate and different concentrations of the test substance in order to assess the inhibitory effects of a test substance on oxygen consumption of activated sludge. These inhibitory effects can be determined for the OUR responses of heterotrophs in carbon removal, for the OUR response of nitrification, and for the OUR response of the whole activated sludge community. The evaluation of the experimental results obtained by the method is performed by calculating the inhibition effect of a test substance at a particular concentration using the percent inhibition of oxygen consumption (I), as given in equation (5.32).

$$I(\%) = \left[\frac{R_B - (R_T - R_{PC})}{R_B}\right] \times 100 \qquad (5.32)$$

where;
   R_T = oxygen consumption rate by test mixture
   R_B = oxygen consumption rate by the blank control
   R_{PC} = oxygen consumption rate by the physicochemical control

When the percent inhibitions (I values) of the known amounts of test material is estimated, the concentration causing 50% (EC_{50}) reduction in the OUR value with respect to blank control (no inhibitor) can be calculated by graphical interpolation as described in ISO 8192 method. The blank control is initiated with the same amounts of activated sludge seed and biodegradable substrate feed as the test dilutions and without any inhibitor addition. The physicochemical control is operated to identify oxygen consumption due to physicochemical phenomena which is done by the aeration of the test substrate and substrate medium, without the addition of activated sludge. ISO 8192 dictates that the EC_{50} values should be determined using the individual OUR values obtained at 30 or 180 minutes after the start of the test (ISO8192, 2007).

Cokgor et al. (2007) have used ISO 8192 procedure for the assessment of metal inhibition using different selected substrates and biomass previously acclimated to the same substrate adjusted to initial concentrations of 400 mg COD/l and 1700 mg VSS/l respectively, corresponding to an initial food to microorganism (S_0/X_0) ratio of around 0.24 mg COD/mg VSS. The EC_{50} levels were determined after 30 minutes of reaction time and were calculated as 33 mg/l for Ni(II) and 60 mg/l for Cr(VI) with peptone-meat extract substrate at 20°C, as given in Figures 5.18 and 5.19.

Although ISO 8192 is a well defined and commonly applied test the assessment of EC_{50} values at single points like 30 and 180 minutes cause discrepancies. Since the OUR profile changes with changing substrate and the different biochemical processes that can take part at different times during experimentation the comparison of OUR values obtained at predefined times might be misleading for the accurate interpretation of the inhibition effects imposed by the test materials.

Figure 5.20 reveals a very good example for the different OUR responses of two systems fed with peptone-meat extract substrate and with and without inhibitor addition for Ni(II). As Figure 5.20 shows the percent inhibition (I)

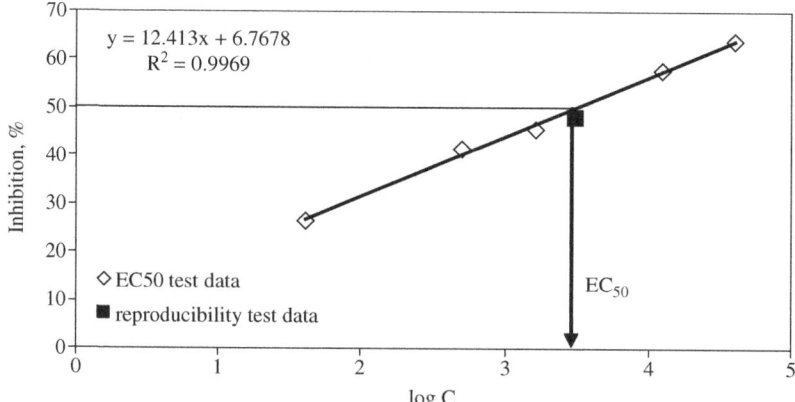

Figure 5.18. Results of the EC_{50} test for nickel using peptone-meat extract substrate (Cokgor et al., 2007)

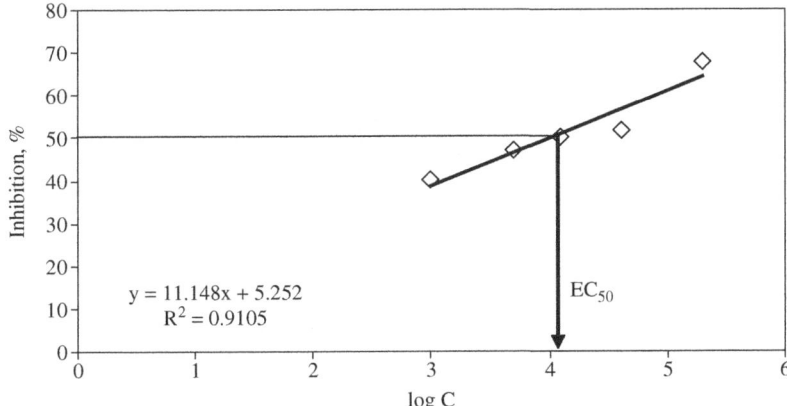

Figure 5.19. Results of the EC_{50} test for hexavalent chromium using peptone-meat extract substrate (Cokgor *et al.*, 2007)

changes with respect to time or in other words with respect to the predominant biochemical processes occurring in the activated sludge system.

The accurate evaluation and understanding of inhibition can only be possible with the description of inhibition both physically and mathematically, in terms of simple parameters that could be incorporated into process rate and mass

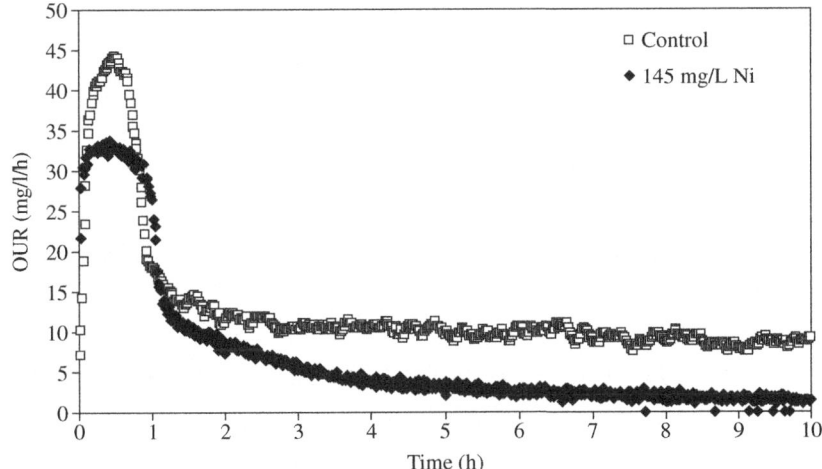

Figure 5.20. OUR profiles obtained for starch-acetate mixture and 145 mg/l Ni(II) addition (Cokgor *et al.*, 2007)

balance expressions defining the biochemical conversions taking place in the activated sludge process. This is only possible by determining the inhibitory effect that is reflected upon the stoichiometric and kinetic coefficients associated with microbial processes. Thus, use of curve fitting techniques to batch respirograms according to the appropriate mechanistic model and experimental assessment of relevant model coefficients are stated to be useful and meaningful for the evaluation of inhibition of activated sludge (Insel et al., 2006).

Figure 5.21 presents the OUR profiles for control and for the individual ASM1 processes where the reaction rates are reduced by 30 and 50% with the rate expressions altered by the introduction of the inhibition expression $K_I/(I + K_I)$ at values of 0.3 and 0.5, respectively. Modeling results have shown that the overall inhibition coefficient of $K_I/(I + K_I)=0.5$ does not necessarily result in 50% reduction in the OUR response of the system, i.e., when hydrolysis is inhibited, 50% OUR reduction can only be obtained at the 40th minute of the experiment when the OUR of growth is reduced only reduced by 40%.

The analogy of the above investigation has also been carried out for ASM3 and similar results have been obtained as given in Figure 5.22. The study of Insel et al. (2006) has clearly shown that 50% reduction in the OUR can be obtained at several times during the test which cannot be predicted without additional information on the stoichiometry and kinetics applicable to the specific experimental conditions.

The results of the study conducted by Insel et al. (2006) has revealed that model interpretation of the OUR profile sets an appropriate basis for the evaluation of inhibition for activated sludge. Modeling may be used to (i) identify the biochemical mechanisms responsible for substrate utilization; (ii) monitor the inhibition effect on individual microbial processes, and (iii) evaluate the level of the impact of inhibition numerically by assessing the change in the values of model coefficients obtained with or without the inhibitor. Modeling is also useful for monitoring the overall impact of the inhibitory compound on every stage of the activated sludge process. However, selection of the appropriate modeling tools, considering different biochemical processes that would be predominant in the system is the crucial issue for the evaluation of inhibition especially when substrates with different biodegradation properties are concerned (Insel et al., 2006).

5.6 MODEL EVALUATION OF EXPERIMENTAL DATA

Activated sludge models are generally regarded as over-parameterized and it is necessary to reduce the order of the model or to simplify with appropriate applications of mathematical techniques. The correct information on the kinetics can be achieved by the splitting of complete experimental data sets in terms of

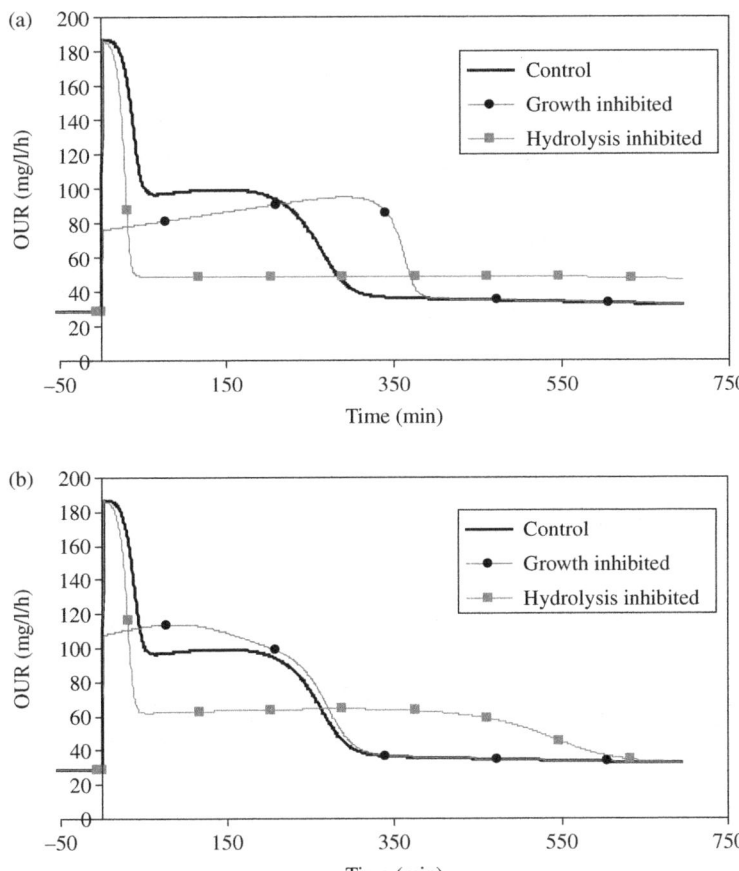

Figure 5.21. OUR profile for control and ASM1 processes with inhibition when (a) $K_I/(I + K_I) = 0.3$ and (b) $K_I/(I + K_I) = 0.5$ (Insel et al., 2006)

time segments (Keesman et al., 1998) and/or estimating only the best parameter combinations (Dochain et al., 1995; Weijers and Vanrolleghem, 1997). Hence it will be possible to arrive at much more accurate predictions on the overall performance of an activated sludge process.

Mechanistic models like ASM1 or ASM3 present a simplified picture of the biochemical processes for substrate utilization and only define growth or storage as the dominant mechanism. Recent understanding of modelling involves not only the simulation of the OUR response but also other relevant experimental

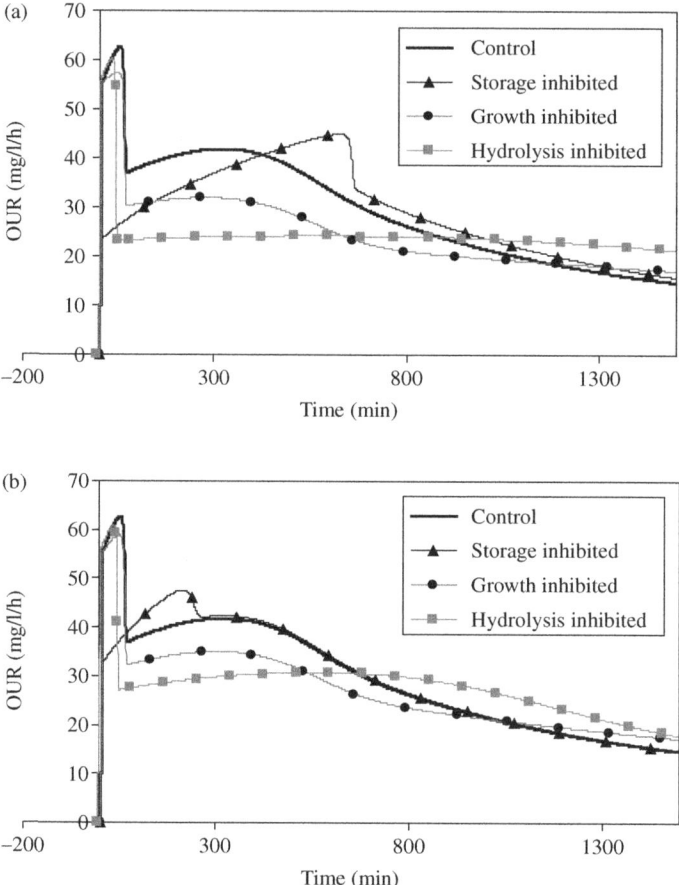

Figure 5.22. OUR profile for control and ASM3 processes with inhibition when (a) $K_I/(I + K_I) = 0.3$ and (b) $K_I/(I + K_I) = 0.5$ (Insel et al., 2006)

data obtained for all related model components, i.e. substrate consumption and storage formation, if any. This approach was illustrated by Karahan et al., (2006) in the modelling of starch utilization, glycogen as storage material and OUR could reveal a mechanism of adsorption, partial storage and simultaneous direct microbial growth. The model could then be calibrated, as shown in Figure 5.23, to yield a more accurate model fit on experimental data using relevant process stoichiometry and kinetics.

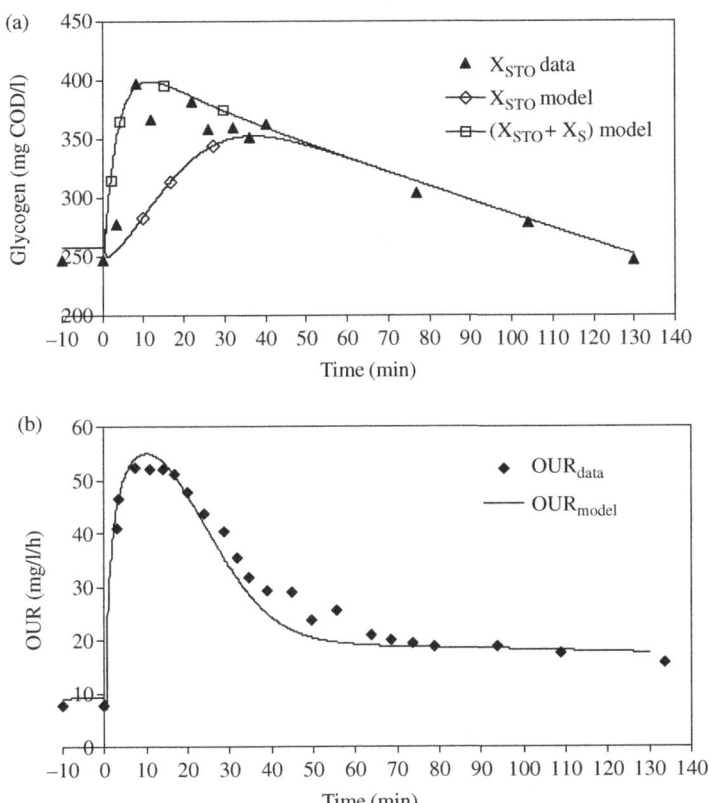

Figure 5.23. Model calibration of (a) glycogen and (b) OUR, of starch utilization for storage and primary growth (Karahan *et al.*, 2006)

Another example of the importance of appropriate modelling approaches has been presented for the evaluation of the experimental results obtained for glucose utilization of *E.coli* under aerobic conditions (Insel *et al.*, 2007). The OUR profile obtained exhibited quite an unusual profile with two peaks as shown in Figure 5.24(a). If the first part of the profile is modelled for microbial growth only, the results would only present a distorted image of growth kinetics. The presence of a lag period until reaching maximum respiration level has also been reported for batch experiments and this observation was attributed to physico-chemical processes of substrate diffusion into floc and dynamics of dissolved oxygen probe (Vanrolleghem *et al.*, 2004). This experimental observation was supported by acetate generation and glycogen storage as other mechanisms

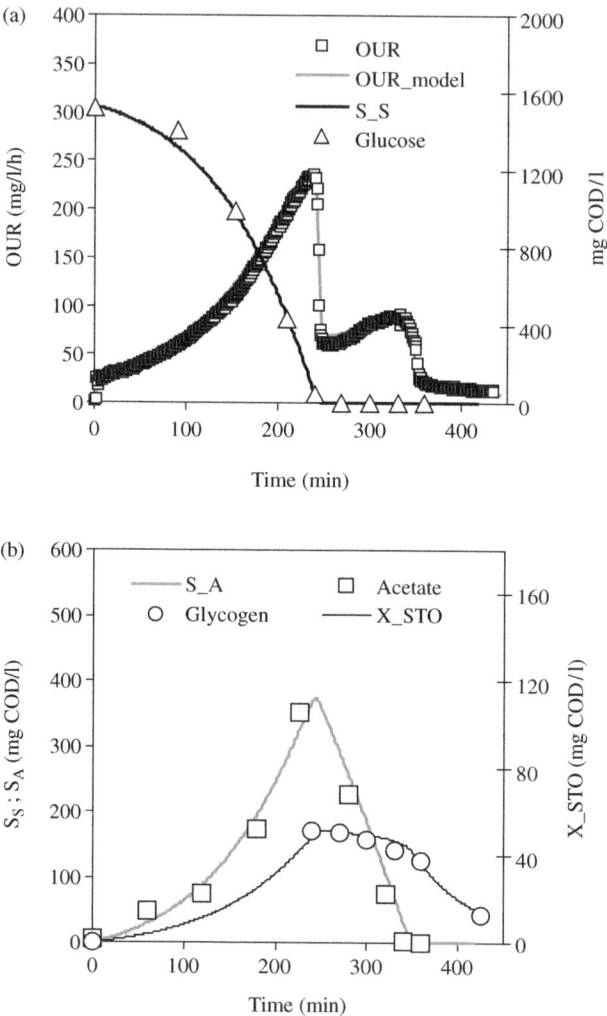

Figure 5.24. Model simulation of (a) glucose, oxygen uptake rate and (b) acetate and glycogen profiles for glucose utilization by cultivated *E. coli* (Insel *et al.*, 2007)

dictating the shape of the OUR profile, which was then modelled together with acetate and glycogen profiles as shown in Figure 5.24(b).

The identifiability of parameters of a model is one of the key issues in modelling. Parameter identifiability covers theoretical and practical identifiability steps. In theoretical identifiably, the model is run under certain conditions

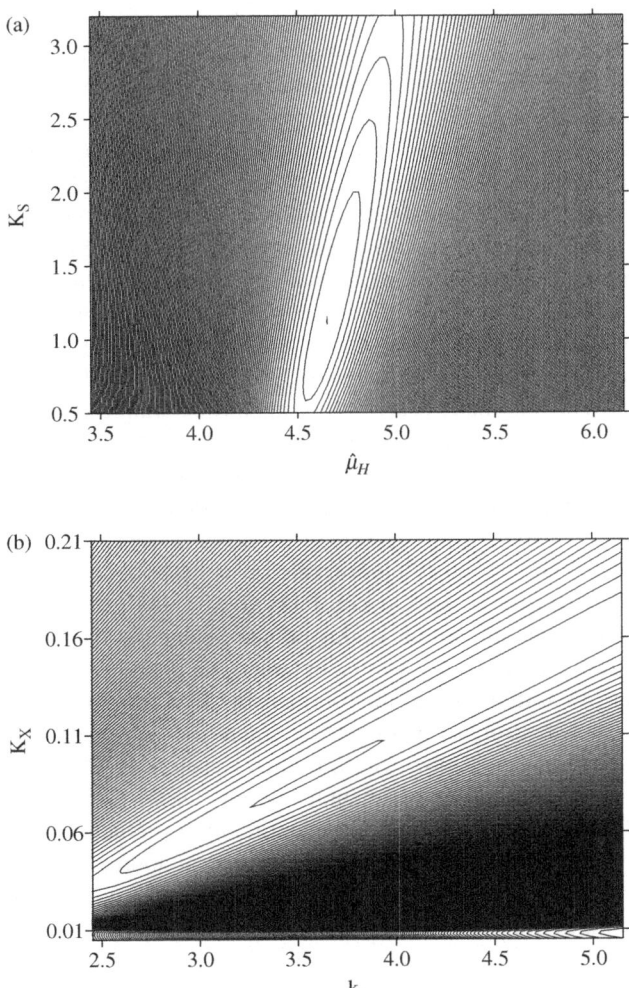

Figure 5.25. Contour plots of objective function for (a) growth and (b) hydrolysis parameters

to gather information on extractable parameters. Mathematical techniques are applied on the models (i.e., linearization, sensitivity analysis) in order to determine parameter subsets under acceptable assumptions (Pohjanpalo, 1978; Walter, 1982). In practical identifiability, the parameters (or subsets) are estimated from the model applied on real experimental data (Vanrolleghem *et al.*, 1995;

Brouwer et al., 1998; Brun et al., 2002). In the real experimental studies measurement noise and data sampling frequency influence the success of parameter estimation. Therefore, the measurement noise should be minimized and data frequency should be selected appropriately.

Data frequency is important for calculating the confidence intervals of the estimated parameters. The correlation of measurements may result in getting contracted confidence ellipsoids, which does not reflect the real case; therefore the measurement should have white noise, ensuring that there is no autocorrelation between the measurements (Dochain and Vanrolleghem, 2001).

The accurate definition of initial experimental conditions is also crucial in parameter estimation since it may expose severe identifiability problems such as parameter correlation, non-informative experiments, especially for (semi-) batch experiments. The automatic search algorithms (i.e., simplex, secant) to find the minimum value of the objective function will also fail if the initial conditions are not well defined. Figure 5.25 illustrates the contour plots of the objective function for growth and hydrolysis parameters simulated for an OUR profile as an example.

Although the simulation results revealed excellent model fit, the parameters especially the hydrolysis process constants show a high degree of uncertainty exhibiting open contours within k_h–K_X domain. Hence, it has been verified that more informative experiments can be obtained via applying batch OUR experiments under low initial food to microorganism (S_0/X_0) ratios for the estimation of growth and hydrolysis parameters, since the parameters have narrow confidence intervals under low (S_0/X_0) ratios (Insel et al., 2002; 2003).

In this sense, Optimal Experimental Design (OED) techniques serve as an important tool to overcome parameter identifiability problems. ORD techniques have been successfully used for the determination of appropriate (initial) experimental conditions and for the adjustment of data frequency (Dochain and Vanrolleghem, 2001; Insel et al., 2003).

REFERENCES

Albek, M., Yetis, U. and Gokcay, C.F. (1997) Effects of Ni(II) on respirometric oxygen uptake. *Appl. Microbiol. Biot.* **48**, 636–641.

Avcioglu, E., Orhon, D. and Sozen, S. (1998) A new method for the assessment of heterotrophic endogenous respiration rate under aerobic and anoxic conditions. *Water Sci. Technol.* **38**(8–9), 95–103.

Avcioglu, E., Karahan-Gul, O. and Orhon, D. (2003) Estimation of stoichiometric and kinetic coefficients of ASM3 under aerobic and anoxic conditions via respirometry. *Water Sci. Technol.* **48**(8), 185–194.

Bagby, M.M. and Sherrard, J.H. (1981) Combined effects of cadmium and nickel on the activated sludge process. *J. Water Pollut. Cont. Fed.* **53**, 1609–1619.

Beun, J.J., Paletta, F., van Loosdrecht, M.C.M. and Heijnen, J.J. (2000) Stoichiometry and kinetics of poly-β-hydroxybutyrate metabolism in aerobic, slow growing, activated sludge cultures. *Biotechnol. Bioeng.* **67**(4), 379–389.

Brouwer, H., Klapwijk, A. and Keesman, K.J. (1998) Identification of activated sludge and wastewater characteristics using respirometric batch experiments. *Water Res.* **32**(4), 1240–1254.

Brun, R., Kuhni, M., Siegrist, H., Gujer, W. and Reichert, P. (2002) Practical identifiability of ASM2d parameters-systematic selection and tuning of parameter subsets. *Water Res.* **36**(16), 4113–4127.

Cokgor, E.U., Sozen, S., Orhon, D. and Henze, M. (1998) Respirometric analysis of activated sludge behaviour I. assessment of the readily biodegradable substrate. *Water Res.* **32**(2), 461–475.

Cokgor, E.U., Wdemir, S., Karahan, O., Insel, G. and Orhon, D. (2007) Critical appraisal of respirometric methods for metal inhibition on activated sludge. *J. Hazard. Mater.* **139**(2), 332–339.

Dircks, K., Beun, J.J., van Loosdrecht, M., Heijnen, J.J., and Henze, M. (2000) Glycogen Metabolism in Aerobic Mixed Cultures. *Biotechnol. Bioeng.* **73**(2), 85–94.

Dochain, D., Vanrolleghem, P.A. and van Daele, M. (1995) Structural identifiability of biokinetic models of activated sludge respiration. *Water Res.* **29**(11) 2571–2579.

Dochain, D. and Vanrolleghem, P.A. (2001) *Dynamical Modelling and Estimation In Wastewater Treatment Processes*. IWA Publishing, London, UK.

Dold, P.L., Ekama, G.A. and Marais, G.v.R. (1980) A general model for the activated sludge process. *Prog. Water Technol.* **12**, 47–77.

Dold, P.L. and Marais, G.v.R. (1986) Evaluation of the general activated sludge model proposed by the IAWPRC Task Group. *Water Sci. Technol.* **18**(6), 63–89.

Ekama, G.A., Dold, P.L., and Marais, G.v.R. (1986) Procedures for determining influent COD fractions and the maximum specific growth rate of heterotrophs in activated sludge systems. *Water Sci. Technol.* **18**, 91–114.

Henze, M., Grady C.P.L. Jr., Gujer, W., Marais, G.v.R. and Matsuo, T. (1987) *Activated sludge model No.1*. IAWPRC Science and Technical Report No. 1, IAWPRC, London, UK.

Henze, M., Gujer, W., Mino, T., Matsuo, T., Wentzel, M.C. and Marais, G.v.R. (1995) *Activated Sludge Model No.2*. IAWPRC Scientific and Technical Report No.2, London: IAWQ.

Germirli, F., Orhon, D. and Artan, N. (1991) Assessment of the initial inert soluble COD in industrial wastewaters. *Water Sci. Technol.* **23**(4–6), 1077–1086.

Gujer, W., Henze, M., Mino, T. and van Loosdrecht, M.C.M. (2000) *Activated Sludge Model No.3*. IWA Scientific and Technical Report No.9 (eds. Activated Sludge Models ASM1, ASM2, ASM2D and ASM3, Henze, M., Gujer, W., Mino, T., van Loosdrecht, M.), London: IWA.

Insel, G., Karahan-Gul, O., Orhon, D., Vanrolleghem, PA. and Henze, M. (2002) Important Limitations in the Modeling of Activated Sludge: Biased Calibration of the Hydrolysis Process. *Water Sci. Technol.* **45**(12), 23–36.

Insel, G., Orhon, D. and Vanrolleghem, P.A. (2003) Identification and modelling of aerobic hydrolysis mechanism-application of optimal experimental design. *J. Chem. Technol. Biot.* **78**(4), 437–445.

Insel, G., Karahan, O., Ozdemir, S., Pala, L., Katipoglu, T., Cokgor, E.U. and Orhon, D. (2006) Unified basis for the respirometric evaluation of inhibition for activated sludge. *J. Environ. Sci. Heal. A.* **41**(9), 1763–1780.

Insel, G., Celikyilmaz, G., Ucisik Akkaya, E., Yesiladali, K., Cakar, P., Tamerler, C. and Orhon, D. (2007) Respirometric evaluation and modeling of glucose utilization by *Escherichia coli* under aerobic and mesophilic cultivation conditions. *Biotechnol. Bioeng.* **96**(1), 94–105.

ISO 8192 (2007) Water Quality-Test for inhibition of oxygen consumption by activated sludge.

Kappeler, J. and Gujer, W. (1992) Estimation of kinetic parameters of heterotrophic biomass under aerobic conditions and characterization of wastewater for activated sludge modelling. *Water Sci.Technol.* **25**(6), 125–139.

Karahan-Gul, O., Artan, N., Orhon, D., Henze, M. and van Loosdrecht, M.C.M. (2002a) Experimental Assessment of Bacterial Storage Yield. *J. Environ. Eng.-ASCE* **128**(11), 1030–1035.

Karahan-Gul, O., Artan, N., Orhon, D., Henze, M. and van Loosdrecht M.C.M. (2002b) Respirometric Assessment of Storage Yield for Different Substrates. *Water Sci. Technol.* **46**(1–2), 345–352.

Karahan, O., van Loosdrecht, M.C.M. and Orhon, D. (2006) Modeling the utilization of starch by activated sludge for simultaneous substrate storage and microbial growth. *Biotechnol. Bioeng.* **94**(1), 43–53.

Keesman, K.J., Spanjers, H. and van Straten, G. (1998) Analysis of endogenous process behaviour in activated sludge. *Biotechnol. Bioeng.* **57**(2), 155–163.

Kelly, C.J., Tumsaroj, N. and Lajoie, C.A. (2004) Assessing wastewater metal toxicity with bacterial bioluminescence in a bench-scale wastewater treatment system. *Water Res.* **38** 423–431.

Krishna, C. and van Loosdrecht, M.C.M. (1999) Substrate flux into storage and growth in relation to activated sludge modeling. *Water Res.* **33**(14), 3149–3161.

Kountz, R.R. and Fourney, C. Jr. (1959) Metabolic energy balances in a total oxidation activated sludge system. *Sewage Indust. Wastes* **31**(8), 819–826.

Marais, G.v.R. and Ekama, G.A. (1976) The activated sludge process Part I – Steady state behaviour. *Water S.A.* **2**(4), 163–199.

McKinney, R.E. (1962) Mathematics of complete-mixing activated sludge. *Proc. Amer. Soc. of Civil Engineers.* **88**(SA3), 87–113.

Orhon, D. and Artan, N. (1994) *Modelling of Activated Sludge Systems*. Technomic Press, Lancaster, PA.

Orhon, D., Artan, N. and Ates, E. (1994) A description of three methods for the determination of the initial inert particulate chemical oxygen demand of wastewater *J. Chem. Technol. Biot.* **61**, 73–80.

Orhon, D., Yildiz, G., Cokgor, E.U. and Sozen, S. (1995) Respirometric evaluation of the biodegradability of confectionary wastewaters. *Water Sci. Technol.* **32**(12), 11–19.

Orhon, D., Ates, E., Sozen, S. and Cokgor, E.U. (1997) Characterization and COD fractionation of domestic wastewaters. *Environ. Pollut.* **95**(2) 191–204.

Orhon, D., Karahan, O. and Sozen, S. (1999a) The Effect of microbial products on the experimental assessment of the particulate inert COD in wastewaters. *Water Res.* **33**(14), 3191–3203.

Orhon, D., Ates Genceli, E. and Ubay Cokgor, E. (1999b) Characterization and modeling of activated sludge for tannery wastewater. *Water Environ. Res.* **71**(1), 50–63.

Orhon, D., Ubay Cokgor, E. and Sozen, S. (1999c) Experimental basis for the hydrolysis of slowly biodegradable substrate in different wastewaters. *Water Sci. Technol.* **39**(1), 87–95.

Orhon, D., Okutman, D. and Insel, G. (2002) Characterization and Biodegradation of Settleable Organic Matter for Domestic Wastewater. *Water S.A.* **28**(3), 299–305.

Orhon, D. and Okutman, D. (2003) Respirometric assessment of residual organic matter for domestic sewage. *Enzyme Microb. Tech.* **32**(5), 560–566

Pearson, E.A. (1966) Kinetics of biological treatment. In *Advances in Water Quality Improvement*. Vol. 1, eds. E.F. Gloyna and W.W. Eckenfelder, Jr. 381–396, Austin, Texas, University of Texas Press.

Pedersen, J. and Sinkjaer, O. (1992) Test of the activated sludge model's capabilities as a prognostic tool on a pilot-scale wastewater treatment plant. *Water Sci. Technol.* **25**(6), 185–194.

Pohjanpalo, H. (1978) System identifiability based on the power series expansion of the solution. *Math. Biosci.* **41**, 21–33.

Ricco, G., Tomei, M.C., Ramadori, R. and Laera, G. (2004) Toxicity assessment of common xenobiotic compounds on municipal activated sludge: comparison between respirometry and microtox. *Water Res.* **38**, 2103–2110.

Sollfrank, U. and Gujer, W. (1991) Characterization of domestic wastewater for mathematical modellin of the activated sludge process. *Water Sci. Technol.* **23**(4), 1057–1066.

Spanjers, H. and Vanrolleghem, P.A. (1995) Respirometry as a tool for rapid characterization wastewater and activated sludge. *Water Sci. Technol.* **31**(2), 105–114.

Sujarittanota, S. and Sherrard, J.H. (1981) Activated sludge nickel toxicity studies. *J. Water Pollut. Control Fed.* **53**, 1314–1322.

Symons, J.M. and McKinney, R.E. (1958) The biochemistry of nitrogen in the synthesis of activated sludge. *Sewage Indust. Wastes* **30**(7), 874–890.

Tunay, O., Zengin, G.E., Kabdasli, I. and Karahan, O. (2004) Performance of magnesium ammonium phosphate precipitation and its effect on biological treatability of leather tanning industry wastewaters. *J. Environ. Sci. Heal. A.* **39**(7), 1891–1902.

Ubay Çokgor, E., Sozen, S., Orhon, D. and Henze, M. (1998) Respirometric analysis of activated sludge behaviour: Assessment of the readily biodegradable substrate. *Water Res.* **32**, 461–75.

van Loosdrecht, M.C.M. and Henze, M. (1999) Maintenance, endogeneous respiration, lysis, decay and predation. *Water Sci. Technol.* **39**(1), 107–117.

Vanrolleghem, P.A., Van Daele, M. and Dochain, D. (1995) Practical identifiability of a biokinetic model of activated sludge respiration. *Water Res.* **29**(11), 2561–2570.

Vanrolleghem, P.A., Sin, G. and Gernaey, K.V. (2004) Transient response of aerobic and anoxic activated sludge activities to sudden substrate concentration changes. *Biotechnol. Bioeng.* **86**(3), 277–290.

Walter, E. (1982) *Identifiability of state space models*. Springer, Berlin, Germany.

Washington, D.R. and Symons, J.M. (1962) Volatile sludge accumulation in activated sludge systems. *J. Water Pollut. Control Fed.* **34**(8), 767–790.

Washington, D.R. and Hettling, L.J. (1965) Volatile sludge accumulation in activated sludge plants. *J. Water Pollut. Control Fed.* **37**, 499–506.

Weijers S.R. and Vanrolleghem P.A. (1997) A procedure for selecting the most important parameters in calibrating the activated sludge model no.1 with full-scale plant data. *Water Sci. Technol.* **36**(5), 69–79.

Wong, K., Zhang, M., Li, X. and Lo, W. (1997) A Luminescence-based scanning respirometer for heavy metal toxicity monitoring. *Biosens Bioelectron.* **12**, 125–133.

Yetis, U., Demirer, G.N. and Gokcay, C.F. (1999) Effect of Cr(VI) on the biomass yield of activated sludge. *Enzyme Microb. Tech.* **25**, 48–54.

6
Management of industrial wastewaters

6.1 NEW TRENDS IN INDUSTRIAL WASTEWATER MANAGEMENT

Considering the diverse nature of industrial activities, geographical differences, various socio-political realities, economical considerations, etc., it is obvious that sound industrial wastewater management requires a skilful, case-specific evaluation based on solid scientific knowledge.

Current understanding of industrial pollution control adopts an integrated management policy that aims to protect the multimedia environment as a whole. Therefore strategies addressing the emissions to air, land and water are developed. On top of that unlike the older approach to wastewater management from industrial sectors, leaving the manufacturing processes untouched as a black box

© 2009 IWA Publishing. *Industrial Wastewater Treatment by Activated Sludge*, by Derin Orhon, Fatos Germirli Babuna and Ozlem Karahan. ISBN: 9781789065282. Published by IWA Publishing, London, UK.

and focusing only on how to treat the effluent in the most efficient and feasible way, the modern point of view concentrates first on the production processes and afterwards on handling the effluents generated. In other words only after realizing *front-of-pipe prevention* measures, the stage is open for *end-of-pipe treatment* applications. The reflections of this current approach, focusing on multimedia effects and highlighting the importance of in-plant control as an essential part of a trustworthy management strategy, can be observed in updated legislation such as European Council Integrated Pollution Prevention and Control (IPPC) Directive; and the US Pollution Prevention Act (PPA) (European Council Directive 96/61/EC; US EPA Pollution Prevention Act of 1990). The *best available techniques* (BAT) concept is extended to cover not only treatment technologies but also production schemes as defined in IPPC Directive (European Commission, 2003). Furthermore due to elevated public sensitivity on environmental and ethical issues, the manufacturers are seeking to improve their images via employing environmentally compatible products and wastes, avoiding wasteful practices, practicing recycling etc. This mode of approach necessitates the adoption of a roadmap constituting the following steps in a stage-wise manner:

(i) waste minimization at source
water use conservation
substitution of chemicals
modifications and/or changes in processes and/or technologies
(ii) reclamation and reuse practices
material reclamation and reuse
wastewater reclamation and reuse
(iii) end-of-pipe treatment

According to an environmental point of view the most favourable actions can be defined as the ones under waste minimization step and the least preferable one as the last item of the list above: End-of-pipe treatment. However depending on case-specific issues, a completely different picture might appear. The cost of fresh water in the region where the plant is located can be considered as an example of case specific issues. It is a well known fact that preventing unnecessary water consumption and segregating the relatively less polluted wastewater fraction for reuse will generate a stronger wastewater, likely to require a higher level of treatment (Orhon *et al.*, 2000; 2001). Therefore the possible adverse effects of such in-plant control measures on the end-of-pipe wastewater quality must be evaluated. For this purpose a feasibility assessment comparing the savings obtained on fresh water demand versus elevated end-of-pipe wastewater treatment costs together with cost of treating reusable streams (where applicable) must be conducted. Consequently according to the results of the feasibility study an action plan must be identified. The cost and availability of environmentally friendly auxiliary chemicals that might be substituted for the currently used ones and the investments allocated for technology modifications are among the other case specific issues.

Depending on the nature of the case, although some steps in the roadmap given above may require intensive investments or application of elaborated technologies, it is very common to get significant improvements with simple changes in housekeeping attitudes that can often be directly implemented at a minimal cost.

Adoption of such a management strategy would create not only many environmental benefits but also some financial gains. However, supervision of skilled professionals especially when establishing the system is required for effective implementation.

6.2 BASIC TOOLS

6.2.1 Assessment at source – process and pollution profiles

It is evident that the application of an approach covering in-plant control as an inseparable part of the management scheme requires a lot more effort devoted to the details of the production processes. Therefore a comprehensive analysis including a wide range of items from the auxiliary chemicals added during each step of the process to the detailed characterization of the segregated wastewater streams must be conducted. In other words, within this new management code of practice, issues such as the data on chemical inputs, the quality of the segregated wastewater streams, etc. can be adopted as useful tools to define either the most appropriate in-plant control alternatives (i.e. wastewater reclamation and reuse, material reclamation and reuse) and/or specific treatment requirements of segregated effluents that will ease the end-of-pipe treatment. By doing so, the respectively unpolluted streams can be segregated for wastewater recovery and reuse. The valuable materials can be recovered from certain effluent streams. Appropriate treatment schemes targeting the removal of specific pollutants can be applied directly to the segregated wastewater streams where the mentioned pollutants exist in the most concentrated form.

What sort of information is required to initiate a sound wastewater management practice for an industrial premise? Despite the fact that almost every type of data gathered on the industry is of a supportive nature, there is certain information which is essential to obtain a trustworthy output. The first one is the data on production capacity. While establishing projections for the future, instead of considering the data on current production level, one should concentrate on the value of production capacity denoting the attainable level of production. On the other hand when used in combination with the current production level, the amount of water usage and/or the amount of discharged wastewater might provide a very beneficial tool in evaluating the situation of the installation in terms of in-plant control measurements. In other words the comparison of the data comprising *the volume of water usage per unit product* (for example 120 m^3 of water / ton dyed cotton fabric), commonly called a *process profile*, obtained from the industry under investigation with the available data of other installations producing the same type of goods contributes an efficacious instrument. Getting a higher level than those obtained by similar producers indicates that more water than required is being used by the industrial facility we are dealing with. Therefore the next step is to check the industrial premises to find out the origin of the problem (i.e. poor housekeeping practices such as running hoses; excessive rinses). In a similar manner, the level of pollutants discharged depending on the current production; *the amount of a certain pollutant parameter per unit product* (for example 40 kg COD / ton dyed cotton fabric), denoting *pollution profile*, can also be compared with the available data. An elevated value might address an issue related to inadequate in-plant control such as the usage of a highly polluting auxiliary, the application of an old process, inadequate dry cleaning efforts etc. A check performed on these values that belong to the whole industry, will provide hints about whether in-plant control requires further attention in this facility or not.

Another very useful tool is the production flowchart. These flowcharts are developed to illustrate every step of the production with the help of boxes and arrows. An example of a production flowchart is given in Figure 6.1.

It is important to note that all sorts of material movements i.e. inputs (from auxiliary chemicals to raw materials) and outputs (from products and by-products to wastes) must be presented while establishing a production flowchart. Depending on the available data, the quantities related to the inputs and outputs have to be tabulated. The next step is to perform a simple material balance to check the reliability of the gathered data. After securing this issue, the production flowchart can be used as a solid source of information. By referring to production flowcharts, segregated wastewater streams originating from the production processes can be defined and quantified. Furthermore they will give an idea about the sort of pollutants these streams might contain. This information together with the data on other water requirements and wastewater discharges (the ones not related to the production processes; e.g. cooling water discharges, domestic wastewaters, boiler make-up) can be considered as a starting point in choosing the proper in-plant control measures.

Industrial wastewaters

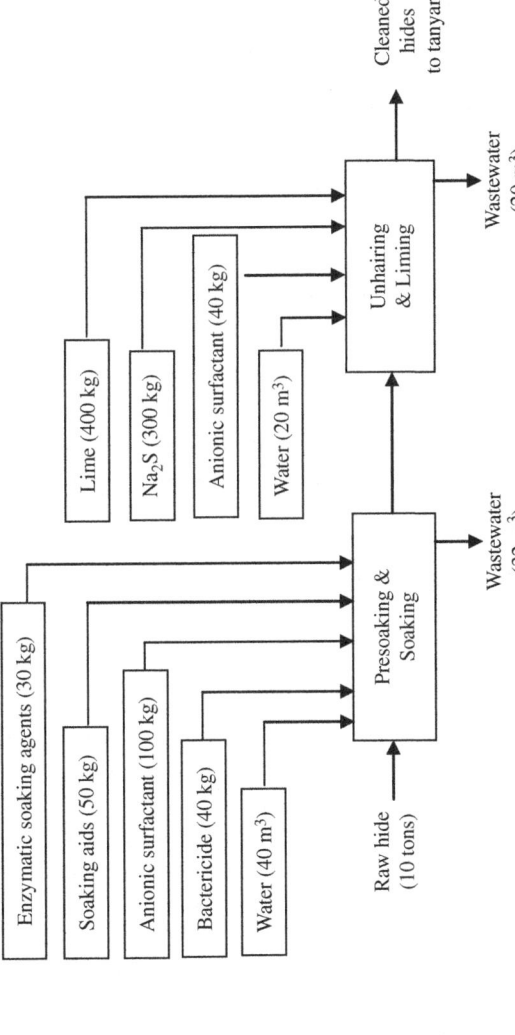

Figure 6.1. Flowchart of beamhouse operations in a shoe upper leather production plant

With the help of the production flowchart one can easily identify the effluent stream on which material reclamation can be applied as it clearly shows all the inputs and segregated wastewater discharges.

The relatively clean discharges (effluent streams obtained from the process stages where no or very small amounts of auxiliaries are added; i.e. rinsings) can further be investigated in terms of wastewater reclamation and reuse. Thus such streams have to be characterized for pollutants.

If the process flowchart indicates the addition of a recalcitrant auxiliary chemical at an upstream position before the discharge of a segregated wastewater stream, a specific treatment scheme can be directly applied on this portion of the effluent. Otherwise this recalcitrant chemical can be substituted with a biodegradable one.

6.2.2 Conventional wastewater characterization

6.2.2.1 Assessment of significant pollutant parameters

Data collected from the characterization of both combined and segregated wastewater streams in terms of conventional parameters will serve as the backbone of a sound management strategy. Apart from providing the necessary information on possible in-plant control practices, conventional wastewater characterization affords a useful database for pre-treatment requirements, for the design of appropriate treatment schemes and for the interpretation of most operational problems related to the treatment plants. However it is important to correctly assess the significant parameters that will be analyzed in the effluents. The significant pollutant parameters must be defined according to the discharge standards applied (all the parameters prescribed in related effluent standards must be covered), the literature data, the auxiliary chemicals added during processes, and the prospective treatment/pre-treatment schemes. In this respect analysing a wastewater stream originating from a dairy industry for heavy metals can be regarded as a useless effort that should be avoided.

Which pollutant parameter one has to use in order to find out the content of organics in an industrial effluent? The use of specific tests targeting the measurement of an organic compound or a group of organics will not produce data of practical importance as wastewaters contain many different organic compounds, probably each in very low concentrations. Occasionally, a specific type of treatment can be exerted to remove a particular compound, i.e. phenol or cyanide. In such cases a specific test can be used. However, since effluents do not contain single organic compounds, but instead a mixture of compounds, each in a different oxidation state, one of the following collective parameters are traditionally used as indirect index values to characterize the overall organic content: Total organic carbon (TOC); biochemical oxygen demand (BOD); and chemical oxygen demand (COD). It should be kept in mind that these collective parameters do not provide the same kind of information. In this respect a brief discussion on the practicality of these collective parameters is presented below:

Although TOC can be considered as a direct and more convenient expression for total organic carbon content than either BOD or COD, it has two main disadvantages: First of all, it does not differentiate between biodegradable and non biodegradable organic compounds. This issue is very important from the biochemical processes point of view as the biological treatment units are only effective in removing the biodegradable fraction of the organic matter. In fact a technique resulting in measuring the total biologically organic carbon exists (Grady *et al.*, 1999). The other significant disadvantage of the TOC test is that; unlike BOD or COD; it is independent of the oxidation state of the organic matter. As the energy available in an organic compound is proportional to its oxidation state, a highly reduced compound will provide more energy as compared to an equal mass of a highly oxidized compound. As a result, a highly reduced compound has a higher pollution potential than that of a highly oxidized one.

The oxidation state of a compound is defined by the number of electrons available for transfer. The number of electrons can be translated directly into the amount of oxygen required to oxidize the compound to carbon dioxide and water, with 8 grams of oxygen being required per electron. This quantity of oxygen is defined as the oxygen demand and is used as a means of pollution potential measurement of a wastewater.

In order for heterotrophs to get energy from complex carbon compounds, an oxidation reaction is necessary, as elaborated in detail in the preceding chapters. Carbon dioxide, hydrogen ion electron and energy are released from this breakdown. The electron and the hydrogen ion must be kept away from the medium for this process to continue. By taking two electrons and two hydrogen ions, half a mole of oxygen produces one water molecule. The released energy is proportional to the used oxygen. Thus, potential oxygen demand can be used as a measure of energy released for metabolism. Consequently, BOD and COD tests are used to measure the carbon energy contained in an effluent by oxygen demand.

The concept of BOD depends on assessing the pollution potential of a wastewater that contains an available organic carbon source for aerobic organotrophic microorganisms by measuring the amount of oxygen utilized during growth of the organisms on a sample of wastewater. Lower BOD values than the actual values are obtained when the microorganisms used as the seed in the test are not acclimated to the substrate in the effluent (which is the case in almost all industrial wastewaters) or when the effluent contains inhibitors that slow down the biological growth (industrial wastewater streams very often carry inhibitors). Furthermore the BOD parameter tends to underestimate the organic concentration as the conditions in the test bottle are too different from those prevailing in the biological reactor: Static culture, low organic matter concentration, low microorganism concentration, diminishing concentrations of dissolved oxygen, etc. (Green et al., 1981). Because of these restrictions and disadvantages the usage of BOD parameter alone, indicating only an unknown fraction of the total organic content, in industrial wastewater management, is not recommended.

COD is a measure of the oxygen equivalent of organic matter content that is susceptible to oxidation by a strong chemical oxidant. COD is often preferred to others as it provides an electron and energy balance between the organic matter, biomass and oxygen utilized with the provision that its biodegradable fraction is ascertained. Although the COD test cannot differentiate between inert and biodegradable organic matter, with the help of the previously described special experimental protocols (Germirli et al., 1991; Orhon et al., 1999c; Orhon and Okutman, 2003) one can get information about the different COD fractions. The biodegradable COD reflects the appropriate electron balance between the substrate, the biomass and the electron acceptor.

Table 6.1 summarizes the outputs of conventional effluent characterization studies obtained from a variety of industrial sectors. Every tabulated value puts forward the average of a statistically representative number (10–30 samples) of different measurements.

6.2.2.2 *Reliability*

Reliable characterization of the combined wastewater that would be the influent of the treatment system should be the prime concern in the site survey. The importance of this step is largely overlooked in most studies. Rather than manufacturing via a single process, industrial premises usually house a variety of production processes functioning in an alternating manner, fluctuating with time. In this respect, wastewater characterization applied to end-of-pipe stream is only meaningful when it relates to the entire set of production processes within the plant with an acceptable accuracy. A few random samples, if not carefully interpreted, may result in a totally misleading picture. Therefore an evaluation based on correct analysis of a statistically valid number of representative wastewater samples is required.

The case study outlined below sets a good example for indicating the danger associated with a superficial wastewater characterization.

Table 6.2 displays measured values for two different industrial wastewater samples evaluated for their compliance with the local discharge to sewer

Table 6.1. Conventional characterization of different industrial wastewaters

Industrial wastewater	Parameter (mg/l)								
	Total COD	Sol. COD	BOD$_5$	Total N	NH$_3$–N	Total P	TSS	VSS	Alkalinity
Textile									
Industrial organized district predominantly textile (Ubay Cokgor et al., 1998)	990	627		54	32	7.9	226	130	655
Denim processing (Germirli Babuna et al., 1998a)	2400	1700		35	5.6	34	500	70	530
Polyester knit fabrics (Germirli Babuna et al., 1998a)	1985	1485		27	1.7	9	213	22	960
Cotton knit fabric (Orhon et al., 1992)	981	535	170	40		14			
Cotton knit fabric (Germirli Babuna et al., 1999a)	2310	2185		14		4.5	135	80	
Cotton knit fabric (Germirli Babuna et al., 1998a)	1470	1165		110	0.5	4	490	160	2350
Cotton PES blend knit fabric (Germirli Babuna et al., 1998a)	2400	1690		20	0.2	7	370	180	1520
Acrylic processing (Germirli Babuna et al., 1999a)	1900	1590		72		4.2	90	43	
Cotton woven fabric (Orhon et al., 1992)	1240	1176	680	144		2.2			

Table 6.1. Continued

Industrial wastewater	Parameter (mg/l)								
	Total COD	Sol. COD	BOD$_5$	Total N	NH$_3$-N	Total P	TSS	VSS	Alkalinity
Dairy processing									
Integrated dairy processing (Ubay Cokgor et al., 1998)	1745	1070		75	41	9.1	400	355	775
Yogurt&Buttermilk processing (Orhon et al., 1993)	1500		1000	63		7.2	191		
Tannery (Orhon et al., 1999b)									
Raw wastewater	4180	1495		250			2070		
Primary effluent	2255	1290		215	160	5.9	770	470	1420
Chemical effluent	1090	1035		197	160	1.8	150	55	1140
Meat processing									
Integrated meat (Ubay Cokgor et al., 1998)	2685	1000		196	85	10	1152	1073	520
Integrated meat (del Pozo et al., 2003)	7230	5500	3180		67	3.3	910	850	
Poultry (Eremektar et al., 1999a)	2690	1700	1595	343	207	30	418		
Confectionery (Orhon et al., 1995)									
Plant I	2840	2500	1840	55		65	260		
Plant II	3630	3020	2700	22		2.5	140		
Plant III	6220	5400	4900	33		8.6	440		
Corn wet mill (Eremektar et al., 1999a)	3800	3230	2800			33	400	315	

Industrial wastewaters 161

Table 6.2. Evaluation of two industrial wastewater samples with respect to local discharge to sewer limitations

Parameters	Sample No. 1	Sample No. 2	Local sewer discharge limitations
COD (mg/l)	5330	600	800
BOD_5 (mg/l)	–	180	250
TSS (mg/l)	245	180	350
Total N (mg/l)	61	31	40
Total P (mg/l)	38	–	10
Detergents (mg/l)	1.5	0.36	5
pH	3.5	8.85	6–10

requirements: The first set depicts a strongly acidic wastewater with a pH of 3.5 and a COD content of 5330 mg/l. Evidently COD largely exceeds the discharge limits together with total N and total P parameters. The second sample has a slightly basic, much weaker character satisfying the applicable discharge limits. In reality, the two sets of analysis belong to the same textile finishing mill dealing with cotton and synthetic knit fabric processing. The samples are collected from the equalization tank at two different dates. A statistical evaluation based on a comprehensive survey involving more than 25 composite samples indicated that the COD of the wastewater could be in fact be characterized as $C_{T1} = 800 \pm 200$ mg/l (Orhon, 1991).

In terms of reliability of characterization there is another issue of practical importance: To decide whether the evaluation for treatment process design should be built upon the average value of the relevant pollutant parameter or not, the average merely indicating that the calculated level is bound to be exceeded with a 50% probability. The correct route to a large extent depends on the nature and the shape of the distribution yielding the average. In cases where the corresponding data are so widespread to involve a risk, it is recommended to adopt a 70 or 80 percentile value for appropriate design.

6.2.2.3 Relationships between major pollutant parameters

The ratio of one major pollutant parameter to another can also be interpreted as an index of treatability.

As previously mentioned in detail, there are strong scientific reasons for not adopting BOD_5 as a design parameter. However, assessment of the BOD_5 content of the wastewater may sometimes be useful when interpreted as the BOD_5/COD ratio. Especially a low BOD_5/COD ratio could be quite informative as a preliminary index for a problematic biodegradation mechanism to be further evaluated with additional testing. For example, a BOD_5/COD ratio of 0.10 (BOD_5 = 100 mg/l; COD = 1000 mg/l) may lead to different interpretations:

- The wastewater contains very slowly biodegradable organic substances.
- The wastewater contains compounds with inhibitory effect (such as heavy metals, toxic organics etc.) towards aerobic microorganisms. The removal of the mentioned inhibitors either with the help of certain in-plant control measures (i.e. substitution of chemicals) or by applying a specific pre-treatment (to a segregated stream or to end-of-pipe effluent) may considerably improve the efficiency of the subsequent biological treatment.
- The majority of the organics in the wastewater is non biodegradable in nature. [According to a literature data, effluents with $BOD_5/COD > 0.4$ may be considered thoroughly biodegradable (Chamarro et al., 2001).]
- Previous to aerobic biological treatment an acclimation period is required. In such cases the usage of BOD_5 parameter as a tool to measure the organic content is meaningless.
- The measurement of BOD_5 or both BOD_5 and COD measurements are wrong. The test or tests should be repeated.

For assessing the extent of particulate matter biodegradation VSS/TSS and particulate COD/VSS ratios can be used. While calculating the magnitude of influent fixed solids that are mostly of inorganic nature and likely to be accumulated in the aeration basin of an activated sludge unit, VSS/TSS ratio can be employed. The difference between the total suspended solids and their volatile fraction may conveniently yield the concentration of the fixed solids ($X_F = X_{TSS} - X_{VSS}$).

The nutritional requirements of the microorganisms are among the major factors dictating the efficiency of biological treatment systems. The impact of this issue is practically determined in terms of COD/N and COD/P ratios measured in the effluent to be treated. If the mentioned ratio is too high compared to the requirements of the microorganisms, as is the case in some industrial wastewaters, energy and biosynthesis reactions may not be fully coupled, with the net result of lower carbon removal efficiency. On the other hand when both or one of the ratios are too low, only very limited N and P removals can be achieved.

A different presentation of conventional wastewater characterization for various industrial sectors/subsectors that highlights the relationships between major pollutant parameters is tabulated in Table 6.3.

As can be seen from Table 6.3, due to the application of different treatment units, different figures are obtained. For example for tannery wastewater,

Table 6.3. Relationships between major pollutant parameters

Industrial wastewater	Parameter							
	Sol.COD/ Tot.COD	BOD$_5$/ Tot.COD	BOD$_5$/ Sol.COD	P.COD/ VSS	VSS/ TSS	NH$_3$-N/ T.N	COD/ T.N	COD/ T.P
Textile								
Denim processing (Germirli Babuna et al., 1998a)	0.71	–	–	10	0.14	0.16	69	71
Polyester knit fabrics (Germirli Babuna et al., 1998a)	0.75	–	–	2.3	0.1	0.06	74	220
Cotton knit fabric (Orhon et al., 1992)	0.54	0.17	0.31	–	–	–	24	70
Cotton knit fabric (Orhon et al., 1992)	0.9	0.47	0.52	–	–	–	38.4	55.3
Cotton knit fabric (Germirli Babuna et al., 1999a)	0.94	–	–	1.6	0.6	–	165	513
Cotton knit fabric (Germirli Babuna et al., 1998a)	0.79	–	–	1.9	0.32	0.004	133	368
Cotton PES blend knit fabric (Germirli Babuna et al., 1998a)	0.7	–	–	3.9	0.48	0.01	120	342
Acrylic processing (Germirli Babuna et al., 1999a)	0.83	–	–	7.2	0.48	–	26.3	452
Cotton woven fabric (Orhon et al., 1992)	0.94	0.55	0.58	–	–	–	8.6	563
Textile								
Industrial organized district predominantly textile (Ubay Cokgor et al., 1998)	0.63	–	–	1.6	0.57	0.6	18	125

Table 6.3. Continued

Industrial wastewater	Parameter							
	Sol.COD/ Tot.COD	BOD$_5$/ Tot.COD	BOD$_5$/ Sol.COD	P.COD/ VSS	VSS/ TSS	NH$_3$-N/ T.N	COD/ T.N	COD/ T.P
Dairy processing								
Integrated dairy processing (Ubay Cokgor et al., 1998)	0.61	–	–	1.9	0.88	0.55	23	191
Yogurt&Buttermilk processing (Orhon et al., 1993)	–	0.66	–	–	–	–	24	208
Tannery								
Raw wastewater (Orhon et al., 1999b)	0.46	–	–	2.8	0.48	0.4	16.7	780
Primary effluent (Orhon et al., 1999b)	0.57	–	–	2.1	0.61	0.75	10.5	382
Chemical effluent (Orhon et al., 1999b)	0.54	–	–	1.0	0.37	0.81	5.5	605
Meat processing								
Integrated meat (Ubay Cokgor et al., 1998)	0.37	–	–	1.5	0.93	0.43	13.7	268
Poultry (Eremektar et al., 1999a)	0.63	0.60	0.94	–	–	0.60	7.8	89.7
Confectionery								
Plant I (Orhon et al., 1995)	0.88	0.65	0.74	–	–	–	51.6	43.6
Plant II (Orhon et al., 1995)	0.87	0.79	0.91	–	–	–	188	723
Plant III (Orhon et al., 1995)	0.83	0.74	0.89	–	–	–	165	1452
Domestic sewage (Orhon et al., 1997)	0.35	0.46	1.18	1.35	0.7	0.8	9.2	52

soluble COD/total COD varies from 0.46 for raw effluents to 0.54 for chemical treatment effluent. Similarly a decrease in particulate COD/VSS ratio from 2.8 to 1.0 with chemical treatment is observed for the wastewaters from the same industrial source.

6.2.3 COD fractions of industrial wastewaters

Assessing the soluble residual COD components of both end-of-pipe and segregated effluent streams can be quoted as an important step in industrial wastewater management since many practical solutions can be derived from such information. First of all, these inert COD fractions either initially present in the wastewater itself and/or microbially generated through the course of the biological processes, are of great importance in meeting the discharge standards. Therefore, with an evaluation based on residual COD levels, one can easily determine if discharge standards can comply with the prescribed biological treatment. Besides, it is possible to address the appropriate type of biological treatment applicable to the wastewater under investigation. The effluent residual COD levels obtained for different industrial wastewaters under aerobic and anaerobic conditions are given in Table 6.4.

According to the figures tabulated in Table 6.4 for the *laying step* of laying chicken industry, the application of aerobic treatment yields a lower effluent residual COD level than a corresponding anaerobic one. Thus an aerobic type of treatment must be prescribed for this industrial effluent. On the contrary, an anaerobic process can be preferred for the treatment of wastewaters from the alcohol distillery.

A similar evaluation can also be adopted for segregated effluent streams. Going through the previously established production flowcharts that clearly illustrate all sorts of inputs including auxiliary chemicals etc, effluent streams susceptible to carrying recalcitrants can be identified. After confirming the inert nature of the mentioned streams by subjecting them to residual COD experiments one can decide on how to manage these streams. According to the so called *minimization at source* philosophy streams with highly recalcitrant character can be directed to pass through a specific treatment alternative before mixing with the others.

The performance of an existing biological treatment system can also be evaluated by referring to inert COD fractions. Table 6.5 outlines the results of COD removal efficiencies achieved in the full scale treatment plants together with inert COD data denoting the attainable levels of COD removals for two industrial sectors. As can be noticed from the table, both of the anaerobic treatment plants are not operating in a proper manner and necessary actions have

Table 6.4. The effluent residual COD levels for different industrial wastewaters under aerobic and anaerobic conditions

Wastewater source (type of treatment)	C_{T1} (mg/l)	S_{I1}/C_{T1} (%)	S_P/C_{T1} (%)	$(S_{I1}+S_P)$ (mg/l)	Reference
Solid waste transport station					
(aerobic)	29900	1	4	1470	Kutluay et al.,
(anaerobic)	29900	1	3	1370	2006
Pulp and paper mill					Germirli Babuna
(aerobic)	21500	21	4	5310	et al., 1998b
Pulp and paper mill					Eremektar et al.,
(anaerobic)	13000	6	11	2240	1998
Antibiotic formulation					Iskender et al.,
(aerobic)	415	78	1	330	2007
Antibiotic formulation					Tezgel et al.,
(aerobic)	400	63	2	260	2007
Brewery					Ince et al.,
(anaerobic)	90000	–	4	3600	1998
Cheese whey					Germirli et al.,
(aerobic)	60000	–	<1	260	1993
Citric acid plant					Germirli et al.,
(aerobic)	29300	6	5	3390	1993
Laying chicken industry*					
(aerobic)	6380	5	<1	350	Germirli Babuna
(anaerobic)	6380	6	5	700	et al., 1999b
Laying chicken industry**					
(aerobic)	9900	7	3	980	Germirli Babuna
(anaerobic)	9900	14	2	1570	et al., 1999b
Alcohol distillery***					
(aerobic)	2350	6	8	340	Eremektar et al.,
(anaerobic)	2350	3	1	85	1999b
Corn wet mill					
(aerobic)	4680	–	9	410	Eremektar et al.,
(anaerobic)	4680	7	7	630	2002

* Chicks step
** Laying step
*** after passing through a chemical treatment

to be taken to improve the COD removal efficiencies up to the levels quoted by inert COD tests.

Furthermore, data on inert COD fractions when enriched with values of other COD components will serve as a solid source of information for modelling

Table 6.5. Evaluation of full-scale treatment plant performance with the help of soluble inert COD fractions under anaerobic conditions

Wastewater source	C_{TI} (mg/l)	Full-scale plant average COD removal efficiency (%)	Inert COD experiments	
			Attainable COD removal efficiency (%)	$(S_{I1}+S_P)$ (mg/l)
Pulp & paper mill (Germirli Babuna *et al.*, 1998b)	13000	68	83	2240
Alcohol distillery∗ (Eremektar *et al.*, 1999b)	2350	80	96	85

∗ after passing through a chemical treatment

studies applied to biological treatment systems. Table 6.6 outlines results of experimental work carried out on COD fractionation of a wide range of industrial wastewaters and domestic sewage. Detailed explanations on how to adopt COD fractionation for modelling purposes are already presented in previous chapters.

6.3 RESOURCE MANAGEMENT

Within the context of the aforementioned stage-wise approach that involves in-plant control measures before an end-of-pipe treatment, segregation and separate evaluation of effluent streams gain importance. A case related to material reclamation and reuse is presented below. The details of another case on water conservation and wastewater recovery and reuse can be found in Chapter 9.

6.3.1 Material reclamation and reuse

This case on material reclamation and reuse emphasizes the significance of considering the segregated effluent qualities while defining a wastewater management strategy. The case handles an industry producing energy drinks with a wastewater generation of 25.4 l/ton product and a COD load of 851 kg/ton product (Oktay *et al.* 2006). Conventional wastewater characterization given in Table 6.7 indicates the generation of a very strong wastewater in terms of organic content.

Costly end-of-pipe treatment alternatives such as the subsequent use of anaerobic and aerobic processes (Austermann-Haun and Rosenwinkel, 1997) or other integrated treatment schemes (Tebai and Hadjivassilis, 1992) can be

Table 6.6. COD fractions of different industrial wastewaters

Industrial wastewater	Parameter (mg/l)							S_{I1}/C_{T1} (%)	C_{S1}/C_{T1} (%)
	C_{T1}	S_{T1}	S_{S1}	S_{I1}	S_{H1}	X_{S1}	X_{I1}		
Textile									
Industrial organized district predominantly textile (Ubay Cokgor et al., 1998)	860	515	144	17	355	229	117	2	85
Denim processing (Germirli Babuna et al., 1998a)	2400	1700	330	100	1270	700	–	4	96
Polyester knit fabric (Germirli Babuna et al., 1998a)	1985	1485	300	415	770	390	110	21	74
Cotton knit fabric (Germirli Babuna et al., 1999a)	2300	1900	420	170	1310	365	35	7	91
Cotton knit fabric (Germirli Babuna et al., 1998a)	1470	1165	330	260	575	288	17	18	81
Cotton PES blend knit fabric (Germirli Babuna et al., 1998a)	2400	1690	165	250	1275	598	112	10	85
Acrylic processing* (Germirli Babuna et al., 1999a)	710	700	86	28	596	–	–	4	96
Integrated dairy processing (Ubay Cokgor et al., 1997)	1410	1075	400	–	675	230	105	–	93

Table 6.6. Continued

Industrial wastewater	Parameter (mg/l)							S_{I1}/C_{T1} (%)	C_{S1}/C_{T1} (%)
	C_{T1}	S_{T1}	S_{S1}	S_{I1}	S_{H1}	X_{S1}	X_{I1}		
Tannery									
Primary diffluent (Orhon et al., 1999b)	2285	1298	434	214	649	724	263	9	79
Chemical effluent (Orhon et al., 1999b)	1090	1090	382	175	534	–	–	16	84
Meat Processing									
Integrated meat (Gorgun et al., 1995)	2600	1140	380	30	730	1155	305	1	87
Integrated meat (del Pozo et al., 2003)	7290	5490	145		4300	2400			94
Poultry (Eremektar et al., 1999a)	2490	1770	308	240	1220	685	35	10	89
Confectionery (Orhon et al., 1995)									
Plant I	3000	2500	2045	45	410	500	–	2	98
Plant II	1800	1495	840	105	550	305	–	6	94
Plant III	8085	6775	5370	45	1360	1310	–	1	99
Corn Wet Mill (Eremektar et al., 1999a)	3800	3192	1710	–	1482	418	190	1	95
Domestic Sewage (Orhon et al., 1997)									
Raw	431	170	42	11	117	229	32	3	90
Settled	300	170	42	11	117	114	16	4	91

* after partial chemical oxidation with H_2O_2

Table 6.7. Conventional wastewater characterization of an industry manufacturing energy drinks

Parameter	Value
Total COD (mg/l)	33000
Soluble COD (mg/l)	27000
TKN (mg/l)	54
Total P (mg/l)	2.5
pH	5.4

prescribed for treating the end-of-pipe effluents, when possible in-plant control practices are not considered. As tabulated in Table 6.8, wastewater characterization conducted on segregated streams shows that only two streams (13 and 14 namely) out of a total of 15 wastewater sources within the whole industry had a significantly high polluting nature. Therefore adopting a wastewater management strategy that segregates these streams having a COD of approximately 80,700 mg/l from the rest of the wastewaters would be necessary.

As stated in the literature (Guyer, 1998), sugar can be recovered by passing the wastewater streams through a membrane process. According to the data summarized in Table 6.9, application of such an in-plant control strategy, not only results in obtaining a valuable material, but also secures an effluent quality that is much lower than the standards without any pre-treatment.

Table 6.8. Segregated streams pollution profile

Source number	Wastewater (l/ton product)	COD	
		(mg/l)	(kg/ton product)
1	0.3125	230	0.072
2	0.3125	40	0.013
3	0.3125	40	0.013
4	0.3125	30	0.009
5	0.3125	750	0.234
6	0.3125	40	0.013
7	1.5	610	0.915
8	1.5	400	0.600
9	1.5	125	0.188
10	1.5	90	0.135
11	1.5	660	0.990
12	1.5	150	0.225
13	5.25	158200	830.6
14	5.25	3200	16.8
15	4	100	0.4

Table 6.9. Effect of sugar recovery on effluent COD and wastewater generation

Options	Effluent COD (mg/l)	Wastewater generation (l/ton product)
Without any in-plant control	33000	25
With in-plant control	250	15

Besides, one should even consider obtaining a feasible output by subjecting this end-of-pipe wastewater to a suitable treatment in order to reuse the effluent.

6.3.2 Auxiliary chemicals

The increased public sensitivity and new legislative framework targeting minimization of pollution at source are forcing the manufacturers to seek environmentally friendly auxiliaries.

In many industrial sectors, some of the various auxiliary chemicals added during manufacturing operations, ends up as constituents of various waste streams. Previously, the attention was given to substituting the chemicals that generate streams with high pollutant loads by the ones that engender less concentrated discharges. Currently, as the research on the possible effects and fates of chemicals in the environment intensifies and new concepts such as *xenobiotics* gain importance, a major shift in this approach is observed. Accordingly, the commonly used chemical auxiliaries are now being substituted by *ecochemicals* having low bio-recalcitrance and toxicity without considering their pollutant loads measured in terms of conventional pollutant parameters. The usage of ecochemicals either eases the treatability of end-of-pipe effluents and/or exerts less toxicity towards the species in the receiving media. Various roadmaps are defined for ecotoxicological assessment and classification of industrial auxiliaries. *TEGEWA Scheme* used in Germany and Danish *SCORE System* are examples of these rating procedures (European Commission, 2003).

An evaluation in terms of toxicity and recalcitrance of some commonly used industrial auxiliary chemicals [namely two dye carriers (Carrier A: *a mixture of isobutanol and tetra propylene benzene sulphonate*; Carrier B: *a mixture of butyl benzoate and dodecyl benzene*), two biocidal finishing agents (Biocide A: *a 2,4,4'-trichloro-2'-hydroxydiphenyl ether*; Biocide B: *a nonionic diphenyl alkane derivative*), two tannin formulations (Tannin A: *a condensation product of aryl sulphonate*; Tannin B: *tannic acid*), and two dispersing agents (Dispersing Agent A: *a naphthalene sulphonic acid derivative*; Dispersing Agent B: *a lignin sulphonate derivative*)] is presented below. Tannins are frequently applied in textile industry during the polyamide dyeing with acid, metal complex or direct dyes at the *after rinsing* stage to increase the fixation rates and to improve wet fastness onto the dyed fabric (e.g. nylon). Naphthalene sulphonic acid derivatives are commonly employed in many industrial sectors covering a wide range of activities from leather tanning operations and textile dyeing mills to production of pharmaceuticals, pesticides, cosmetics, polymers, construction materials (Rivera-Utrilla *et al.*, 2002; Ercole *et al.*, 2005; Shiyun *et al.*, 2002). Lignosulphonates are used in the textile dyeing industry as dispersing agents in conjunction with reactive, disperse and metal complex dyestuffs to dye cotton, cellulose acetate and polyester fabrics respectively (Clapp *et al.*, 1993). Recently, naphthalene sulphonates have been replaced by the more biodegradable (ecologically more-friendly) lignin sulphonates as alternative dye assisting chemicals in many EU countries. Biocides are frequently applied in various sectors including pharmaceutical, paint, personal health care, tannery and textile industries. Most of them are known to have a recalcitrant and toxic nature (Okamura *et al.*, 2000; Thomas *et al.*, 2003). The dye carriers are used in batch dyeing of synthetic fibers to promote absorption and diffusion of hydrophobic disperse dyestuffs under relatively low-temperature ($\leq 100\ ^\circ C$) conditions. They are important for dyeing hybrid fibers of wool and polyester, as wool cannot withstand dyeing under high temperatures (European Commission, 2003). Most of them have a toxic and recalcitrant nature.

The organic contents of segregated discharges carrying these auxiliary chemicals are outlined in Table 6.10.

As evident from the table almost all of the listed auxiliaries have either negligible or low BOD_5 values leading to very low BOD_5/COD ratios roughly indicating the recalcitrant nature of the formulations.

Total and residual COD contents of segregated discharges carrying these chemicals are tabulated in Table 6.11.

The rows related to natural and synthetic tannin formulations represent an example that emphasizes the importance of biodegradability while substituting an auxiliary with another one (Germirli Babuna *et al.*, 2007a). The COD contents (C_{T1}) given in the table for tannin formulations, reveal that application of synthetic tannin as an auxiliary chemical must be preferred as in this case less COD is introduced to the segregated effluent. However a correct conclusion can be extracted from the data on initially inert soluble COD. Despite the fact that natural tannin formulation exerted a COD contribution more than two fold of the one related to its alternative, only 2 % of this organic input yielding 25 mg/l COD can be considered as initially inert. Therefore the segregated wastewater stream containing natural tannin did not necessitate any pre-treatment to

Table 6.10. Organic contents of segregated effluent discharges carrying commonly used industrial auxiliary chemicals

Auxiliary	COD	BOD$_5$
	(mg/l)	
Carrier A	4800	negligible
(Arslan-Alaton, et al., 2006a; 2007)		
Carrier B	4600	negligible
(Arslan-Alaton, et al., 2006a; 2007)		
Biocide A	240	negligible
(Arslan-Alaton, et al., 2006a; b)		
Biocide B	200	negligible
(Arslan-Alaton, et al., 2006a; b)		
Natural tannin	1100	86
(Koyunluoglu et al., 2006)		
Synthetic tannin	465	negligible
(Koyunluoglu et al., 2006)		
Lignin sulphonate	615	90
(Germirli Babuna et al., 2007c)		
Naphtalanesulphonate	1180	40
(Germirli Babuna et al., 2007b)		

improve its biodegradability. When the segregated discharges containing both of the tannin formulations are comparatively evaluated in terms of the lowest achievable COD levels after biotreatment, natural tannin is observed to have a residual COD of 100 mg/l. Synthetic tannin formulation containing segregated discharge on the other hand is monitored to yield a residual COD of 190 mg/l of which 135 mg/l comes from initially inert soluble COD (S_{I1}) and the rest from inert metabolic products (S_P). Therefore, contrary to the previous immature finding, a lump sum appraisal points out natural tannin as the best alternative.

According to the figures given in Table 6.11, quite a high percentage of the COD generated by naphthalene sulphonic acid derivative carrying discharge, namely 87%, must be regarded as biorecalcitrant in nature (Germirli Babuna et al., 2007b). This significantly high fraction supports the necessity of applying a partial chemical pre-treatment to the segregated wastewater stream with naphthalene sulphonic acid derivative prior to letting it mixed with the other wastewater sources that will consequently pass through a conventional biological treatment.

The data presented in Table 6.11 indicates that the initially inert soluble COD of the segregated discharge containing the lignosulphonate derivative is determined as 320 mg/l, corresponding to 52% of the total COD. Due to the formation of soluble metabolic products, the effluent quality cannot be improved

Table 6.11. Inert COD contents of segregated industrial effluents carrying auxiliary chemicals

Auxiliary chemical	C_{TI}	S_{II}	$S_{II}+S_P$	S_{II}/C_{TI}
		(mg/l)		(%)
Natural tannin (Germirli Babuna et al., 2007a)	1100	25	100	2
Synthetic tannin (Germirli Babuna et al., 2007a)	465	135	190	29
Ligninsulphonate (Germirli Babuna et al., 2007c)	615	320	340	52
Naphtalanesulphonate Germirli Babuna et al.,2007b)	1180	1027	1034	87

better than 340 mg/l of COD when a biological treatment is prescribed for the segregated stream carrying this lignosulphonate derivative.

The ED_{50} values (causing 50% inhibition) of the segregated effluents carrying auxiliary chemicals are tabulated in Table 6.12. According to the figures listed for acute toxicities towards water flea *Daphnia magna*, Carrier A is observed to be three times more toxic than Carrier B. On the other hand both of the carriers are monitored to impart serious toxicities on marine microalgae *Phaeodactylum tricornutum* supporting the fact that the toxicity responses of different test organisms differed from each other. Although previously known as an *eco* chemical, due to the represented findings it's not possible to categorize Carrier B with such an ascription. In terms of activated sludge inhibition both of the carriers shows a similar behaviour. The biocide formulations are highly and almost equally toxic towards both *Phaeodactylum tricornutum* and *Daphnia magna*. Both biocides are observed to cause 50–60 % inhibition in activated sludge systems. Both of the tannin formulations impart very high acute toxicities on *Phaeodactylum tricornutum*. Likely activated sludge inhibition rates are obtained for natural tannin and synthetic tannin. Lignin sulphonate formulation exerted a high acute toxicity towards marine microalgae *Phaeodactylum tricornutum*. Relatively less toxicity towards *Phaeodactylum tricornutum* is observed for naphthalene sulphonate formulation.

In conclusion, either due to their toxic and/or inhibitory behaviour towards the test organisms and/or their elevated recalcitrant nature at least a partial chemical treatment must be prescribed for the segregated discharges containing the presented auxiliary chemicals.

It should be kept in mind that there exists a possibility of obtaining more toxic byproducts than the original effluent especially under partial pre-treatment conditions (Sharma et al., 2007). In addition the literature does not provide

Table 6.12. Results of the acute toxicity tests and activated sludge (AS) inhibition

Auxiliary	Daphnia magna	Phaeodactylum tricornutum	AS Inhibition (ISO 8192, 2007)
		ED_{50} (% v/v)	
Carrier A (Arslan-Alaton, et al., 2007)	3	< 1	10
Carrier B (Arslan-Alaton, et al., 2007)	9	< 1	12
Biocide A (Arslan-Alaton, et al., 2006b)	8	3–5	50–60
Biocide B (Arslan-Alaton, et al., 2006b)	7	3–5	50–60
Natural tannin (Germirli Babuna et al., 2007a; Arslan-Alaton & Koyunluoglu 2007)	ND	< 1	42
Synthetic tannin (Germirli Babuna et al., 2007a; Arslan-Alaton & Koyunluoglu 2007)	ND	< 1	45
Lignin sulphonate (Germirli Babuna et al., 2007c)	ND	6–8	ND
Naphtalane sulphonate (Germirli Babuna et al., 2007b)	ND	30–35	ND

ND = not determined

adequate information on byproduct toxicity (Hao et al., 2000). Therefore a careful evaluation is required to assure that one pollution problem is not being substituted by another one while applying partial pre-treatment to segregated recalcitrant and/or toxic industrials effluents.

6.3.3 Xenobiotics

Due to economical constrains the application of biodegradable and non-toxic auxiliary chemicals in industrial facilities is limited on a worldwide basis. Instead, countless types of chemicals having xenobiotic characteristics are employed in manufacturing processes. By definition a xenobiotic is a foreign chemical which is found in an organism but is not normally produced or expected to be present in it. Treatment of many xenobiotics via conventional processes is proved to be inadequate because of their toxic and/or recalcitrant nature. The xenobiotic emissions even at low concentrations pose major problem for the receiving media. Besides, while buying goods consumers start to question the environmental and ethical issues. In this context xenobiotics must be handled according to the minimization at source principle. Thus initial efforts should be given to substitute them with the ecochemicals as much as possible. Nevertheless, if the usage of a certain xenobiotic is an unavoidable part of a manufacturing process, the effluent stream originating from this very process (or even operation) must be segregated from the other wastewaters. Only after passing through a specially prescribed partial pre-treatment can this segregated effluent be directed towards a conventional end-of-pipe treatment plant. Biological treatment employing either specially adopted or genetically engineered microorganisms, a type of chemical oxidation, advanced systems etc. can be applied as effective treatment units.

In future, research activities are expected to focus on developing new synthetic auxiliaries that match in the naturally existing catabolic potential of the microorganisms or in other words are *benign by design* (Rieger *et al.*, 2002).

6.4 REFERENCES

Arslan Alaton, I., Insel, G., Eremektar, G., Germirli Babuna, F. and Orhon, D. (2006a) Effect of textile auxiliaries on the biodegradation of dyehouse effluent in activated sludge. *Chemosphere*. **62**(9), 1549–1557.

Arslan Alaton, I., Okay S.O., Eremektar, G. and Germirli Babuna, F. (2006b) Toxicity assessment of raw and ozonated textile biocides. International Magazine for Textile Design Processing and Testing, *AATCC Rev.* **6**(5), 43–48.

Arslan Alaton, I., Iskender, G., Ozerkan, B., Germirli Babuna, F. and Okay, O. (2007) Effect of chemical treatment on the acute toxicity of two commercial textile dye carriers. *Water Sci. Technol.* **55**(10), 253–260.

Arslan Alaton, I. and Koyunluoglu, S. (2007) Ozonation of two commercially important biocidal finishing agents. *Ozone Sci. Eng.* **29**, 335–342.

Austermann-Haun, U. and Rosenwinkel, K.H. (1997) Two examples of anaerobic pre-treatment of wastewater in the beverage industry. *Water Sci. Technol.* **36**(2–3), 311–319.

Chamarro, E., Marco, A. and Esplugas, S. (2001) Use of fenton reagent to improve organic chemical biodegradability, *Water Res.* **35**(4), 1047–1051.

Clapp, C.E., Hayes, M.H.B. and Swift, R.S. (1993) Isolation, fractionation, functionalities and concepts of structure of soil organic macromolecules. In *Organic Substances in Soil and Water*, ed. A. J. Beck, K. C. Jones, M. B. H. Hayes and U. Mingelgrin, Royal Society of Chemistry, Cambridge.

del Pozo, R., Tas, D.O., Dulkadiroglu, H., Orhon, D. and Diez, V. (2003) Biodegradability of slaughterhouse wastewater with high blood content under anaerobic and aerobic conditions. *J. Chem. Technol. Biot.* **78**(4), 384–391.

Ercole, C., Botta, A.L., Sulpizii, M., Veglio, F. and Lepidi, A. (2005) Microbial desulphonation and β-naphthol formation from 2-naphthalenesulphonic acid. *Process Biochem.* **40**, 2297–2303.

Eremektar, G., Germirli Babuna, F. and Orhon, D. (1998) Inert COD fractions in two-stage treatment of a pulp and paper mill effluent, *J. Chem. Technol. Biot.* **72**, 7–14.

Eremektar, G., Ubay Cokgor, E., Ovez, S., Germirli Babuna, F. and Orhon D. (1999a) Biological treatability of poultry processing plant effluent-a case study. *Water Sci. Technol.* **40**, 323–329.

Eremektar, G., Germirli Babuna, F. and Ince, O. (1999b) Fate of inert COD fractions in two-stage treatment of a strong wastewater. *J. Environ. Sci. Heal. A.* **34**(6), 1329–1340.

Eremektar, G., Karahan Gul, O., Germirli Babuna, F., Ovez, S., Uner, H. and Orhon, D. (2002) Biological treatability of a corn wet mill effluent. *Water Sci. Technol.* **45**(12), 339–346.

European Commission. (2003) *Integrated Pollution Prevention and Control (IPPC) Reference Document on Best Available Techniques for the Textiles Industry.* 586 pages, Brussels.

European Council Directive (1996). *Integrated Pollution Prevention and Control (IPPC).* 96/61/EC.

Germirli, F., Orhon, D. and Artan, N. (1991) Assessment of the initial inert soluble COD in industrial wastewaters. *Water Sci. Technol.* **23**(4–6), 1077–1086.

Germirli, F., Orhon, D., Artan, N., Ubay, E. and Gorgun, E. (1993) Effect of two-stage treatment on the biological treatability of strong industrial wastes. *Water Sci. Technol.* **28**(2), 145–154.

Germirli Babuna, F., Orhon, D., Ubay Cokgor, E., Insel, G. and Yapraklı, B. (1998a) Modelling of activated sludge for textile wastewaters. *Water Sci. Technol.* **38**(4–5), 9–17.

Germirli Babuna, F., Ince, O., Orhon, D. and Simsek, A. (1998b) Assessment of inert COD in pulp and paper mill wastewater under anaerobic conditions. *Water Res.* **32**(11), 3490–3494.

Germirli Babuna, F., Soyhan, B., Eremektar, G. and Orhon, D. (1999a) Evaluation of treatability for two textile mill effluents. *Water Sci. Technol.* **40**, 145–152.

Germirli Babuna, F., Cekyay, E., Eremektar, G. and Orhon, D. (1999b) Pollution loads and inert COD in laying chickens industry. *Water Sci. Technol.* **40**, 207–213.

Germirli Babuna, F., Yilmaz, Z., Okay, O., Arslan Alaton, I. and Iskender, G. (2007a) Ozonation of synthetic versus natural tannin: inert COD and toxicity towards *Phaeodactylum tricornutum. Water Sci. Technol.* **55**(10), 45–52.

Germirli Babuna, F., Camur, S., Arslan Alaton, I., Okay, O. and Iskender, G. (2007b) The application of ozonation for the detoxification and biodegradability improvement of a textile auxiliary: naphthalene sulphonic acid. *Proceedings of the 10th International Conference on Environmental Science and Technology,* Kos Island, Greece, A-402–409.

Germirli Babuna, F., Oructut, N., Arslan Alaton, I., Iskender, G. and Okay, O. (2007c) Reducing the toxicity and recalcitrance of a textile xenobiotic through ozonation. *International Conference on Environmental Survival and Sustainability.* 19–24 February 2007, Near East University, Nicosia, Northern Cyprus. (provisionally accepted for publication in J. Environ. Sci. Heal. A)

Gorgun, E., Ubay Cokgor, E., Orhon, D., Germirli, F. and Artan, N. (1995) Modelling biological treatability for meat processing effluent. *Water Sci. Technol.* **32**(12), 43–52.

Grady, C.P.L., Daigger, G.T. and Lim, H.C. (1999) *Biological Wastewater Treatment:* 2nd edition, Marcal Dekker Inc., New York.

Green, M., Sheler, G. and Moraine, R. (1981) Chemical and biochemical oxygen demand as indicators of biodegradable substrate concentration. *J. Water Pollut. Control Fed.* 655–658.

Guyer H.H. (1998) *Industrial Processes and Waste Stream Management*. John Wiley & Sons Inc., 592 pages, Canada.

Hao, O.J., Kim, H. and Chiang, P.C. (2000) Decolorization of wastewater. *Crit. Rev. Env. Sci. Tec.* **30**(4), 449–505.

Ince, O., Germirli Babuna, F., Kasapgil, B. and Anderson, G.K. (1998) Experimental determination of the inert soluble COD fraction of a brewery wastewater under anaerobic conditions. *Environ. Technol.* **19**, 437–442.

Iskender, G., Sezer, A., Arslan Alaton, I., Germirli Babuna, F. and Okay, O.S. (2007) Treatability of Cefazolin antibiotic formulation effluent with O_3 and O_3/H_2O_2 processes. *Water Sci. Technol.* **55**(10), 217–225.

Koyunluoglu, S., Arslan Alaton, I., Eremektar, G. and Germirli Babuna, F. (2006) Pre-ozonation of commercial textile tannins: effect on biodegradability and toxicity. *J. Environ. Sci. Heal. A.* **41**(9), 1873–1886.

Kutluay, G., Iskender, G., Germirli Babuna, F. and Orhon, D. (2007) Treatment options for effluents from a solid waste transport station, *Desalination* **211**, 96–101.

Oktay, S., Iskender, G., Germirli Babuna, F., Kutluay, G. and Orhon, D. (2007) Improving the wastewater management for a beverage industry with in-plant control. *Desalination* **211**, 138–143.

Okamura, H., Aoyama, I., Liu, D., Maguire, R.J., Pacepavicius, G.J. and Lau, Y.L. (2000) Fate and ecotoxicity of a new antifouling compound: Irgarol 1051 in the aquatic environment. *Water Res.* **34**, 3523–3530.

Orhon, D. (1991) Design of pre-treatment. In *Pre-treatment of Industrial Wastewaters*, (O.Tunay, D.Orhon, and A. Bederli, eds.), pp.131–173, ISO-SKATMK, Istanbul, (in Turkish).

Orhon D., Artan N., Buyukmurat, S. and Gorgun, E. (1992) The effect of residual COD on the biological treatability of textile wastewater. *Water Sci. Technol.* **26**(3–4), 815–825.

Orhon, D., Gorgun, E., Germirli, F. and Artan, N. (1993) Biological treatability of dairy wastewaters. *Water Res.* **27**(4), 625–633.

Orhon, D., Yildiz, G., Ubay Cokgor, E. and Sozen, S. (1995) Respirometric evaluation of the biodegradability of confectionary wastewater. *Water Sci. Technol.* **40**(1), 1–11.

Orhon D., Tasli, R. and Sozen, S. (1999a). Experimental basis of activated sludge treatment for industrial wastewaters-the state of the art. *Water Sci. Technol.* **32**(12), 11–19.

Orhon, D., Ates Genceli, E., Sozen, S. and Ubay Cokgor, E. (1997) Characterization and COD fractionation of domestic wastewaters. *Environ. Pollut.* **92**(2), 191–204.

Orhon, D. and Ubay Cokgor, E. (1997) COD fractionation in wastewater characterization – the state of the art. *J. Chem. Technol. Biot.* **68**, 283–293.

Orhon, D., Ates Genceli, E. and Ubay Cokgor, E. (1999b) Characterization and modeling of activated sludge for tannery wastewater. *Water Environ. Res.* **71**(1), 50–63.

Orhon, D., Karahan, O. and Sozen, S. (1999c). The Effect of microbial products on the experimental assessment of the particulate inert COD in wastewaters. *Water Res.* **33**(14), 3191–3203.

Orhon, D., Sozen, S., Kabdaslı, I., Germirli Babuna, F., Karahan, O., Insel, G., Dulkadiroglu, H., Dogruel, S., Kıran, N., Baban, A. and Kemerdere Kaya, N.

(2000b) Recovery and reuse in the textile industry – A case study at a wool and blends finishing mill. In: *Chemical Water and Wastewater Treatment VI*, ed. H. H. Hahn, E. Hoffman and H. Odegaard, pp 305–315, Springer Verlag, Berlin.

Orhon D., Germirli Babuna F., Kabdasli, I, Insel, G., Karahan, O., Dulkadiroglu, H., Dogruel, S., Sevimli, F. and Yediler, A. (2001). A scientific approach to wastewater recovery and reuse in the textile industry. *Water Sci. Technol.* **43**(11), 223–231.

Orhon, D. and Okutman, D. (2003). Respirometric assessment of residual organic matter for domestic sewage. *Enzyme Microb. Tech.* **32**(5), 560–566

Rieger, P.G., Meier, H.M., Gerle, M., Vogt, U., Groth, T. and Knackmuss, H.J. (2002) Xenobiotics in the environment: present and future strategies to obviate the problem of biological persistence. *J. Biotechnol.* **94**, 101–123.

Rivera-Utrilla, J., Sanchez-Polo, M., Mondaca, M.A. and Zaror, C.A. (2002) Effect of ozone and ozone/activated carbon treatments on genotoxic activity of naphthalene-sulfonic acids. *J. Chem. Technol. Biot.* **77**, 883–890.

Sharma, K.P., Sharma, S., Subhasini Sharma Singh, P.K., Kumar, S., Grover, R. and Sharma, P.K. (2007) A comparative study on characterization of textile wastewaters (untreated and treated) toxicity by chemical and biological tests. *Chemosphere.* **69**, 48–54.

Shiyun, Z., Xuesong, Z. and Daotang, L. (2002) Ozonation of naphthalene sulfonic acids in aqueous solutions. Part I: elimination of COD, TOC and increase of their biodegradability. *Water Res.* **36**, 1237–1243.

Tebai, L. and Hadjivassilis, I. (1992) Soft drinks industry wastewater treatment. *Water Sci. Technol.* **25**(1), 45–51.

Tezgel, T., Germirli Babuna, F., Arslan-Alaton, I., Iskender, G. and Okay, O. (2007) Pretreatment of Ceftriaxone formulation effluents: drawbacks and benefits. *International Conference on Environmental Survival and Sustainability*. 19–24 February 2007, Near East University, Nicosia, Northern Cyprus. (provisionally accepted for publication in J. Environ. Sci. Heal. A)

Thomas, K.V., Mc Hugh, M., Hilton, M. and Waldock, M. (2003) Increased persistance of antifouling paint biocides when associated with paint particles. *Environ. Pollut.* **123**, 153–161.

Ubay Cokgor, E., Sozen, S., Orhon, D. and Henze, M. (1998) Respirometric analysis of activated sludge behaviour-1. assessment of the readily biodegradable substrate. *Water Res.* **32**(2), 461–475.

US EPA Pollution Prevention Act of 1990.

7
Continuous flow activated sludge technology

7.1 INTRODUCTION

The striking feature of the activated sludge technology is perhaps the long transition period which was spent fully understanding the process, after its discovery and its promotion as a treatment method. A sound conceptual basis was indeed very much needed for appropriate modelling of the process that translated into a rational approach for design. The early design practice depended only on a few crude measurements such as BOD_5 on the wastewater to be treated, and used arbitrary parameters like volumetric loadings for determining reactor volume. At the early stages, the operation of activated sludge plants encountered major difficulties. The quest for improvement however had to rely on experience and good judgement. New ideas were developed by trial and error and resulted in a number of process modifications. The conceptual understanding

© 2009 IWA Publishing. *Industrial Wastewater Treatment by Activated Sludge*, by Derin Orhon, Fatos Germirli Babuna and Ozlem Karahan. ISBN: 9781789065282. Published by IWA Publishing, London, UK.

of the activated sludge process improved along the way, mostly as an auxiliary effort. Almost 50 years after its discovery, McKinney (1962) introduced *reactor hydraulics* as an important dimension for design. Pearson (1967) applied for the first time the concept of mass balance for the interpretation of substrate removal and microbial growth. The *sludge age, θ_X* was then defined as the major parameter for design (Jenkins and Garrison, 1968). A major milestone for a better system design was the adoption of COD as a substrate parameter and the recognition of COD fractions with different biodegradation characteristics (Eckhoff and Jenkins, 1967; Grady and Williams, 1975; Dold et al., 1980).

In recent years, the capability of activated sludge was greatly expanded to focus on nutrient removal. In this context, mechanistic models providing insight into the process mainly explored nutrient removal from domestic sewage (Henze et al., 1987, 1995; Orhon and Artan, 1994). It was only after the promotion of respirometric techniques for the experimental evaluation of biodegradation characteristics that a unified basis could be established for the design of biological treatment systems for industrial wastewaters, the same way as domestic sewage (Orhon, 1998; Orhon et al., 1999). The available information today enables the definition of, as outlined in the following sections, major system functions of the activated sludge process, such as *reactor biomass excess sludge, oxygen utilized, nutritional requirements,* etc., using basic design parameters, relevant biochemical reactions and mass balance so that they can be integrated with a rational design approach both for organic carbon and nutrient removal.

7.2 BASIC PARAMETERS

A limited number of key parameters have emerged from years of practice with the activated sludge system, as the most popular biological treatment technology both for domestic sewage and industrial wastewater. Extensive modelling studies have only confirmed the merit of these parameters as indispensable tools for appropriate design. The most important ones, namely, *the sludge age, the heterotrophic net yield coefficient, the food to microorganism ratio* and *the mean hydraulic retention time* are described below, mainly to emphasize their role in establishing relevant relationships between process kinetics and system operation.

7.2.1 The sludge age

The sludge age, θ_X is the most vital parameter in the design and operation of the activated sludge systems. From a process kinetics standpoint, it defines the inverse of the net growth rate:

$$\frac{1}{\theta_X} = \mu_H - b_H \tag{7.1}$$

The above equation is expressed in terms of the heterotrophic growth. However, the resulting sludge age is rate limiting and it is applicable to all microbial communities, i.e., nitrifiers, etc., maintained in the biological reactor. In system operation, the sludge age reflects the mean solids retention time, SRT. Accordingly, it is conveniently defined as the ratio of the total biomass in the reactor to the daily amount of total excess sludge that needs to be wasted in order to maintain the activated sludge reactor at steady state:

$$\theta_X = \frac{V_R X_T}{P_{XT}} \tag{7.2}$$

where, V_R = reactor volume
X_T = total biomass concentration
P_{XT} = daily amount of excess biomass

For carbon removal systems, the selection of an appropriate sludge age value is not so much affected by kinetic considerations, but it is mainly focused on sustaining a flocculent biomass that settles well. For domestic sewage, the experience suggests that good biomass settling is secured above a minimum sludge age of three days. A much higher sludge age value in the range of 8–10 days is generally selected for industrial wastewaters.

7.2.2 The heterotrophic net yield coefficient

The heterotrophic net yield coefficient, Y_{NH} is also a very significant parameter because it relates substrate utilization directly to biomass generation and oxygen demand. It can be easily defined using process kinetics as the ratio between the overall biomass growth rate and substrate utilization rate, as shown in Figure 2.3:

$$Y_{NH} = \frac{r_X}{r_S} \tag{7.3}$$

Both r_S and r_X can be defined in terms of the heterotrophic growth rate, μ_H:

$$r_X = (\mu_H - b_H) X_H$$

$$r_S = Y_H \mu_H X_H$$

Using the above expressions and replacing μ_H from equation (7.1) yields:

$$Y_{NH} = \frac{Y_H}{1 + b_H \theta_X} \quad (7.4)$$

This equation shows that Y_{NH} decreases as a function of θ_X depending on the relative impact of endogenous respiration with respect to microbial growth. Variation of Y_{NH} with θ_X is illustrated in the following example.

Example 7.1 Variation of Y_{NH} with θ_X

Evaluate the variation of Y_{NH} with θ_X. Calculate (a) Y_{NH} for $\theta_X = 10$ days; (b) the sludge age value that would reduce the Y_{NH} / Y_H ratio to 0.3.

$$Y_H = 0.60 \text{ g cell COD/g COD}$$
$$b_H = 0.15/d$$

Using equation (7.4):

$$Y_{NH} = \frac{Y_H}{1 + b_H \theta_X} = \frac{0.6}{1 + 0.15 \times \theta_X}$$

The relationship between Y_{NH} and θ_X is plotted in Figure 7.1.

(a) For $\theta_X = 10$ days,

$$Y_{NH} = \frac{0.6}{1 + 0.15 \times 10} = 0.24 \text{g cell COD/g COD}$$

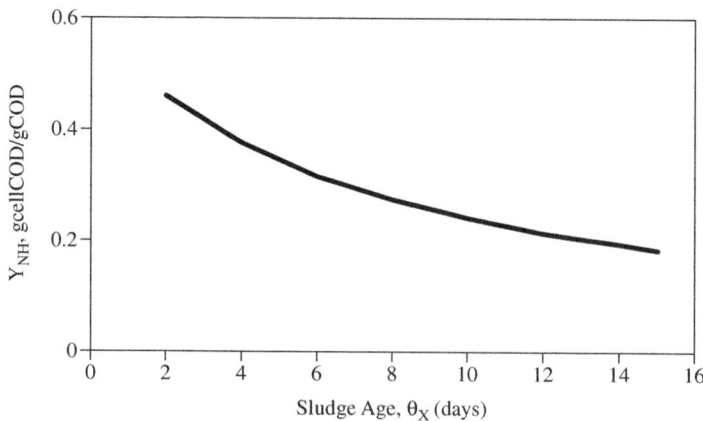

Figure 7.1. Variation of Y_{NH} with the sludge age

Continuous flow technology 185

(b) If the Y_{NH}/Y_H ratio is 0.3, then,

$$\frac{Y_{NH}}{Y_H} = 0.3 = \frac{1}{1 + 0.15 \times \theta_X}$$

and, $\theta_X = 15.6$ days

7.2.3 The food to microorganism ratio

The food to microorganism ratio, (F/M) simply indicates the amount of substrate supplied per day per unit amount of biomass in the biological reactor. When COD is used as a substrate parameter, the F/M ratio should be expressed in terms of the influent biodegradable COD:

$$\frac{F}{M} = \frac{QC_{S1}}{V_R X_H} \tag{7.5}$$

where, Q = wastewater flow rate
C_{S1} = influent biodegradable COD
X_H = active heterotrophic biomass

This parameter is generally used as a design parameter: A selected F/M value determines the required amount of biomass in the reactor and the reactor volume is computed for a suitable biomass concentration. This approach may not be accurate mainly because the F/M ratio cannot be selected independently from the adopted sludge age. In fact from growth kinetics, the F/M ratio may be expressed in terms of μ_H with the assumption that all biodegradable substrate is depleted in the biological reactor:

$$\frac{F}{M} = \frac{1}{Y_H}\mu_H \tag{7.6}$$

Replacing μ_H from equation (7.1) and expressing Y_H in terms of Y_{NH} yields:

$$\frac{1}{\theta_X} = Y_{NH}\left(\frac{F}{M}\right) \tag{7.7}$$

or $$\frac{F}{M} = \frac{1 + b_H \theta_X}{Y_H \theta_X} \tag{7.8}$$

This fundamental relationship shows that F/M cannot be independently adjusted for an activated sludge system operated at a given sludge age as it varies as a function of the sludge age, the main design parameter for activated sludge systems. It also indicates that adoption of VSS as an overall biomass parameter may lead to wrong interpretations if the F/M ratio is to be used as a design parameter as suggested in many guidelines.

7.2.4 The mean hydraulic retention time

The mean hydraulic retention time, θ_h is a useful parameter relating the wastewater flow rate to the selected reactor volume:

$$\theta_h = \frac{V_R}{Q} \tag{7.9}$$

It is extensively used in design and as one of the main parameters in all mass balance equations. It should be noted however that θ_h is merely an average value and it does not give any indication regarding the mixing characteristics in the biological reactor, i.e. the applicable residence time distribution.

7.3 ASSESSMENT OF SYSTEM FUNCTIONS

Assessment of activated sludge operation and performance includes a check list of a number of fundamental functions that need to be analysed for appropriate design and operation of activated sludge systems:

(1) Amount of biomass maintained in the reactor
(2) Excess biomass generated
(3) Effluent quality
(4) Amount of oxygen utilized
(5) Recycle ratio
(6) Nutritional requirements

Evaluation of the first five functions has the same basis for both domestic sewage and industrial wastewaters. Analysis of nutritional balance is a specific requirement for industrial wastewaters, which are mostly nutrient deficient.

Assessment of these functions relies on mass balance for the specific process components involved. Mass balance is usually established around the inlet and outlet streams. Other control sections may also be selected as required for specific parameters such as the recirculation ratio, R. For activated sludge systems, mass balance generally involves a number of simplifying assumptions.

(1) No biochemical reactions take place in the settling tank
(2) Biomass storage in the settling tank is negligible
(3) The influent stream contains no biomass
(4) Complete settling is achieved ($X_e \approx 0$) so that the waste stream includes all excess biomass
(5) The biological reactor has the mixing characteristics of a single CSTR

The process diagram used for mass balance for carbon removal is schematically given in Figure 7.2.

Mass balance requires adoption of a mechanistic model for description and incorporation of different biochemical processes involved. In the following sections, the *endogenous decay* model is selected for the evaluations. In essence, interpretation of specific system functions for design purposes requires only two major model coefficients, the *heterotrophic yield coefficient*, Y_H and the *endogenous decay rate*, b_H together with the selected sludge age, θ_X. The adopted mechanistic model, although not explicitly included in the assessment process, should be used to support and verify every step of the evaluation.

7.3.1 Reactor biomass

Unlike the traditional approach which involves a single overall biomass parameter usually characterised by VSS, the recent mechanistic models, including the endogenous decay model adopted for the evaluation, differentiate at least three biomass components, namely active heterotrophic biomass X_H, influent inert particulate COD, X_{I1} and inert particulate microbial products, X_P.

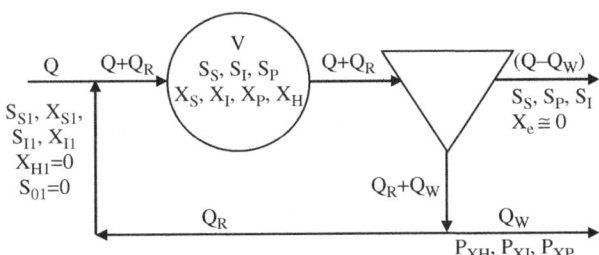

Figure 7.2. Schematic process diagram of the activated sludge process for organic carbon removal

Possible inclusion of the remaining portion of the slowly biodegradable COD, X_S entrapped in biomass is neglected with the assumption that system design secures total utilization of all available substrate. This assumption needs to be verified by model simulation.

7.3.1.1 Active heterotrophic biomass

The amount of active heterotrophic biomass M_{XH} cannot be arbitrarily selected as usually attempted in the practice. It is set as a function of (i) sludge age, θ_X and (ii) influent biodegradable COD load, QC_{S1}. It is best derived from the basic mass balance for biodegradable substrate expressed in terms of the specific substrate removal rate, q which indicates by definition the amount of substrate removed per unit amount of active biomass per day (g COD/g cell COD. day):

$$QC_{S1} - QC_S - qV_R X_H = 0 \qquad (7.10)$$

Based on process kinetics, q equals μ_H/Y_H and also approximates the F/M ratio with the assumption of complete removal of biodegradable substrate ($C_S \approx 0$):

$$q = \frac{\mu_H}{Y_H} = \frac{QC_{S1}}{V_R X_H} \qquad (7.11)$$

This equation may be rearranged to define M_{XH}:

$$M_{XH} = V_R X_H = Y_{NH} QC_{S1} \theta_X \qquad (7.12)$$

The above equation may be used to derive another meaningful parameter, m_{XH} which defines the amount of active biomass sustained in the reactor per unit wastewater flow:

$$m_{XH} = \frac{M_{XH}}{Q} = Y_{NH} C_{S1} \theta_X \qquad (7.13)$$

Evaluation of m_{XH} allows a general evaluation of reactor active biomass which can be coupled to any selected wastewater flow in the design procedure. The active biomass concentration in the reactor may then be calculated as:

$$X_H = Y_{NH} C_{S1} \frac{\theta_X}{\theta_h} \qquad (7.14)$$

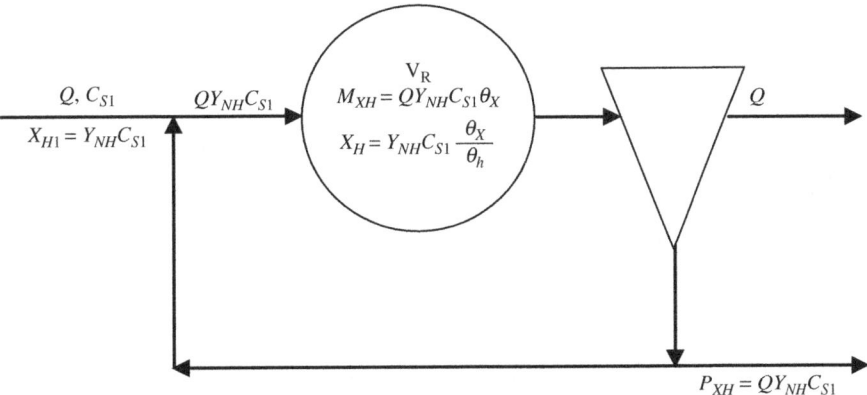

Figure 7.3. Active biomass accumulation in the biological reactor

Equations (7.12–7.14) provide the basis for an alternative evaluation which drastically simplifies the substrate/biomass interaction in the activated sludge systems: In fact, the term $Y_{NH} C_{S1}$ may be interpreted as a fictitious active biomass concentration in the influent stream, X_{H1} equivalent to available substrate, C_{S1} as shown in Figure 7.3. The system accumulates this biomass concentration by a factor θ_X/θ_h. More simply, the influent generates an equivalent active biomass load of $Y_{NH} Q C_{S1}$ which is accumulated in the reactor by a factor of θ_X to satisfy the selected sludge age for the system. As will be shown in the following sections, this accumulation holds true for all particulate components.

7.3.1.2 Inert particulate microbial products

The following mass balance equation describes the generation of particulate inert microbial products, X_P:

$$-P_{XP} + V_R f_{EX} b_H X_H = 0 \tag{7.15}$$

where, P_{XP} = the daily rate of particulate inert microbial products generation.

The general definition of θ_X equally applies to X_P:

$$P_{XP} = \frac{V_R X_P}{\theta_X} \tag{7.16}$$

The value of X_P can be computed by substituting the value of P_{XP} in equation (7.15):

$$X_P = f_{EX} b_H X_H \theta_X \tag{7.17}$$

or,

$$\frac{X_P}{X_H} = f_{EX} b_H \theta_X \qquad (7.18)$$

The last expression shows that the X_{XP}/X_H ratio increases with the sludge age. The sum of the two biomass fractions X_P and X_H can be conveniently expressed as follows:

$$X_H + X_P = X_H(1 + f_{EX} b_H \theta_X) \qquad (7.19)$$

Accordingly in design calculations, X_H is first computed and then multiplied by $(1 + f_{EX} b_H \theta_X)$ to easily obtain the level of $X_H + X_P$ sustained in the reactor. A similar approach also applies for the mass of particulate inert microbial products, M_{XP}:

$$M_{XP} = M_{XH} f_{EX} b_H \theta_X \qquad (7.20)$$

and, $\quad M_{XH} + M_{XP} = QY_{NH} C_{S1} \theta_X (1 + f_{EX} b_H \theta_X) \qquad (7.21)$

7.3.1.3 Inert particulate COD of influent origin

The corresponding mass balance for this component is quite simple, because it does not include a reaction term. This fraction is basically entrapped and accumulated in the biomass and leaves the biological reactor with the sludge wasting:

$$QX_{I1} - P_{XI} = 0 \qquad (7.22)$$

where, P_{XI} = the daily rate of influent inert particulate COD wastage

Expressing P_{XI} as a function of θ_X:

$$P_{XI} = \frac{V_R X_I}{\theta_X} \qquad (7.23)$$

and, substituting it into equation (7.22) yields:

$$X_I = X_{I1} \frac{\theta_X}{\theta_h} \qquad (7.24)$$

and, $\quad M_{XI} = V_R X_I = QX_{I1} \theta_X \qquad (7.25)$

Continuous flow technology 191

These two equations provide additional clarification to the accumulation of influent COD fractions in the reactor setting an example of a non-reactive process component: The mass is accumulated by θ_X and the concentration by θ_X / θ_h. Similarly the X_I/X_H ratio may be expressed as:

$$\frac{X_I}{X_H} = \frac{X_{I1}}{Y_{NH}C_{S1}} \tag{7.26}$$

This expression shows that X_I/X_H ratio remains independent of the sludge age and it is set by $Y_{NH}C_{S1}$ defining the active biomass equivalent of available biodegradable COD in the influent. In other words the ratio in the influent is also maintained in the reactor volume.

7.3.1.4 Total biomass in the reactor

The total biomass in the reactor, M_{XT} is calculated by adding the amount of three main particulate biomass components:

$$M_{XT} = M_{XH} + M_{XP} + M_{XI} \tag{7.27}$$

The value of M_{XT} is obtained by combining equations (7.21) and (7.25):

$$M_{XT} = Q[Y_{NH}C_{S1}(1 + f_{EX}b_H\theta_X) + X_{I1}]\theta_X \tag{7.28}$$

A similar general expression holds true for the total biomass concentration, X_T:

$$X_T = X_H + X_P + X_I \tag{7.29}$$

The total biomass concentration can be calculated using equations (7.19) and (7.24):

$$X_T = [Y_{NH}C_{S1}(1 + f_{EX}b_H\theta_X) + X_{I1}]\frac{\theta_X}{\theta_h} \tag{7.30}$$

These expressions indicate that the viability of activated sludge is significantly affected by the sludge age selected for system operation. The variation of biomass viability (X_H/X_T) as a function of the sludge age is illustrated in the following example.

Example 7.2 Variation of X_H/X_T ratio with the sludge age

Evaluate the variation of the X_H/X_T ratio with the sludge age for an activated sludge system treating an industrial wastewater with the following characteristics. Calculate the corresponding X_H/X_T ratio for $\theta_X = 5$ days and $\theta_X = 15$ days.

$C_{S1} = 1250$ mg/l
$X_{I1} = 150$ mg/l
$Y_H = 0.64$ g cell COD/gCOD
$b_H = 0.15$/day
$f_{EX} = 0.2$

(a) Using equation (7.14)

$$X_H \theta_h = Y_{NH} C_{S1} \theta_X$$

Similarly from equation (7.30)

$$X_T \theta_h = [Y_{NH} C_{S1}(1 + f_{EX} b_H \theta_X) + X_{I1}] \theta_X$$

Then,

$$\frac{X_H}{X_T} = \frac{Y_{NH} C_{S1} \theta_X}{[Y_{NH} C_{S1}(1 + f_{EX} b_H \theta_X) + X_{I1}] \theta_X}$$

or

$$\frac{X_H}{X_T} = \frac{1}{(1 + f_{EX} b_H \theta_X) + \frac{X_{I1}}{Y_{NH} C_{S1}}}$$

where, $Y_{NH} = \dfrac{Y_H}{1 + b_H \theta_X}$

Using the data for industrial wastewater:

$$\frac{X_H}{X_T} = \frac{1}{(1 + 0.2 \times 0.15 \times \theta_X) + \frac{150 \times (1 + 0.15 \times \theta_X)}{0.64 \times 1250}}$$

and

$$\frac{X_H}{X_T} = \frac{1}{1.18 + 0.058 \times \theta_X}$$

The variation of $\frac{X_H}{X_T}$ with the sludge age, θ_X is plotted in Figure 7.4.

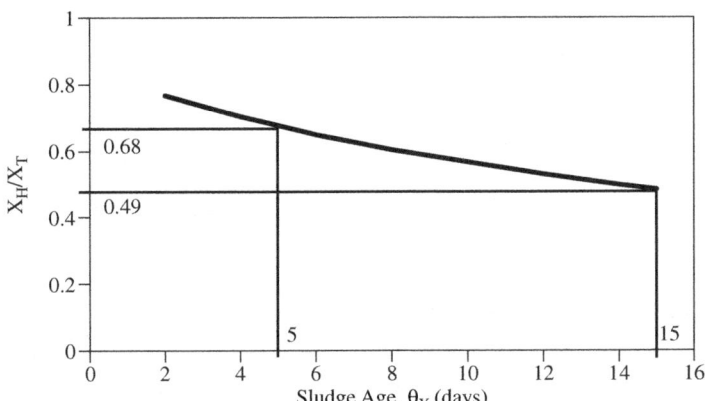

Figure 7.4. Variation of X_H/X_T with the sludge age

(b) $\theta_X = 5$ days;

$$Y_{NH} = \frac{0.64}{1+0.15\times 5} = 0.37 \text{ g cell COD/g COD}$$

$$X_H\theta_h = Y_{NH}C_{S1}\theta_X$$

$$X_H\theta_h = 0.37\times 1250\times 5 = 2312 \text{ mg cell COD.day/l}$$

$$X_T\theta_h = [0.37\times 1250\times(1+0.2\times 0.15\times 5)+150]\times 5$$

$$X_T\theta_h = [462.5\times(1+0.15)+150]\times 5$$

$$X_T\theta_h = 681.3\times 5 = 3406 \text{ mg COD.day/l}$$

$$\frac{X_H}{X_T} = \frac{2312}{3406} = 0.68 \text{ g cell COD/g COD}$$

or $\quad \dfrac{X_H}{X_T} = \dfrac{1}{(1+f_{EX}b_H\theta_X)+\dfrac{X_{I1}}{Y_{NH}C_{S1}}}$

$$\frac{X_H}{X_T} = \frac{1}{(1+0.2\times 0.15\times 5)+\dfrac{150\times(1+0.15\times 5)}{0.64\times 1250}}$$

$$\frac{X_H}{X_T} = \frac{1}{1.18 + 0.058 \times 5} = 0.68 \text{ g cell COD/g COD}$$

$\theta_X = 15$ days;

$$\frac{X_H}{X_T} = \frac{1}{1.18 + 0.058 \times 15} = 0.49 \text{ g cell COD/g COD}$$

7.3.2 Excess sludge production

The amount of biomass to be wasted for secure steady state operation must also be evaluated in terms of major particulate COD components of the biological reactor. The daily rate of active heterotrophic biomass generation, P_{XH} may be calculated combining the following basic mass balance equation for X_H with equation (7.12):

$$-P_{XH} + V_R X_H (\mu_H - b_H) = 0 \qquad (7.31)$$

$$\text{and} \quad P_{XH} = Q Y_{NH} C_{S1} \qquad (7.32)$$

The same value may also be obtained using the concept of equivalent active biomass load of influent biodegradable substrate. In this expression, P_{XH} is simply the fictitious influent biomass concentration, $Y_{NH} C_{S1}$ multiplied by the wastewater flow rate, Q.

Similarly, the rate of particulate inert microbial products generation to be wasted on a daily basis, P_{XP} can be computed directly from the corresponding mass balance equation (7.15):

$$P_{XP} = f_{EX} b_H V_R X_H \qquad (7.33)$$

or, using the value of $V_R X_H$ in equation (7.12):

$$P_{XP} = Q Y_{NH} C_{S1} f_{EX} b_H \theta_X \qquad (7.34)$$

Consequently, excess sludge production through generation of active and endogenous biomass fractions may be conveniently expressed as:

$$P_{XH} + P_{XP} = Q C_{S1} Y_{NH} (1 + f_{EX} b_H \theta_X) \qquad (7.35)$$

The contribution to the amount of excess sludge from inert particulate COD of influent origin can be directly computed from the mass balance equation (7.22):

$$P_{XI} = QX_{I1} \tag{7.36}$$

It is possible to define X_{I1} in terms of the influent biodegradable COD fraction, C_{S1}:

$$X_{I1} = \frac{f_{XI}}{f_S} C_{S1}$$

where, $\quad f_{XI} = X_{I1}/C_{T1}$ and $f_S = C_{S1}/C_{T1}$

then

$$P_{XI} = Q \frac{f_{XI}}{f_S} C_{S1} \tag{7.37}$$

The total daily excess sludge, P_{XT} may then be calculated in terms of its components:

$$P_{XT} = QY_{NH}C_{S1}(1 + f_{EX}b_H\theta_X) + Q\frac{f_{XI}}{f_S}C_{S1} \tag{7.38}$$

P_{XT} may also be defined introducing an overall sludge yield coefficient, Y_N:

$$P_{XT} = Y_N Q C_{S1} \tag{7.39}$$

where,

$$Y_N = Y_{NH}(1 + f_{EX}b_H\theta_X) + \frac{f_{XI}}{f_S} \tag{7.40}$$

Y_N is a useful design parameter defining the amount of sludge produced per unit amount of available biodegradable substrate:

$$Y_N = \frac{P_{XT}}{QC_{S1}} \tag{7.41}$$

The above expressions indicate that total excess sludge, P_{XT} varies as a function of the sludge age and the particular inert COD fraction in the industrial wastewater. It is important to remember that P_{XT} cannot give the exact value for the total amount of sludge to be discharged as it does not include the inorganic particulate matter accumulated in the reactor either from the influent stream or from the inorganic residue of endogenous decay. Unless otherwise determined for industrial discharges where the inorganic solids are important, the conversion factor of $i_{SS,COD} = 0.9$ g TSS/g COD suggested by Gujer et al., (2000) for domestic sewage may also be adopted for industrial wastewaters. The unit of the total excess sludge can be converted to total suspended solids, P_{SS} (kg TSS/day), using the above conversion factor, $i_{SS,COD}$. The same conversion can also be used for the total biomass M_{SS} (kg TSS), and the total biomass concentration, X_{SS} (mg TSS/l). The effect of sludge age and inert particulate COD in the influent stream on excess sludge production is illustrated in Example 7.3.

Example 7.3 Variation of excess sludge generation with θ_X and X_{I1}

For an industrial wastewater with characteristics given below,
  (a) *evaluate the variation of total excess sludge per unit wastewater flow with the sludge age;*
  (b) *evaluate the variation of the P_{XI}/P_{XT} with the sludge age;*
  (c) *calculate the overall yield coefficient, Y_N for $\theta_X = 5$ days;15 days*

$C_{S1} = 1250$ mg/l
$C_{T1} = 1500$ mg/l
$X_{I1} = 150$ mg/l
$Y_H = 0.64$ g cell COD/gCOD
$b_H = 0.15$/day
$f_{EX} = 0.2$

(a) Equation (7.38) is used to define the amount of total excess sludge per unit wastewater flow:

$$\frac{P_{XT}}{Q} = Y_{NH}C_{S1}(1 + f_{EX}b_H\theta_X) + \frac{f_{XI}}{f_S}C_{S1}$$

or $\quad \dfrac{P_{XT}}{Q} = Y_{NH}C_{S1}(1 + f_{EX}b_H\theta_X) + X_{I1}$

For the industrial wastewater given in the example

$$f_{XI} = \frac{X_{I1}}{C_{T1}} = \frac{150}{1500} = 0.10$$

$$f_S = \frac{C_{S1}}{C_{T1}} = \frac{1250}{1500} = 0.83$$

and, $\quad \dfrac{P_{XT}}{Q} = \dfrac{0.64}{1+0.15 \times \theta_X} \times 1250 \times (1 + 0.2 \times 0.15 \times \theta_X) + \dfrac{0.1}{0.83} \times 1250$

$$\frac{P_{XT}}{Q} = \frac{950 + 46.5 \times \theta_X}{1 + 0.15 \times \theta_X}$$

The variation of $\frac{P_{XT}}{Q}$ with the sludge age is plotted in Figure 7.5.

(b) From equation (7.36):

$$\frac{P_{XI}}{Q} = X_{I1} = 150$$

Since $\quad \dfrac{P_{XT}}{Q} = \dfrac{950 + 46.5 \times \theta_X}{1 + 0.15 \times \theta_X}$

Then, $\quad \dfrac{P_{XI}}{P_{XT}} = \dfrac{150 + 22.5 \times \theta_X}{950 + 46.5 \times \theta_X}$

Since P_{XI} is not a function of the sludge age, the $\frac{P_{XI}}{P_{XT}}$ ratio exhibits as shown in Figure 7.6 the same decreasing trend as previously evaluated for $\frac{P_{XT}}{Q}$.

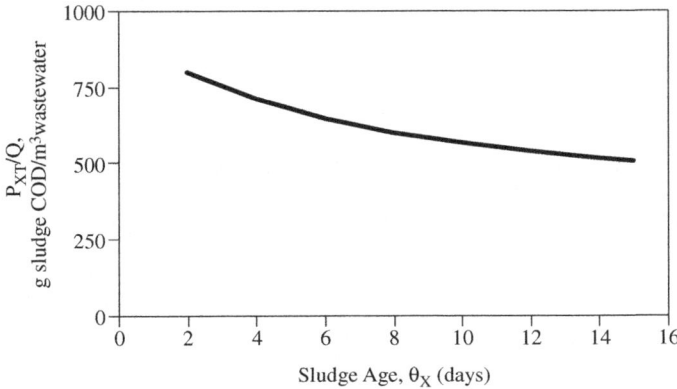

Figure 7.5. Variation of $\frac{P_{XT}}{Q}$ with the sludge age

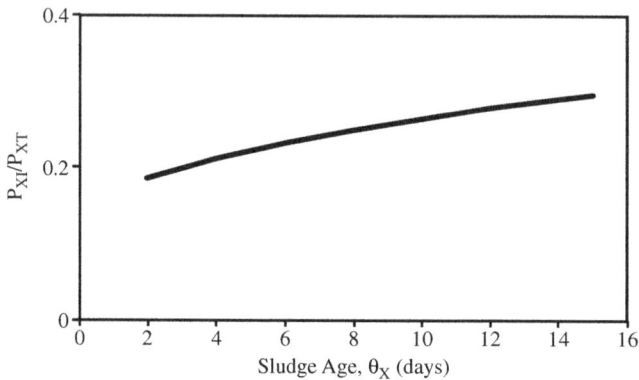

Figure 7.6. The variation of the $\frac{P_{XI}}{P_{XT}}$ ratio with the sludge age

(c) From equation (7.40):

$$Y_N = Y_{NH}(1 + f_{EX}b_H\theta_X) + \frac{X_{I1}}{C_{S1}}$$

For $\theta_X = 5$ days

$$Y_{NH} = \frac{0.64}{1 + 0.15 \times 5} = 0.37 \text{ g cell COD/g COD}$$

and,

$$Y_N = 0.37 \times (1 + 0.20 \times 0.15 \times 5) + \frac{150}{1250}$$

$$Y_N = 0.55 \text{ g cell COD/g COD}$$

For $\theta_X = 15$ days

$$Y_{NH} = \frac{0.64}{1 + 0.15 \times 15} = 0.20 \text{ g cell COD/g COD}$$

and,

$$Y_N = 0.20 \times (1 + 0.20 \times 0.15 \times 15) + \frac{150}{1250}$$

$$Y_N = 0.41 \text{ g cell COD/g COD}$$

7.3.3 Effluent quality

The level of effluent COD primarily depends upon the fate of soluble COD components in the reactor. In the biological treatment of industrial wastewaters, the mean hydraulic retention time, θ_h of the system is usually long enough to secure complete removal of soluble biodegradable COD fractions, S_S and S_H so that the effluent soluble COD practically involves (i) inert soluble COD from the influent, S_I and (ii) soluble residual microbial products, S_P. The particulate component of the effluent COD is essentially composed of biomass escaping from final settling.

The level of soluble inert microbial products in the effluent can be readily calculated from the following mass balance expression:

$$-QS_P + V_R f_{ES} b_H X_H = 0 \tag{7.42}$$

Substituting the value of $V_R X_H$ in equation (7.12) into the above expression yields:

$$S_P = Y_{NH} C_{S1} f_{ES} b_H \theta_X \tag{7.43}$$

The value of f_{ES} must be specifically determined for each industrial wastewater as indicated in the preceding chapters. In the absence of experimental data, a default value of $f_{ES} = 0.05$ may be adopted for calculating the magnitude of S_P. A typical approach for calculating the expected total effluent COD is defined in Example 7.4. For industrial wastewaters, it is advisable to check the effluent quality by model simulation.

Example 7.4 Effect of S_P on effluent quality

For an industrial wastewater with characteristics given below,
    (a) evaluate the variation of S_P with the sludge age
    (b) calculate the total effluent COD, C_{TE} for $\theta_X = 5$ days; $\theta_X = 15$ days

$C_{S1} = 1250$ mg/l
$S_{II} = 100$ mg/l
$Y_H = 0.64$ g cell COD/gCOD
$b_H = 0.15$/day
$f_{ES} = 0.05$
$f_X = 1.42$ g cell COD/g COD
effluent VSS = $X_{VSSE} = 40$ mg/l

(a) From Equation (7.43)

$$S_P = Y_{NH} C_{SI} f_{ES} b_H \theta_X$$

For the industrial wastewater in the example, S_P may be expressed as:

$$S_P = \frac{0.64}{1 + 0.15 \times \theta_X} \times 1250 \times 0.05 \times 0.15 \times \theta_X$$

$$S_P = \frac{6 \times \theta_X}{1 + 0.15 \times \theta_X}$$

The variation of S_P with the sludge age, θ_X is plotted in Figure 7.7.

(b) The effluent may be assumed to have only two soluble COD components, S_{I1} and S_P, since soluble biodegradable COD fractions are completely removed during biological treatment of industrial wastewaters. The other component of the effluent COD comes from biomass VSS escaping final settling:

$$C_{TE} = S_I + S_P + f_x X_{VSSE}$$

$$f_x X_{VSSE} = 1.42 \times 40 = 57 \text{ mg/l COD}$$

For $\theta_X = 5$ days

$$S_P = \frac{6 \times 5}{1 + 0.15 \times 5} = 17 \text{ mg/l COD}$$

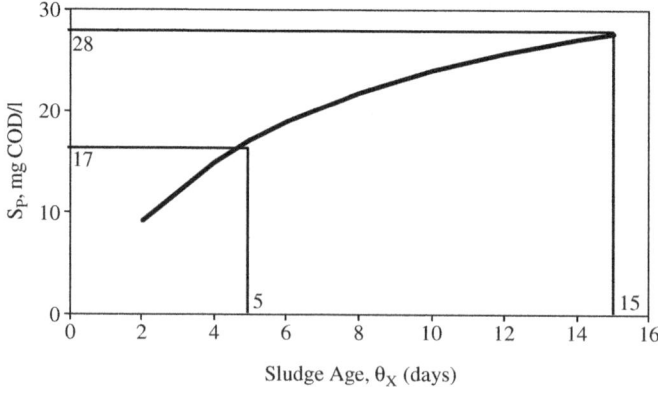

Figure 7.7. Variation of S_P with the sludge age

and, $$C_{TE} = 100 + 17 + 57 = 174 \text{mg/l}$$

For $\theta_X = 15$ days

$$S_P = \frac{6 \times 15}{1 + 0.15 \times 15} = 28 \text{ mg/l COD}$$

and, $$C_{TE} = 100 + 28 + 57 = 185 \text{ mg/l}$$

7.3.4 Oxygen consumption

The evaluation of oxygen utilization is basically the same as in the traditional modelling approach with the exception of the effect of microbial products generation. It essentially accounts for the electron acceptor requirements of microbial growth and endogenous respiration: One of the major parameters for oxygen consumption is the oxygen utilization rate, OUR, which defines the amount of oxygen utilization per unit amount of reactor volume per unit time. From a process kinetics standpoint, it is equivalent to the overall process rate for oxygen, r_{SO}:

$$OUR = r_{SO} = \left(\frac{1 - Y_H}{Y_H}\right)\hat{\mu}_H X_H + (1 - f_E)b_H X_H \quad (7.44)$$

where, $f_E = f_{EX} + f_{ES}$

The oxygen utilization rate, OUR is now significantly explored and constitutes the major parameter for the respirometric evaluation of biodegradation (Spanjers et al., 1998; Orhon et al., 2007). As the main experimental tool for the evaluation and modelling of different aspects of substrate biodegradation, the use of OUR has been extensively elaborated in the preceding chapters.

The total oxygen requirement can be derived from the expression which defines OUR, giving the same parameter per unit reactor volume:

$$OR = OUR \times V_R = QC_{S1}[1 - Y_{NH}(1 + f_E b_H \theta_X)] \quad (7.45)$$

It should be noted that the oxygen consumption considered above relates to organic carbon removal. In single sludge systems where nitrification takes place the oxygen utilization related to heterotrophic activity, OR_H should be differentiated from the oxygen requirement of nitrifiers, OR_A. Based on expression (7.43), the amount of oxygen required per unit amount of available biodegradable COD in the influent can be evaluated as follows:

$$\frac{OR}{QC_{S1}} = 1 - Y_{NH}(1 + f_E b_H \theta_X) \tag{7.46}$$

The calculated oxygen utilization, OR only defines the requirement corresponding to the metabolic activities of the biomass. This requirement should be calculated for the most critical operation temperature of the activated sludge system. It should be satisfied with the fraction of oxygen supplied by aeration that is available to the microorganisms. The oxygen transfer efficiency of aeration, e, however is usually defined under standard conditions. Therefore, OR should first be re-calculated as the oxygen requirement under standard conditions, SOR. Then the required air supply to the biological reactor should be computed based upon SOR and the oxygen transfer efficiency, e, which depends on operating conditions and the type of selected aeration equipment. The aeration rate must satisfy two different operating conditions: It should be lower than a critical upper value to avoid disruption of flocs and deterioration of settling properties. It should also secure the minimum mixing requirements of the biomass in the reactor. Detailed coverage of aeration system design is provided in different papers and textbooks (Groves *et al.*, 1992; U.S. EPA, 1989; Orhon and Artan, 1994; Grady *et al.*, 1989).

7.3.5 Recycle ratio

The recycle ratio, R is another significant parameter that needs to be set accurately for the operation of the activated sludge systems:

$$R = \frac{Q_R}{Q} \tag{7.47}$$

where, Q_R = the flow rate of biomass recycled.

The magnitude of R can be assessed in terms of a mass balance for total biomass around the settling tank:

$$(Q + Q_R)X_T - Q_R X_{RT} = P_{XT} \tag{7.48}$$

where, X_{SS} = the suspended solids concentration in the reactor
X_{RSS} = the suspended solids concentration in the recycle flow

R is defined by dividing both sides of equation (7.43) by VX_T and rearranging as:

$$R = \frac{1 - \frac{\theta_h}{\theta_X}}{\frac{X_{RSS}}{X_{SS}} - 1} \qquad (7.49)$$

In activated sludge systems, R is the main operating instrument for adjusting the biomass balance between the biological reactor and the settling tank. When the level of X_{RSS}, and consequently X_{RSS}/X_{SS} ratio, drops due to deterioration of settling properties, biomass accumulation in the settling tank can be avoided by increasing the recycle ratio. At high θ_X values when the θ_h/θ_X ratio becomes too low, Equation (7.47) is simplified into:

$$R = \frac{X_{SS}}{X_{RSS} - X_{SS}} \qquad (7.50)$$

In practice, the total biomass concentration in the sludge recycle stream is estimated on the basis of settling properties of activated sludge, as commonly measured by the *sludge volume index*, SVI. This experimental parameter indicates the volume of a unit amount of mixed liquor suspended solids, MLSS, after 30 minutes of settling:

$$SVI = \frac{V_S}{X_{SS}} \qquad (7.51)$$

where, SVI = sludge volume index, (ml/g)
X_{SS} = MLSS concentration, (g/l)
V_S = the volume of settled sludge, (ml/l)

With the assumption that the SVI experiment approximates biomass settling, the value of X_{RSS} can be correlated with the SVI level:

$$X_{RSS} \approx \frac{10^6}{SVI} \qquad (7.52)$$

An appropriate MLSS concentration can be estimated based on a selected design value for SVI. Well settling sludge should have an SVI value of less than 100 g/ml. Sludges with SVI values higher than 120 g/ml are considered as *bulking sludge*. The recommended range for the MLSS concentration is 3000–4000 mg/l. MLSS values over 5000 mg/l are not usually recommended mainly because they often lead to overloading of final settling tanks.

7.3.6 Nutrient balance

Nutritional requirements of active biomass have significant impact on the operation and performance of activated sludge systems. Biological treatment of industrial wastewaters mainly concerns organic carbon removal. As previously explained, two simultaneous biochemical processes are responsible for organic carbon removal: (i) oxidation to CO_2 through energy reactions; (ii) incorporation into biomass through biosynthesis reactions. Nutrients such as nitrogen and phosphorus are essentially removed during biosynthesis in proportions determined by the elementary composition of the active biomass. The relative magnitude of nutrient removal may be evaluated as a function of COD/N and COD/P ratios in the industrial wastewater to be treated. The chemical composition of microorganisms defines a biochemical threshold for these ratios. Partial nutrient removal occurs if these ratios are lower than the threshold level. This is the case for domestic sewage, which is usually characterised by lower COD/N and COD/P ratios, where complete COD removal is coupled with a fraction of available nutrients in the wastewater. Conversely if these ratios are higher than the microbial requirements, energy and biosynthesis reactions may not be fully coupled and removal efficiency of organic carbon is reduced. From a process kinetics standpoint, nutrients rather than organic carbon act as the rate limiting substrate. In this case, external addition of nutrients is required to balance the metabolic reactions and to increase system performance to the desired level of efficiency.

7.3.6.1 Nutrient removal

The nutrient removal potential of an activated sludge system under selected operating conditions may be easily evaluated using the basic reaction kinetics. The sludge age and the characteristics of the industrial wastewater to be treated are the essential elements of this evaluation. Table 7.1 gives the matrix format of the reaction kinetics for organic carbon removal including nitrogen (S_{NH}) and phosphorus (S_{PO}) as model components. Nitrogen for example is removed through microbial growth and released back to the solution through endogenous decay reduced for particulate endogenous residues. This mechanism in incorporated into the reaction matrix by means of a stoichiometric coefficient, i_{XN}, which defines the organic nitrogen fraction of the biomass (g N/g cell COD). A similar mechanism also applies to phosphorus with i_{XB} the phosphorus fraction of biomass (g P/g cell COD) acting as the corresponding stoichiometric coefficient. A default value of

Table 7.1. Reaction kinetics for nitrogen and phosphorus associated with organic carbon removal

Component →	1	2	3	4	5	6	7	Process rate
Process ↓	S_S	X_S	X_H	X_P	S_O	S_{NH}	S_{PO}	$ML^{-3}T^{-1}$
1 Growth	$-\dfrac{1}{Y_H}$		1		$-\dfrac{(1-H_Y)}{Y_H}$	$-i_{XN}$	$-i_{XP}$	$\hat{\mu}_H \left(\dfrac{S_S}{K_S + S_S} \right) X_H$
2 Decay		$1-f_{EX}$	-1	f_{EX}		$i_{XN}(1-f_{EX})$	$i_{XP}(1-f_{EX})$	$b_H X_H$
3 Hydrolysis	1	-1						$k_h \left(\dfrac{X_S/X_H}{K_x + (X_S/X_H)} \right) X_H$
Parameter, ML^{-3}	COD	COD	Cell COD	COD	O_2	NH_3-N	PO_4-P	

0.086 g N/g cell COD was suggested for i_{XN} and 0.02 g P/g cell COD for i_{XP} in ASM1 and ASM2, respectively. (Henze *et al.*, 1987; 1995).

The overall nitrogen removal rate, r_N per unit reactor volume can be derived as follows from the reaction matrix in Table 7.1:

$$r_N = -i_{XN}\mu_H V_R X_H + i_{XN}(1 - f_{EX})b_H V_R X_H \qquad (7.53)$$

This expression indicates that the amount of nitrogen removed per day, P_N may be correlated by P_{XHP} defined in equation (7.35) with i_{XN}:

$$P_N = i_{XN} P_{XHP} = i_{XN} Q Y_{NH} C_{S1}(1 + f_{EX} b_H \theta_X) \qquad (7.54)$$

Consequently, the amount of nitrogen removed per unit volume of wastewater treated, N_X may be calculated as:

$$N_X = \frac{P_N}{Q} = i_{XN} Y_{NH} C_{S1}(1 + f_{EX} b_H \theta_X) \qquad (7.55)$$

It should be noted that N_X has a concentration unit and conveniently represents the fraction of the influent total nitrogen concentration that is removed by heterotrophic biomass. Similarly, the amount of phosphorus removed per unit volume of wastewater treated, PO_X may be computed as follows:

$$PO_X = i_{XP} Y_{NH} C_{S1}(1 + f_{EX} b_H \theta_X) \qquad (7.56)$$

7.3.6.2 *Nutrient requirements*

Nutrient deficiency is a common feature for most industrial wastewaters. Therefore, the magnitude of nitrogen and phosphorus available in the wastewater should be checked and compared with the biochemical threshold levels for COD/N and COD/P ratios required for efficient removal of organic carbon. The stoichiometric coefficients for nitrogen and phosphorus, i_{XN} and i_{XP} together with the heterotrophic yield, Y_H provide the basic stoichiometry for evaluating the respective threshold levels. In fact, complete removal of a unit amount of biodegradable COD generates Y_H amount of active biomass through microbial growth and requires $i_{XN} Y_H$ amount of nitrogen available in the wastewater to be incorporated into biomass. Therefore the COD/N ratio should be equal or lower than $1/(i_{XN} Y_H)$ on a biodegradable COD basis to satisfy the minimum nitrogen requirement. Expressed in terms of total influent COD, this ratio becomes:

$$\frac{COD}{N} \leq \frac{1}{f_S i_{XN} Y_H} \qquad (7.57)$$

where, $f_S = C_{S1}/C_{T1}$

For phosphorus the same ratio may be expressed as:

$$\frac{COD}{P} \leq \frac{1}{f_S i_{XP} Y_H} \qquad (7.58)$$

It should be noted that the above ratios are calculated based on Y_H and do not take into account nutrient release through endogenous respiration. This approach is justifiable in view of the fact that the level of nutrients available should secure in the first place microbial growth which should proceed independently from by-products of endogenous decay. A numerical evaluation of nutritional requirements of an industrial wastewater is presented in Example 7.5:

Example 7.5 Nutritional requirements
For an industrial wastewater with characteristics given below, calculate
  (a) *the minimum levels of nitrogen and phosphorus that should be available in the influent to sustain effective organic carbon removal.*
  (b) *nitrogen and phosphorus removals achieved in an activated sludge system operated at a sludge age of 10 days for organic carbon removal.*

$C_{S1} = 1250$ mg/l
$C_{T1} = 1500$ mg/l
$C_{N1} = 85$ mg/l
$C_{PO1} = 10$ mg/l
$Y_H = 0.64$ g cell COD/g COD
$b_H = 0.15$/day
$f_{EX} = 0.2$
$i_{XN} = 0.08$ g N/g cell COD
$i_{XP} = 0.02$ g P/g cell COD

(a) Minimum nitrogen requirement:

$$f_S = \frac{C_{S1}}{C_{T1}} = \frac{1250}{1500} = 0.83$$

From equation (7.57)

$$\frac{COD}{N} = \frac{1}{f_{si_{XN}}Y_H} = \frac{1}{0.83 \times 0.08 \times 0.64} = 24$$

$$\frac{COD}{N} = \frac{1500}{N} = 24$$

$$\Delta N = 63 \text{ mg N/l}$$

Since the influent contains $C_{N1} = 85$ mg/l > 63 mg/l, it is not nitrogen deficient for full COD removal.

Minimum phosphorus requirement:

From equation (7.58)

$$\frac{COD}{P} = \frac{1}{f_{si_{XP}}Y_H} = \frac{1}{0.83 \times 0.02 \times 0.64} = 94$$

$$\frac{COD}{P} = \frac{1500}{P} = 94$$

$$\Delta P = 16 \text{ mg P/l}$$

A minimum of 16 mg/l phosphorus should be presented in the influent for sustaining full removal of biodegradable COD. Since the influent phosphorus content $C_{P01} = 10$ mg/l < 16 mg/l, the industrial wastewater is deficient for phosphorus.

(b) $Y_{NH} = \dfrac{Y_H}{1 + b_H \theta_X} = \dfrac{0.64}{1 + 0.15 \times 10} = 0.26$ g cell COD/g COD

From equation (7.55):

$$N_X = i_{XN} Y_{NH} C_{S1}(1 + f_{EX} b_H \theta_X)$$

$$N_X = 0.08 \times 0.26 \times 1250 \times (1 + 0.2 \times 0.15 \times 10)$$

$$N_X = 34 \text{ mg N/l}$$

Since the nitrogen level in the influent is $C_{N1} = 85$ mg/l, the nitrogen removal achieved at $\theta_X = 10$ days:

$$S_N = C_{N1} - N_X$$

$$S_N = 85 - 34 = 51 \text{ mg N/l}$$

From equation (7.56):

$$PO_X = i_{XP} Y_{NH} C_{S1} (1 + f_{EX} b_H \theta_X)$$

$$PO_X = 0.02 \times 0.26 \times 1250 \times (1 + 0.2 \times 0.15 \times 10)$$

$$PO_X = 8.5 \text{ mg P/l}$$

At this sludge age, the influent phosphorus level appears to be sufficient, considering phosphorus release through endogenous respiration.

7.4 PROCESS DESIGN FOR ORGANIC CARBON REMOVAL

Design of activated sludge systems for industrial wastewaters is generally based on empirical engineering experience and expertise. A number of different sources provide information for detailed engineering involved in the design of activated sludge systems (Benefield and Randall, 1980; Archievala, 1981; ATV, 1991; WEF and ACSE, 1991; Eckenfelder and Grau, 1992; Orhon and Artan, 1994; Grady *et al.*, 1999; Metcalf & Eddy, 2003). This section does not intend to introduce another design version but to focus on the fundamental basis for the evaluation of the activated sludge treatment of industrial wastewaters. It attempts to describe and underline the basic relationships between mass balance and design parameters used in operation. Modelling is not directly involved in the design but it is an integral part of the procedure. It is implicit in establishing the basic stoichiometry for major process components. Furthermore, verification of results by model evaluation and simulation is always strongly recommended as part of conceptual design.

7.4.1 The concept of pre-treatment

The first step before process design is to make sure that the characteristics of the influent is optimally suitable for biological treatment. For most industrial wastewaters this is not the case and a *pre-treatment* step needs to be incorporated as a corrective measure to safeguard the efficient operation of the activated sludge process at the following stage. Generally, pre-treatment refers to the necessary level of preliminary treatment applied to individual industrial polluters in order to comply with discharge to sewer conditions. After pre-treatment, different waste streams are usually collected in a joint treatment plant. Different aspects of pre-treatment practice are well documented in the literature (WEF, 1994; U.S. EPA, 1996).

Pre-treatment as applied to individual biological treatment plants for industrial wastewaters essentially relies on the following principles:

(i) Protection of the biological process from wastewater flow and pollutant load transients;
(ii) More effective removal of certain specific pollutants, such as inorganic solids, oil and grease, etc., before biological treatment
(iii) Removal of pollutants likely to exert inhibitory and toxic effects on biological treatment
(iv) Reduction of the organic carbon load of biological treatment.

The production scheme in most industrial plants involves a sequence of batch operations with different rates of wastewater generation. There may be short-term surges of wastewater discharges after specific batch operations. Some plants only work with one of two shifts a day and the daily wastewater discharge becomes restricted to 8–16 working hours of the day. Furthermore, a treatment plant that would operate only during a selected period of the day or a single shift may be evaluated as a more feasible option. All these cases require *equalization* as the major component of pre-treatment. The equalization step does not only provide a uniform flow regime for the influent of the biological treatment unit but it also homogenizes fluctuations in pollutant loads and pH variations.

Certain pollutants are more effectively removed at the pre-treatment stage. For example, *the pumice stone* used for stone bleaching in denim processing can be fully removed by plain settling. Similarly, the *oil and grease* content of the edible oil industry effluent is best removed by flotation or an equivalent physico-chemical process before biological treatment. Wastewater pH has to be adjusted to the optimal range of 7.5–9.0 for biochemical reactions. A sequence of coagulation, flocculation and chemical settling is often used for the removal *of heavy metals* and *other particulate chemicals* for eliminating their adverse effects on biological processes. Extensive information on different processes, incorporated as pre-treatment before biological units, is provided in different sources (WEF, 1994; U.S. EPA, 1996).

Generally, plain or chemical settling is used to reduce the COD (organic carbon) load of the following biological treatment stage. The feasibility of preliminary settling much depends on the particle size distribution of the influent COD. Settling and filtration can be conveniently used as useful size indexes for

Table 7.2. Effect of COD size distribution on the selection of required pre-treatment

COD fractions	Industrial wastewaters		
	Sample 1	Sample 2	Sample 3
Total COD (mg/l)	1000	1000	1000
Settled COD (mg/l)	200	900	950
Filtered COD (mg/l)	100	100	900

COD fractionation. For this purpose, experimental assessment of *settled COD*, measured in the supernatant after two-hour settling and *filtered COD*, determined in the filtrate after 0.45 µm filtration may be quite helpful in evaluating the potential of plain or chemical settling as pre-treatment. Filtered COD provides a conservative approximation for chemical settling, which may be slightly more efficient as a result of partial COD adsorption on chemical flocs. The use of total COD, settled COD and filtered COD fractions as indicators of pre-treatment potential is illustrated with the data in Table 7.2, giving the COD size distribution characteristics of three different industrial wastewater samples: In Sample 1, 80% of the COD is settleable and influent COD can be reduced to 200 mg/l by plain settling; in Sample 2, plain settling remains quite ineffective as it removes only 10% of the COD but presumably, the colloidal nature of the remaining COD would favour chemical settling after coagulation and flocculation, which would secure more than 90% COD removal; the results for Sample 3 provide a clear indication that pre-treatment either by plain settling or by chemical settling will not be needed as 90% of the COD is filterable.

Removal of the COD load to the extent possible prior to biological treatment represents the traditional approach which somewhat overlooks the trade between liquid and solid waste streams and the cost of sludge treatment. A more rational approach today should involve the feasibility of the selected treatment scheme considering both wastewater treatment and sludge disposal. It should also be noted that organic carbon in biological treatment is an essential ingredient for nutrient removal.

7.4.2 Conceptual design procedure

The procedure for conceptual design basically involves calculation of essential entities and functions of the system as elaborated in the previous sections. These entities and functions then serve as the basis for detailed engineering.

Full insight and understanding of all the production steps in the industrial facility is an essential prerequisite for appropriate design. In this comprehensive survey, a complete *pollution profile* of the plant must be established with identification of different wastewater streams. Wastewater quantity and quality should be determined for each individual stream and the composite effluent. This evaluation would indicate if one or a few individual streams need to be segregated for separate treatment or pre-treatment before joint biological treatment.

Conventional characterization should always be accompanied by experimental assessment of influent COD fractions and model coefficients indicating the specific biodegradation characteristics of the wastewater to be studied. These coefficients provide the necessary stoichiometric and kinetic feedback to the selected model. It should be noted that only a few of these coefficients such as Y_H, b_H, etc., are actually incorporated into design expressions but the entire specific information related to process stoichiometry and kinetics is needed to make the selected model functional and relevant for design verification, calibration of the experimental data and prediction of system performance under different operating conditions.

Table 7.3 compiles necessary information on COD fractionation for a typical industrial wastewater. It also includes similar information for a typical domestic sewage for the main purpose of serving as a basis for comparison. Similarly, selected values for stoichiometric and kinetic coefficients taking part in the endogenous decay model adopted for evaluating organic carbon removal are listed in Table 7.4 both for the industrial wastewater and domestic sewage. These values will be used in all examples throughout the text, where necessary.

After collection of all necessary characteristics defining the industrial wastewater to be treated, the design starts by selecting an appropriate value for the sludge age, θ_X that will be adopted for the operation of the activated sludge system. It involves a stepwise evaluation approach as schematically given in Figure 7.8, based on specific expressions and relationships defined for major design parameters and functions. In this framework, the recommended

Table 7.3. COD fractionation for a typical industrial wastewater and domestic sewage

Parameter	Industrial wastewater		Domestic sewage	
	mg/l	% C_{T1}	mg/l	% C_{T1}
Total COD, C_{T1}	1500		500	
Total Soluble COD, S_{T1}	1125	75	175	35
Total Particulate COD, X_{T1}	375	25	325	65
Total Biodegradable COD, C_{S1}	1250	83	425	85
Readily Biodegradable COD, S_{S1}	225	15	50	10
Rapidly Hydrolysable COD, S_{H1}	800	53	100	20
Slowly Hydrolysable COD, X_{S1}	225	15	275	55
Total Inert COD, C_{I1}	250	17	75	15
Soluble Inert COD, S_{I1}	100	7	25	5
Particulate Inert COD, X_{I1}	150	10	50	10

Table 7.4. Stoichiometric and kinetic coefficients for a typical industrial wastewater and domestic sewage

Parameter	Unit	Industrial wastewater	Domestic sewage
$\hat{\mu}_H$	day^{-1}	4.0	6.0
K_S	mg COD/l	15.0	5.0
k_{hS}	day^{-1}	2.0	3.5
K_{XS}	mg COD/mg COD	0.2	0.1
k_{hX}	day^{-1}	1.0	1.5
K_{XX}	mg COD/mg COD	0.2	0.2
b_H	day^{-1}	0.15	0.15
Y_H	mg cell COD/mg COD	0.64	0.64
f_{EX}		0.20	0.20
f_{ES}		0.05	0.05
$i_{SS,COD}$	mg TSS/mg COD	0.90	0.90
f_{FS}	mg TSS/mg COD	0	0.07
i_{XN}	mg N/mg COD	0.08	0.08
i_{XP}	mg P/mg COD	0.02	0.02
SVI	ml/g TSS	120	100

procedure for the design of an activated sludge system for organic carbon removal from the typical industrial wastewater defined in Tables 7.3 and 7.4 is illustrated in Example 7.6.

Example 7.6 Activated sludge design for organic carbon removal

Design an activated sludge system for organic carbon removal, for an industrial wastewater with characteristics given below and in Tables 7.4 and 7.5.

Wastewater flow, $Q = 100$ m^3/day
Total Influent COD, $C_{TI} = 1500$ m^3/day
Total influent nitrogen, $C_{NI} = 85$ mg/l
Total influent phosphorus, $C_{POI} = 10$ mg/l
Make all assumptions as necessary

(1) Select θ_X:
A θ_X value of 8 days is selected for design, based on experience to ensure satisfactory settling properties of biomass

(2) Calculate Y_{NH}:

$$Y_{NH} = \frac{Y_H}{1 + b_H \theta_X} = \frac{0.64}{1 + 0.15 \times 8} = 0.29 \text{ g cell COD/g COD}$$

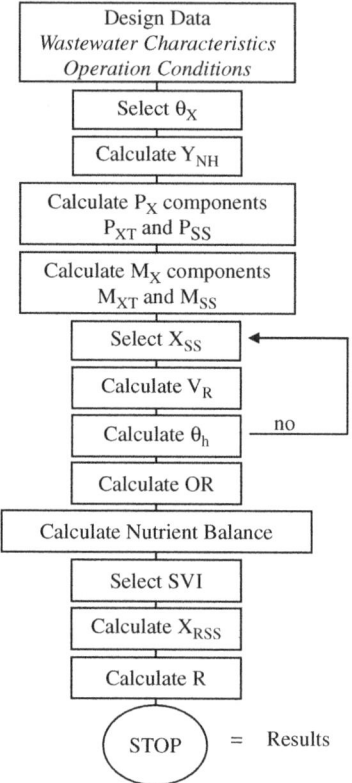

Figure 7.8. Design algorithm for organic carbon removal

(3) Calculate P_X components:

$$Q = 100 \text{ m}^3/day$$

$$C_{SI} = 1250 \text{ mg/l} = 1.25 \text{ kg/m}^3$$

From equation (7.32)

$$P_{XH} = QY_{NH}C_{SI}$$

$$P_{XH} = 100 \times 0.29 \times 1.25 = 36.25 \text{ kg cell COD/day}$$

$$b_H = 0.15 \text{ day}^{-1}$$

$$f_{EX} = 0.20$$

From equations (7.32) and (7.35)

$$P_{XHP} = P_{XH} + P_{XP} = P_{XH}(1 + f_{EX}b_H\theta_X)$$

$$P_{XHP} = 36.25 \times (1 + 0.20 \times 0.15 \times 8) = 44.95 \text{ kg COD/day}$$

$$X_{II} = 150 \text{ mg/l} = 0.15 \text{ kg/m}^3$$

From equation (7.36)

$$P_{XI} = QX_{II} = 100 \times 0.15 = 15 \text{ kg COD/day}$$

Calculate P_{XT}

$$P_{XT} = P_{XH} + P_{XP} + P_{XI}$$

$$P_{XT} = 44.95 + 15 = 59.95 \text{ kg COD/day}$$

Calculate P_{SS}

$$\text{Assume } i_{SS,COD} = 0.90 \text{ kg SS/kg COD}$$

$$P_{SS} = 59.95 \times 0.90 = 53.95 \text{ kg SS/day}$$

(4) Calculate M_X components:

$$M_{HX} = V_R X_H = P_{XH}\theta_X$$

$$M_{HX} = 36.25 \times 8 = 290 \text{ kg cell COD}$$

$$M_{XHP} = M_{XH} f_{EX} b_H \theta_X$$

$$M_{XHP} = 290 \times 0.20 \times 0.15 \times 8 = 69.6 \text{ kg COD}$$

$$M_{XI} = P_{XI}\theta_X = 15 \times 8 = 120 \text{ kg COD}$$

Calculate M_{XT}:

$$M_{XT} = M_{XH} + M_{XP} + M_{XI}$$

$$M_{XT} = 290 + 69.6 + 120 = 479.6 \text{ kg COD}$$

M_{XT} could also be calculated as:

$$M_{XT} = P_{XT}\theta_X = 59.95 \times 8 = 479.6 \text{ kg COD}$$

$$\text{Viability} = \frac{M_{XH}}{M_{XT}} \times 100 = \frac{290}{479.6} \times 100 = 60\%$$

60% of the total biomass is composed of active heterotrophs. Calculate M_{SS}

$$M_{SS} = M_{XT} i_{SS,COD} = 479.6 \times 0.9 = 431.6 \text{ kg SS}$$

(5) Select X_{SS}:
X_{SS} is selected as 4000 mg SS/l < 5000 mg/l

(6) Calculate V_R:

$$X_{SS} = 4000 \text{ mg/l} = 4 \text{ kg/m}^3$$

$$V_R = \frac{M_{SS}}{X_{SS}} = \frac{431.6}{4} = 107.9 \, m^3$$

$$\text{Select } V_R = 110 \, m^3$$

(7) Calculate θ_h:

$$\theta_h = \frac{V_R}{Q} = \frac{110}{100} = 1.1 \text{ days} = 26.4 \text{ hrs}$$

(8) Calculate OR:
From equation (7.45)

$$OR = QC_{S1}[1 - Y_{NH}(1 + f_E b_H \theta_X)]$$

$$f_E = f_{ES} + f_{EX} = 0.05 + 0.20 = 0.25$$

$$OR = 100 \times 1.25 \times [1 - 0.29 \times (1 + 0.25 \times 0.15 \times 8)]$$

$$OR = 125 \times (1 - 0.377) = 77.9 \text{ kg } O_2/\text{day}$$

Considering the long hydraulic retention time, mixing requirements should be checked and aeration should be provided to satisfy mixing requirements.

(9) Calculate nutrient balance:
  (9.1) Nitrogen requirement
$$C_{NI} = 85 \text{ mg/l}$$

From equation (7.57) the threshold COD/N ratio

$$\frac{COD}{N} = \frac{1}{f_S i_{XN} Y_H}$$

$$f_S = \frac{C_{S1}}{C_{T1}} = \frac{1250}{1500} = 0.83$$
$$i_{XN} = 0.08 \text{ g N/g cell COD}$$
$$Y_H = 0.64 \text{ g cell COD/g COD}$$

$$\frac{COD}{N} = \frac{1}{0.83 \times 0.08 \times 0.64} = 24$$

$$\frac{C_{T1}}{C_{N1}} = \frac{1500}{85} = 17.6$$

$$\frac{C_{T1}}{C_{N1}} < \frac{COD}{N}$$

Influent nitrogen is sufficient for effective carbon removal.

  (9.2) Phosphorus requirement:
$$C_{PO1} = 10 \text{ mg/l}$$

From equation (7.58) the threshold COD/P ratio

$$\frac{COD}{P} = \frac{1}{f_S i_{XP} Y_H}$$

$$i_{XP} = 0.02 \text{ g P/g cell COD}$$

$$\frac{COD}{P} = \frac{1}{0.83 \times 0.02 \times 0.64} = 94$$

$$\frac{C_{T1}}{C_{PO1}} = \frac{1500}{10} = 150$$

$$\frac{C_{T1}}{C_{PO1}} > \frac{COD}{P}$$

Influent phosphorus is not sufficient. Addition of at least 5 mg/l phosphorus is necessary to sustain effective organic carbon removal.

(10) Select SVI:

It is assumed that the SVI of activated sludge could be maintained under 120 ml/g.

(11) Calculate X_{RSS}:
From equation (7.52)

$$X_{RSS} = \frac{10^6}{120} \cong 8300 \text{ mg SS/l}$$

(12) Calculate R:
From equation (7.49)

$$R = \frac{1 - \frac{\theta_h}{\theta_X}}{\frac{X_{RSS}}{X_{SS}} - 1} = \frac{1 - \frac{1.1}{8}}{\frac{8300}{4000} - 1} = \frac{0.86}{1.075}$$

$$R = 0.8$$

Table 7.5. Summary of activated sludge design for the selected industrial wastewater and domestic sewage

Parameter	Unit	Industrial wastewater	Domestic sewage
Q	m³/day	100	100
θ_X	days	8	8
P_{XT}	kg COD/day	59.95	20.3
P_{SS}	kg SS/day	53.95	18.3
Y_N	kg SS/kg COD	0.43	0.42
M_{XT}	kg COD	479.6	162.6
M_{SS}	kg SS	431.6	146.4
X_{SS}	mg/l	4000	4000
V_R	m³	110	36
θ_h	days	1.1	0.37
θ_h	hrs	26.4	8.9
OR	kg O₂/day	77.87	26.5
Nutrient Balance		P deficient	Excess N & P
SVI	ml/g	120	100
X_{RSS}	mg/l	8300	10.000
R		0.8	0.64

Similar design calculations are also performed for the domestic sewage described in Tables 7.3 and 7.4. Results are listed in Table 7.5 to visualize the effect of wastewater characterization on system design.

7.5 BIOLOGICAL NITROGEN REMOVAL

Nitrogen discharge is a major concern, especially in sensitive areas, regardless of the source. Appropriate techniques for effective nitrogen removal are well documented for domestic sewage (WPCF, 1983; Bohnke, 1989; Orhon and Artan, 1994). They rely on the same biochemical principles and equally apply to industrial wastewaters. In a typical domestic sewage the influent COD/N ratio is almost constant and mostly varies in a narrow range around 10 g COD/g N. For most industrial wastewaters this ratio remains quite low so that a security check should be performed for nitrogen deficiency. Consequently the main problem for these types of wastewaters is not nitrogen removal but supply of convenient nitrogen forms for sustaining effective operation of biological treatment for COD removal. However, certain specific wastewaters, such as tannery effluents, are characterized by nitrogen contents significantly higher than the level associated with domestic sewage, often in excess of 100–150 mg/l. For these wastewaters, an acceptable nitrogen removal performance without additional external carbon requires full understanding and manipulation of the nitrogen removal potential of the activated sludge configurations and especially pre-denitrification systems. The influent nitrogen level is not the only factor in this evaluation. It should be considered together with a number of other operating parameters including the influent COD/N ratio.

7.5.1 Activated sludge design for nitrification

In activated sludge systems treating wastewaters with excess nitrogen, aside from the part that is incorporated into biomass nitrification needs to be accounted for as it occurs naturally, under suitable operating conditions without specific engineering adjustments. If the wastewater temperature, the selected sludge age and the level of dissolved oxygen are favourable, a nitrifying autotrophic biomass develops in the reactor and oxidizes available ammonia to nitrate. The occurrence of nitrification should be evaluated as an integral part of system design in terms of appropriate parameters.

7.5.1.1 Aerobic sludge age

A particular activated sludge configuration may include a number of different aerobic, anoxic and anaerobic phases as dictated by the specific mode of operation. Nitrification is a strictly aerobic process. In single sludge systems nitrifiers can only be sustained together with heterotrophs if the aerobic sludge age, θ_{XA}, i.e. the sludge age fraction corresponding to the aerobic volume, is favourable. The basic mass balance for autotrophic growth at steady-state may be formulated as follows:

$$-P_{XA} + V_R X_A \left(\hat{\mu}_A \frac{S_{NH}}{K_{NH} + S_{NH}} - b_A \right) = 0 \qquad (7.59)$$

where, P_{XA} = the daily amount of autotrophic biomass wasted
X_A = autotrophic biomass
K_{NH} = half saturation coefficient for nitrification

By definition:

$$\theta_{XA} = \frac{V_R X_A}{P_{XA}}$$

Then,

$$\frac{1}{\theta_{XA}} = \hat{\mu}_A \frac{S_{NH}}{K_{NH} + S_{NH}} - b_A \qquad (7.60)$$

The effluent ammonia concentration, S_{NH} may be calculated as a function of the selected sludge age, θ_{XA} by rearranging the above expression:

$$S_{NH} = \frac{K_{NH}(1 + b_A \theta_{XA})}{\hat{\mu}_A \theta_{XA} - (1 + b_A \theta_{XA})} \qquad (7.61)$$

This expression should only be considered as a conservative approximation, because it is derived from a mass balance which defines the aeration tank where nitrification occurs, as a completely mixed ideal reactor (CSTR). In fact, biological reactors with different mixing characteristics as compared to CSTRs are likely to provide a more effective nitrification. The basic mass balance equation (7.60) is quite useful in indicating that lowering the sludge age increases the resulting S_{NH} value. This equation implies the existence of a minimum aerobic sludge age, θ_{XM} for which the reactor S_{NH} level equals the available

Continuous flow technology 221

ammonia in the influent, $S_{NH} = S_{NH1}$. From a practical standpoint, θ_{XM} defines the threshold level where nitrification stops and nitrifiers are washed out from the system. It should be remembered that the threshold value for the sludge age, θ_{XM}, has to be calculated for the $\hat{\mu}'_A$ specifically calculated for the most adverse conditions of operation, reflecting the negative effect of oxygen limitation, temperature and inhibitors:

$$\frac{1}{\theta_{XM}} = \hat{\mu}'_A \frac{S_{NH1}}{K_{NH} + S_{NH1}} - b_A \qquad (7.62)$$

Since K_{NH} is usually small enough to be neglected as compared to S_{SH1}, equation (7.62) may be approximated as:

$$\theta_{XM} \approx \frac{1}{\hat{\mu}'_A - b_A} \qquad (7.63)$$

In this context, in single sludge systems operating for organic carbon removal and nitrification, the major concern is to ensure a sludge age lower than θ_{XM} where all nitrifiers will be washed out of the system, with a safety margin. Therefore, the adopted design procedure should first determine θ_{XM} and then apply a safety factor, f_{SA} for selecting a higher value for the design sludge age, θ_{XAD} that would secure consistent nitrification:

$$\theta_{XAD} = f_{SA}\theta_{XM} \qquad (7.64)$$

7.5.1.2 Nitrification parameters

After kinetic assessment of nitrification and the resulting ammonia concentration, S_{NH}, the nitrogen mass balance may be used to identify two important parameters associated with nitrification that are essential for system design, namely, the *specific nitrogen removal*, N_X and the *nitrification capacity*, N_{OX}. As previously defined by means of equation (7.55), the first parameter, N_X defines the amount of nitrogen incorporated into biomass per unit volume of wastewater treated:

$$N_X = i_{XN} Y_{NHE} C_{S1} \qquad (7.65)$$

where, $\qquad Y_{NHE} = Y_{NH}(1 + f_{EX} b_H \theta_{XA})$

The second parameter N_{OX} defines the amount of ammonia nitrogen oxidized or nitrate nitrogen formed per unit volume of wastewater treated. The nitrification capacity, N_{OX} may be calculated from the basic mass balance for ammonia nitrogen:

$$N_{OX} = C_{N1} - i_{NXI}X_{I1} - i_{NSI}S_{I1} - N_X - S_{NH} \tag{7.66}$$

In this expression, the non biodegradable nitrogen compounds by-passing biological conversions are expressed in terms of fractions (i_{NSI}; i_{NXI}) of their soluble and particulate counterparts in the wastewater.

7.5.1.3 Autotrophic biomass

No analytical technique is so far available for the determination of the autotrophic fraction of the active biomass f_A. However, f_A can be estimated indirectly for the amount of active biomass produced for heterotrophic and autotrophic microorganisms:

$$f_A = \frac{P_{XA}}{P_{XA} + P_{XH}} \tag{7.67}$$

Using the same principles as applied to heterotrophic growth, P_{XA} may be expressed in terms of N_{OX}:

$$P_{XA} = Y_{NA}QN_{OX} \tag{7.68}$$

where,

$$Y_{NA} = \frac{Y_A}{1 + b_A \theta_{XA}}$$

The autotrophic fraction of active biomass, f_A may then be calculated using equations (7.68) and (7.34):

$$f_A = \frac{Y_{NA}Q_A}{Y_{NA}Q_A + Y_{NH}C_{S1}} \tag{7.69}$$

Assessment of this ratio enables direct calculation of the amount of autotrophic biomass in the reactor, M_{XA} and the autotrophic biomass concentration, X_A.

7.5.1.4 Autotrophic oxygen demand

While the level of autotrophic biomass remains quite low as compared to the heterotrophic fraction in the reactor, the amount of oxygen consumed by nitrification often constitutes an important dimension in the total oxygen demand of the system.

The oxygen requirement rate for nitrification, OR_A can be calculated from system stoichiometry:

$$OR_A = (4.57 - Y_{NA})QN_{OX} \qquad (7.70)$$

and,

$$OR_T = OR_H + OR_A \qquad (7.71)$$

7.5.1.5 Alkalinity consumption

Alkalinity consumption due to nitrification may be calculated from the following expression:

$$S_{ALK} = S_{ALK1} - 7.14 N_{OX} \qquad (7.72)$$

where, S_{ALK} = remaining alkalinity in wastewater (mg/l $CaCO_3$)

The alkalinity check is an important part of system design in nitrification because pH in the biological reactor tends to lose its stability below alkalinity values of around 50 mg/l $CaCO_3$. Therefore, a minimum residual alkalinity of above 100 mg/l $CaCO_3$ is generally recommended for safe operation (Orhon and Artan 1994).

7.5.1.6 Design Procedure

An appropriate design for nitrification should always take into account the effect of temperature on the process kinetics. Critical wastewater temperatures under summer and winter conditions need to be determined as significant design parameters. Under low temperature conditions, the designed system should be able to (i) sustain a nitrifying biomass and (ii) secure an effluent ammonia concentration below the applicable effluent discharge limitation. This is accomplished by selecting an aerobic sludge age corrected for temperature effects. It is preferable to use a temperature correction coefficient, θ that is experimentally determined for the wastewater to be treated. The effect of high wastewater temperature should also be considered when evaluating the oxygen requirement of nitrification under summer conditions.

After the selection of the design sludge age, θ_{XAD}, and the verification of the effluent ammonia concentration, S_{NH}, the amount of nitrogen incorporated into biomass, N_X can be calculated as a function of the excess heterotrophic sludge produced at the selected θ_{XAD}. Then, the amount of oxidized nitrogen per unit volume of wastewater treated, N_{OX} is computed from mass balance. N_{OX} is the central parameter for nitrification, which allows verifying alkalinity, calculating

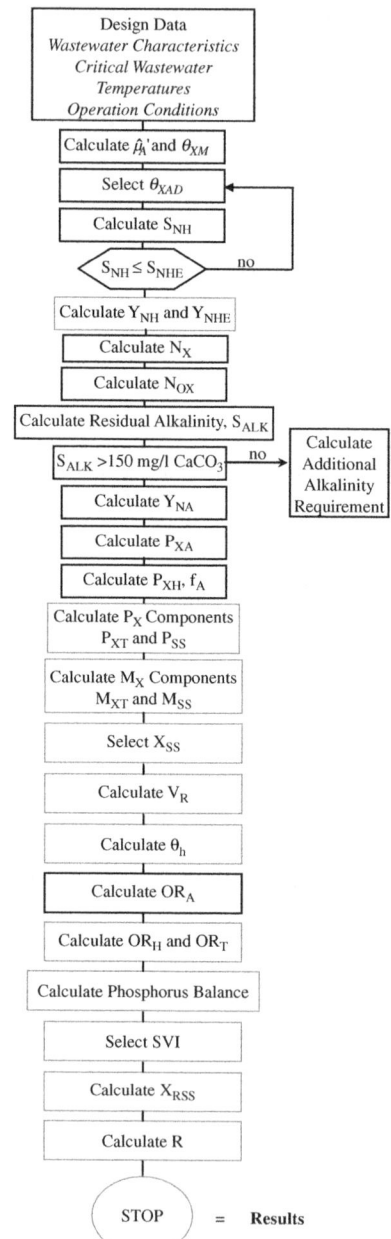

Figure 7.9. Design algorithm for combined organic carbon removal and nitrification

the autotrophic biomass production rate, P_{XA} and assessing the oxygen requirement for the autotrophs, OR_A. The remaining part of the design proceeds as defined for organic carbon removal. The stepwise design approach using expressions and relationships derived for major design parameters and functions is outlined in Figure 7.9. In this context, the recommended design procedure for combined nitrification and organic carbon removal from the typical industrial wastewater defined in Tables 7.3 and 7.4 is illustrated in Example 7.7:

Example 7.7 Design of a combined COD removal-nitrification system
Design an activated sludge system for combined organic carbon removal and nitrification for an industrial wastewater with characteristics given below and in Tables 7.3 and 7.4, to achieve en effluent ammonia nitrogen concentration below 10 mg/l N. Assume 22°C and 14°C reflect summer and winter design temperatures respectively.

Wastewater flow, $Q = 100$ m³/day
Total influent biodegradable COD, $C_{S1} = 1250$ mg/l
Total influent nitrogen, $C_{TKN1} = 180$ mg/l
Influent alkalinity, $S_{ALK1} = 460$ mg/l $CaCO_3$
Dissolved oxygen concentration, $S_O = 2.0$ mg/l
$Y_A = 0.24$ g cell COD/g N
$\hat{\mu}_A$ at $20°C = 0.40$ /day
$K_{NH} = 1.0$ mg/l
$K_{OA} = 0.4$ mg/l
$b_A = 0.10$ /day
Temperature coefficient, $\theta = 1.1$

It should be noted that the example uses the same industrial wastewater previously analysed for organic carbon removal in Example 7.6. In this example, the influent nitrogen concentration, C_{TKN1} is increased to 180 mg/l to better visualize the impact of nitrification on system design. The selected wastewater characteristics with high nitrogen content, approximate chemically-settled tannery effluent.

(1) Calculate $\hat{\mu}'_A$ and θ_{XM}:
Under summer conditions:

$$\hat{\mu}'_A = \hat{\mu}_A \left(\frac{S_O}{K_{OA} + S_O} \right) \theta^{T-20}$$

$$= 0.4 \times \left(\frac{2.0}{0.4 + 2.0}\right) \times 1.1^{22-20}$$

$$= 0.4 \times 0.83 \times 1.21 = 0.40 \, day^{-1}$$

$$\theta_{XM} = \frac{1}{\hat{\mu}'_A - b_A} = \frac{1}{0.4 - 0.1} = 3.4 \, days$$

Under winter conditions:

$$\hat{\mu}'_A = 0.4 \times \left(\frac{2.0}{0.4 + 2.0}\right) \times 1.1^{14-20}$$

$$= 0.4 \times 0.83 \times 0.56 = 0.19 \, day^{-1}$$

$$\theta_{XM} = \frac{1}{\hat{\mu}'_A - b_A} = \frac{1}{0.19 - 0.10} = 11.1 \, days$$

(2) Select θ_{XAD}:
Select a safety factor, $f_{SA} = 1.2$ to comply with both summer and winter conditions, and calculate θ_{XAD}.

$$\theta_{XAD} = f_{SA} \theta_{XM} = 1.2 \times 11.1$$

$$\theta_{XAD} = \theta_{XA} = 13 \, days$$

(3) Calculate S_{NH}:

$$S_{NH} = \frac{K_{NH}(1 + b_A \theta_{XA})}{\hat{\mu}'_A \theta_{XA} - (1 + b_A \theta_{XA})}$$

Under summer conditions:

$$S_{NH} = \frac{1.0 \times (1 + 0.1 \times 13)}{0.4 \times 13 - (1 + 0.1 \times 13)} = 0.8 \, mg/l$$

Under winter conditions:

$$S_{NH} = \frac{1.0 \times (1 + 0.1 \times 13)}{0.19 \times 13 - (1 + 0.1 \times 13)} = 13.5 \, mg/l$$

Since $S_{NH} > 10$ mg/l, a higher θ_{XA} must be selected.
Select $\theta_{XA} = 15$ days

$$S_{NH} = \frac{1.0 \times (1 + 0.1 \times 15)}{0.19 \times 15 - (1 + 0.1 \times 15)} = 7.14 \text{ mg/l}$$

$S_{NH} < 10$ mg/l, selected θ_{XA} of 15 days is acceptable.

(4) Calculate Y_{NH}:

$$Y_{NH} = \frac{Y_H}{1 + b_H \theta_{XA}}$$

$Y_H = 0.64$ g cell COD/g COD
$b_H = 0.15$ day^{-1}

$$Y_{NH} = \frac{0.64}{1 + 0.15 \times 15} = 0.20 \text{ g cell COD/g COD}$$

(5) Calculate Y_{NHE}:

$$Y_{NHE} = Y_{NH}(1 + f_{EX} b_H \theta_{XA})$$

$$f_{EX} = 0.2$$

$$Y_{NHE} = 0.20 \times (1 + 0.2 \times 0.15 \times 15)$$
$$Y_{NHE} = 0.29 \text{ g COD/g COD}$$

(6) Calculate N_X:

$$N_X = i_{XN} Y_{NHE} C_{S1}$$

$$i_{NX} = 0.08 \text{ g N/g cell COD}$$

$$N_X = 0.08 \times 0.29 \times 1250 = 29 \text{ mg/l}$$

(7) Calculate N_{OX}:

$$N_{OX} = C_{N1} - i_{NXI} X_{I1} - i_{NSI} S_{I1} - N_X - S_{NH}$$

228 Industrial Wastewater Treatment by Activated Sludge

$$X_{I1} = 150 \text{ mg/l}$$
$$S_{I1} = 100 \text{ mg/l}$$

Assume: $i_{NSI} = 0.02$ g N/g COD
$i_{NXI} = 0.03$ g N/g COD

$$C_{NI1} = i_{NXI}X_{I1} + i_{NSI}S_{I1} = 0.03 \times 150 + 0.02 \times 100 = 6.5 \text{ mg N/l}$$

$$N_{OX} = 180 - 6.5 - 29 - 7.14 = 137 \text{ mg N/l}$$

This gives N_{OX} under winter conditions.
Under summer conditions:

$$S_{NH} = \frac{1.0 \times (1 + 0.1 \times 15)}{0.4 \times 15 - (1 + 0.1 \times 15)} = 0.71 \text{ mg N/l}$$

and
$$N_{OX} = 180 - 6.5 - 29 - 0.71 \cong 144 \text{ mg N/l}$$

(8) Calculate Residual Alkalinity:

$$S_{ALK} = S_{ALK1} - 7.14 N_{OX}$$

Use critical summer conditions

$$S_{ALK} = 460 - 7.14 \times 144 = -568 \text{ mg/l CaCO}_3$$

A minimum additional alkalinity of 670 mg/l $CaCO_3$ must be supplied to secure nitrification.

(9) Calculate Y_{NA} and P_{XA}:

$$Y_{NA} = \frac{Y_A}{1 + b_A \theta_{XA}} = \frac{0.24}{1 + 0.1 \times 15} \approx 0.1 \text{g cell COD/g N}$$

$$P_{XA} = Y_{NA} Q N_{OX}$$

$$P_{XA} = 0.1 \times 100 \times 0.144$$

$$P_{XA} = 1.44 \text{ kg cell COD/day}$$

(10) Calculate P_{XH} and f_A:

$$P_{XH} = Y_{NH}QC_{S1}$$

$$P_{XH} = 0.20 \times 100 \times 1.25 = 25 \text{ kg cell COD/day}$$

$$f_A = \frac{P_{XA}}{P_{XA} + P_{XH}} = \frac{1.44}{1.44 + 25} = 0.055 = 5.5\%$$

(11) Calculate OR_A, OR_H, and OR_T:

$$OR_A = (4.57 - Y_{NA})QN_{OX}$$

For summer conditions:

$$OR_A = (4.57 - 0.1) \times 100 \times 0.144$$
$$= 64.4 \text{ kg } O_2/day$$

$$OR_H = QC_{S1}[1 - Y_{NH}(1 + f_E b_H \theta_{XA})]$$

$$f_E = f_{EX} + f_{ES} = 0.20 + 0.05 = 0.25$$

$$OR_H = 100 \times 1.25 \times [1 - 0.20 \times (1 + 0.25 \times 0.15 \times 15)]$$

$$OR_H = 85.9 \text{ kg } O_2/day$$

$$OR_T = OR_H + OR_A$$

$$OR_T = 85.9 + 64.4 = 150.3 \text{ kg } O_2/day$$

(12) Calculate P_X components:

$$P_{XH} + P_{XP} = P_{XH}(1 + f_{EX} b_H \theta_{XA})$$

$$P_{XH} = 25 \text{ kg cell COD/day}$$

$$P_{XH} + P_{XP} = 25 \times (1 + 0.2 \times 0.15 \times 15)$$
$$= 36.25 \text{ kg COD/day}$$

$$X_{I1} = 150 \text{ mg/l} = 0.15 \text{ kg/m}^3$$

$$P_{XI} = QX_{I1} = 100 \times 0.15 = 15 \text{ kg COD/day}$$

$$P_{XT} = P_{XH} + P_{XP} + P_{XA} + P_{XI}$$

$$P_{XT} = 36.25 + 1.44 + 15 = 52.7 \text{ kg COD/day}$$

Assume $i_{SS,COD} = 0.9$ kg SS/kg COD

$$P_{SS} = 52.7 \times 0.9 = 47.4 \text{ kg SS/day}$$

(13) Calculate M_X components:

$$M_{XH} = P_{XH}\theta_{XA} = 25 \times 15 = 375 \text{ kg COD}$$

$$M_{XP} = M_{XH} f_{EX} b_H \theta_{XA}$$

$$M_{XP} = 375 \times 0.2 \times 0.15 \times 15$$
$$= 168.7 \text{ kg COD}$$

$$M_{XI} = P_{XI}\theta_{XA} = 15 \times 15 = 225 \text{ kg COD}$$

$$M_{XA} = P_{XA}\theta_{XA} = 1.44 \times 15 = 21.6 \text{ kg COD}$$

$$M_{XT} = M_{XH} + M_{XP} + M_{XI} + M_{XA}$$

$$M_{XT} = 375 + 168.7 + 225 + 21.6$$
$$= 790.3 \text{ kg COD}$$

$$M_{XT} = P_{XT}\theta_{XA} = 52.7 \times 15 \cong 790 \text{ kg COD}$$

$$M_{SS} = 790.3 \times 0.9 = 711 \text{ kg SS}$$

$$\text{Viability} = \frac{M_{XH}}{M_{XT}} \times 100 = \frac{375}{790} \times 100 = 47\%$$

(14) Select $X_{SS} = 4000$ mg SS/l

(15) Calculate V_R:

$$V_R = \frac{M_{SS}}{X_{SS}} = \frac{711}{4} = 177.8 \ m^3$$

Select $V_R = 180 \ m^3$

(16) Calculate θ_h:

$$\theta_h = \frac{V_R}{Q} = \frac{180}{100} = 1.8 \text{ days} = 43 \text{ hrs}$$

(17) Select SVI = 120 ml/g

(18) Calculate X_{RSS}:

$$X_{RSS} = \frac{10^6}{120} \cong 8300 \text{ mg SS/l}$$

(19) Calculate R:

$$R = \frac{1 - \dfrac{\theta_h}{\theta_X}}{\dfrac{X_{RSS}}{X_{SS}} - 1} = \frac{1 - \dfrac{1.1}{8}}{\dfrac{8300}{4000} - 1} = \frac{0.86}{1.075} = 0.82$$

Table 7.6. Summary of activated sludge design for the selected industrial wastewater for combined nitrification and organic carbon removal

Parameter	Unit	Nitrification and organic carbon removal	Organic carbon removal
Q	m³/day	100	100
θ_X	days	15	8
P_{XT}	kg COD/day	52.7	59.95
P_{SS}	kg SS/day	47.4	53.95
M_{XT}	kg COD	790	479.6
M_{SS}	kg SS	711	431.6
X_{SS}	mg SS/l	4000	4000
V_R	m³	180	110
θ_h	days	1.8	1.1
OR_H	kg O₂/day	85.9	77.9
OR_A	kg O₂/day	64.4	–
OR_T	kg O₂/day	150.3	77.9
X_{RSS}	mg SS/l	8300	8300
R		0.82	0.8

Although the same wastewater is treated as in Example 7.6, inclusion of nitrification and the corresponding high sludge age changes the basic design of the activated sludge system. Design calculations are summarized in Table 7.6 together with similar results obtained for the same industrial wastewater treated for organic carbon removal in Example 7.6.

7.5.2 Activated sludge design for nitrogen removal

7.5.2.1 Process configurations

The following functions are required for nitrogen removal:

(i) an aerobic volume for nitrification
(ii) an anoxic volume for denitrification
(iii) presence of the necessary amount of oxidized nitrogen – nitrate – in the anoxic volume to ensure the desired effluent total nitrogen concentration
(iv) presence of the required amount of biodegradable COD in the anoxic zone to remove the available nitrate.

These functions indeed offer a large number of combinations for different process configurations. They were largely exploited for deriving different activated sludge alternatives, often developed with an empirical approach without the fundamental basis to support a particular choice. Extensive descriptions of these alternatives are presented elsewhere (Christensen and Harremoes, 1977; Ekama and Marais, 1984a; Orhon and Artan, 1994).

The level of understanding reached today on the microbial mechanism of nitrogen removal gives a clear indication that the functions listed above are best fulfilled with a single biomass including an autotrophic fraction and a heterotrophic fraction with the ability to switch and sustain specific metabolic activities under a sequence of aerobic/anoxic conditions. The process configurations designed to operate with a single biomass are called *single sludge systems*. These systems are engineered with the view in mind that organic carbon can also be removed in the aerobic volume and if so arranged, this may upset the required COD/N balance in the anoxic zone. Therefore, the main concern in the design of the single sludge activated process is to budget and optimize the available biodegradable COD since addition of an external carbon source may prove quite costly. In this context, a single sludge process for nitrogen removal may have two different configurations: A *pre-denitrification system* and a *post-denitrification system*.

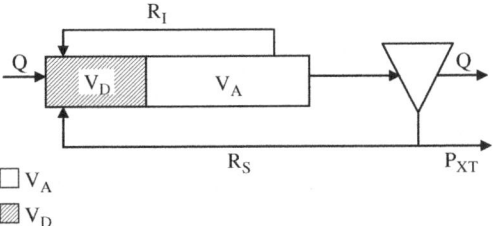

Figure 7.10. Schematic process diagram for pre-denitrification

The *pre-denitrification system* basically consists of a sequence of two volumes, the first operated under anoxic conditions and the following under aerobic conditions without an intermediate settling, as shown in Figure 7.10. The wastewater is introduced to the first anoxic volume which benefits from the entire internal carbon potential available in the influent. The following aerobic volume receives the denitrified effluent and the aerobic sludge age is so selected to sustain nitrification. The oxidized nitrogen generated through nitrification at this stage is re-circulated back to the anoxic volume. Provision of the necessary amount of nitrate for denitrification usually requires internal recycling aside from recycling of the settled activated sludge.

The major drawback of the pre-denitrification system is the hydraulic limitation imposed on the extent of internal recycling which puts a constraint on the magnitude of the amount of nitrate to be introduced and removed in the anoxic zone. This is particularly a problem for industrial wastewaters with high nitrogen content. One of the ways to overcome the problem is to select a process scheme with double anoxic zones including two anoxic and two aerobic volumes as given in Figure 7.11. The first anoxic/aerobic volume couple essentially works as a basic pre-denitrification system with internal recycle. Nitrate not totally utilized in this section is removed in the second part of the reactor serving as the electron acceptor for endogenous respiration and this way, utilizing the endogenous carbon source. A large number of process alternatives have been developed using this principle some including step aeration, others multiple anoxic zones. They are covered as part of the sequencing batch reactor technology process (SBR), in the following chapter as this process offers a better explanation for the mechanistic basis of different operation alternatives (Artan and Orhon, 2005).

The single sludge activated sludge system operated as a *post denitrification process* includes two separate volumes. As opposed to pre-denitrification, the first volume is aerobic and the second one is anoxic as indicated in Figure 7.12. In this flow scheme, the anoxic part of the reactor receives the full amount of

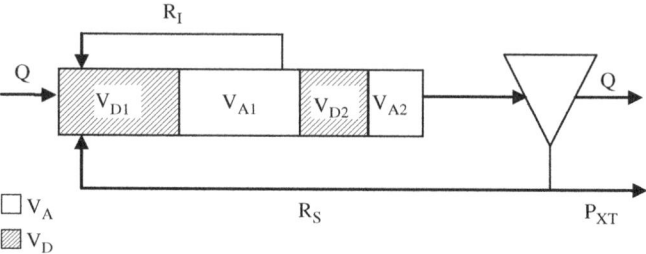

Figure 7.11. Schematic process diagram for an activated sludge system for dual anoxic zones for nitrogen removal

oxidized nitrogen generated by means of nitrification, but no biodegradable COD which would be depleted through aerobic growth of heterotrophs in the first volume. At this stage endogenous respiration acts as the sole energy source and it is likely to provide minor nitrogen removal. In this context, this configuration becomes a viable but costly alternative as it relies on external carbon addition which can potentially achieve full removal of oxidized nitrogen.

Nitrogen removal can also be obtained using a *separate sludge* configuration consisting of two sets of reactors, each with individual settling and sludge recycle. Each set are operated to sustain different types of microbial communities. The first reactor is usually aerated for combined carbon removal and nitrification. The second reactor is kept anoxic for denitrification at the expense of external carbon addition. Methanol is the most feasible external carbon source commonly utilized for this purpose. When the influent COD/N ratio is suitable, a portion of the wastewater flow may be diverted to the anoxic tank to reduce and if possible eliminate the external carbon requirement. The separate sludge systems, highly promoted during the early stages of biological nitrogen removal

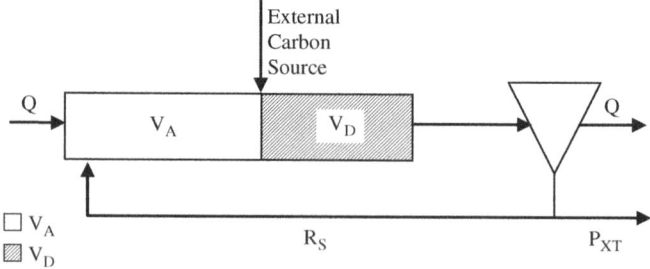

Figure 7.12. Schematic process diagram of single sludge post denitrification with external carbon source

practice, have lost their viability in view of the potential, flexibility and cost attraction offered by single sludge activated sludge. Today, the single sludge process is by far the preferred biological nitrogen alternative owing to a number of reasons:

(i) no need for the expense of external carbon source
(ii) lower overall oxygen demand
(iii) no need for additional intermediate settling
(iv) better sludge settling properties.

7.5.2.2 The overall sludge age

In a single sludge activated sludge system designed for biological nitrogen removal, the overall sludge age, has two main components: (i) the aerobic sludge age, θ_{XA} which controls the autotrophic growth and secures nitrification; (ii) the anoxic sludge age, θ_{XD} associated with the anoxic volume fraction where nitrate reduction takes place:

$$\theta_X = \theta_{XA} + \theta_{XD} \tag{7.73}$$

Heterotrophic growth takes place both in the entire reactor volume and it can be evaluated accurately by means of process modelling in the anoxic volume. For design purposes, the reduced heterotrophic activity can be approximated by a single overall correction factor, η which seemingly reduces the anoxic reactor volume as indicated by the following mass balance for heterotrophs:

$$(V_A + \eta V_D) X_H (\mu_H - b_H) - P_{XH} = 0 \tag{7.74}$$

where, V_A = the aerobic reactor volume
V_D = the anoxic reactor volume

and
$$\theta_X = \frac{\theta_A}{1 - \frac{V_D}{V}} \tag{7.75}$$

Setting

$$V_A + \eta V_D = c V_R$$

$$c = 1 - (1 - \eta)\frac{V_D}{V} \tag{7.76}$$

Substituting the value of c in equation (7.76), the mass balance for the heterotrophic growth may be re-expressed as:

$$(\mu_H - b_H) = \frac{1}{\theta_{XH}} = \frac{P_{XH}}{cV_R} = \frac{1}{c\theta_X} \qquad (7.77)$$

The above equation indicates that all design parameters connected with heterotrophic growth have to be re-adjusted with $c\theta_X$ to account for reduced heterotrophic activity under anoxic conditions. Expressions defining major parameters to be used in the design of single sludge systems for biological nitrogen removal are given as part of design calculations in Example 7.8.

7.5.2.3 Denitrification potential

A significant parameter in the design of single sludge systems for denitrification is the nitrate equivalent of biodegradable COD that may be potentially removed as the energy source in the anoxic phase. This represents the denitrification potential, N_{DP} of the wastewater for the selected process. N_{DP} is not meaningful by itself. It has a practical value if evaluated together with available nitrate, N_A in the anoxic volume. N_A is introduced into the anoxic zone from the aerobic part of the reactor. The N_{DP}/N_A ratio should be envisaged as a useful design instrument representing the balance that can be maintained between substrate utilization and nitrate removal in the anoxic volume. It is the refined version of the COD/N ratio, directly applicable to design.

Calculation of N_{DP} should take into account the amount of biodegradable COD utilized in biosynthesis for microbial growth. Model evaluation provides the exact level of the N_{DP} corresponding to each process alternative for design. N_{DP} can also be estimated on the basis of basic stoichiometry. In a simplified way, N_{DP} may be approximated as a function of the total biodegradable COD in the wastewater, C_{S1} with the assumption that substrate utilization occurs in proportion to the V_D/V_R ratio.

$$N_{DP} = \frac{V_D}{V_R}(1 - Y_{NHE})\frac{C_{S1}}{2.86} \qquad (7.78)$$

Direct use of this expression is not totally compatible with currently used mechanistic models which all differentiate COD fractions with different biodegradation characteristics (Henze et al., 1987; Gujer et al., 2000). Accordingly, the expression has been refined recently in a way that separately identifies N_{DP} components associated with readily biodegradable COD, S_{S1}, slowly biodegradable COD, X_{S1} and endogenous carbon (Sozen et al., 2002). N_{DP} may also be

Continuous flow technology

defined in terms of the oxygen requirement rate per unit volume of wastewater treated, or as it applies to denitrification, the total oxygen equivalent of the total electron acceptor demand associated with the removal biodegradable COD (Orhon and Artan, 1994). Using this approach and the same simplifying assumption of the overall rate correction for heterotrophic activities under anoxic conditions, N_{DP} may be expressed as:

$$N_{DP} = \frac{\eta V_D}{c V_R} \frac{OR_H}{2.86 Q} \tag{7.79}$$

where, OR_H = the oxygen requirement rate of heterotrophs

For pre-denitrification systems, expressions (7.78) and (7.79) could be improved by differentiating the readily biodegradable COD fraction, S_{S1} from the rest and assuming that it is totally depleted within the anoxic zone fully contributing to nitrate removal:

$$N_{DP} = \frac{(1 - Y_{NHE})}{2.86} \left[S_{S1} + \frac{V_D}{V_R}(C_{S1} - S_{S1}) \right] \tag{7.80}$$

$$\text{or} \quad N_{DP} = \frac{OR_{HS}}{2.86 Q} + \frac{\eta V_D}{c V_R} \frac{(OR_H - OR_{HS})}{Q 2.86} \tag{7.81}$$

where, OR_{HS} = oxygen requirement of the readily biodegradable COD fraction.

7.5.2.4 Oxygen requirement

In activated sludge systems designed for denitrification, the oxygen requirement is reduced as a function of the amount of organic carbon removed under anoxic conditions, using nitrate instead of oxygen as the final electron acceptor. The reduction in the oxygen requirement due to denitrification, OR_D may be expressed in terms of nitrate removed per unit volume of wastewater treated, S_D:

$$OR_D = 2.86 Q N_D \tag{7.82}$$

Then, the total oxygen requirement OR_T may be defined as:

$$OR_T = OR_H + OR_A - OR_D \tag{7.83}$$

where $\quad OR_H = Q C_{S1}[1 - Y_{NH}(1 + f_E c b_H \theta_X)]$

and $\quad OR_A = (4.57 - Y_{NA}) Q N_{OX} \tag{7.84}$

7.5.2.5 System design for pre-denitrification

It should first be remembered that designing industrial wastewater treatment commonly relies on technologies selected *a priori* without too much concern on the suitability of the selected alternative for the expected removal performance. Literature is replete with studies which merely mention the process with a *black box* approach and report successful performance without scientific justification. In short, a pre-selected process often restricts design rather than design exploring and defining the most appropriate process. While this approach may be somewhat tolerated for conservatively designed carbon removal systems, it may end up with the wrong solution for nutrient removal. An appropriate design procedure should fully benefit from fundamentals of process kinetics for providing a *tailor made* solution for the specific wastewater studied. For nitrogen removal, a pre-denitrification flow scheme should always be given priority as the most feasible alternative to be tested for compliance with the effluent quality requirements. Other alternatives such as *systems with multiple anoxic zones, step aeration, post-denitrification*, etc., should only be considered if pre-denitrification proves to be ineffective.

The adopted procedure should start by selecting a design value for the aerobic sludge age, θ_{XA} suitable for sustaining effective nitrification within the aerobic part of the reactor. Then, an anoxic volume ratio, V_D/V_R is selected for defining the total sludge age, θ_X of the system. A V_D/V_R ratio lower than 0.5 is recommended to prevent possible deterioration of the settling properties of the biomass. The total sludge age enables calculation of all the essential parameters of nitrification, including the amount of oxidized nitrogen per unit volume of wastewater treated, N_{OX}.

The design for denitrification starts by calculating the denitrification potential, N_{DP} as a function of the tested V_D/V_R ratio. A preliminary check could be made by comparing N_{DP} with N_{OX}. If $N_{DP} > N_{OX}$, the wastewater characteristics are potentially suitable for pre-denitrification and nitrogen removal efficiency will depend on the magnitude of nitrate that can be introduced into the anoxic zone by recycling. The case where N_{OX} is found significantly higher than N_{DP} may be taken as a sign of shortage of internal carbon and partial addition of external carbon with post-denitrification may be necessary.

As previously mentioned, a pre-denitrification system requires internal recycling, R_I aside from regular sludge recycling, R_S and the total recycle ratio determines the amount of nitrate introduced into the anoxic volume per unit flow of wastewater treated, also defined as the *available nitrate*, N_A. If $N_{DP} > N_A$, for a selected R_T, then N_A becomes fully consumed in the anoxic zone.

The resulting effluent nitrate concentration, S_{NO} may be calculated from mass balance:

$$S_{NO} = N_{OX} - N_A \tag{7.85}$$

Since
$$N_A = R_T S_{NO}$$

Then
$$S_{NO} = \frac{N_{OX}}{1 + R_T} \tag{7.86}$$

The level of S_{NO} may be reduced to meet the expected effluent requirement. In most cases, it is not practically feasible to operate the system with $R_T > 5$. If $N_{DP} < N_A$, the expression (7.85) becomes

$$S_{NO} = N_{OX} - N_{DP} \tag{7.87}$$

In this case R_T is adjusted to only supply $N_A = N_{DP}$ in the anoxic zone. The rest of the design procedure remains the same as the one implemented for organic carbon removal and nitrification. Denitrification imparts alkalinity to the solution. The overall alkalinity balance of the system may be expressed as follows:

$$\Delta Alk[mmoles/l] = \frac{1}{14}(N_D - 2N_{OX} - N_X + C_{ND1}) \tag{7.88}$$

where N_D = the amount of oxidized nitrogen removed per unit volume of wastewater. Since,

$$N_D = N_{OX} + S_{NO1} - S_{NO}$$

and $\quad N_{OX} = C_{ND1} + S_{NH1} - S_{NH} - N_X$

Equation (7.88) becomes

$$\Delta Alk = \frac{1}{14}(S_{NO1} - S_{NO} + S_{NH} - S_{NH1}) \tag{7.89}$$

Using equation (7.89), the alkalinity concentration sustained in the reactor, S_{ALK} may be calculated from the expression below:

$$S_{ALK} = S_{ALK1} + \frac{1}{14}(S_{NO1} - S_{NO} + S_{NH} - S_{NH1}) \tag{7.90}$$

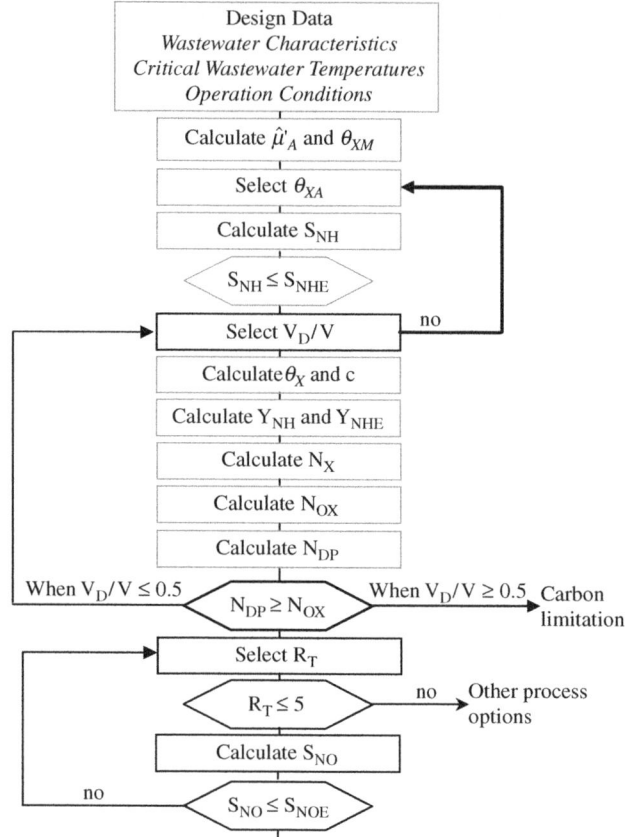

Figure 7.13. Recommended design procedure for a pre-denitrification system

It is desirable to keep the level of S_{ALK} above 100–150 mg/l $CaCO_3$ by alkalinity addition if necessary. The recommended design procedure for a pre-denitrification system, as explained above is shown in Figure 7.13. This procedure is further illustrated in the following design example.

Example 7.8 Design of a single sludge pre-denitrification system
Design a single sludge pre-denitrification system for an industrial wastewater with characteristics summarized in example 7.7. The system is required to denitrify under summer conditions considering that the applicable design temperature is set as 20° C.

Continuous flow technology 241

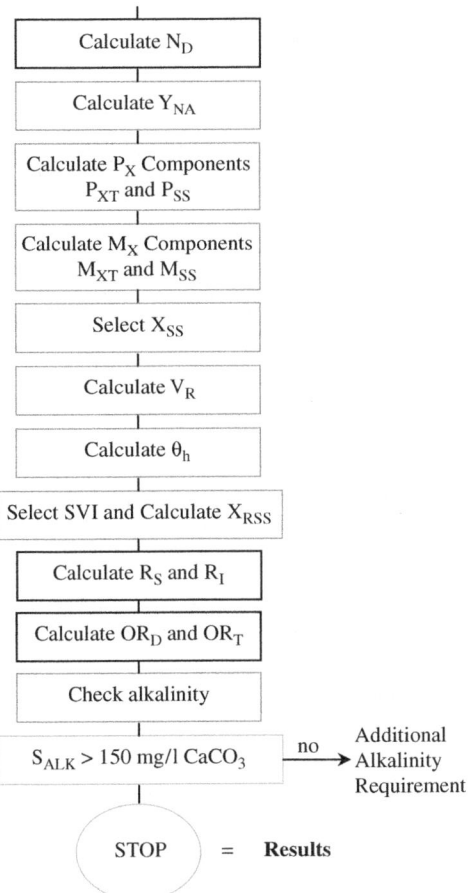

Figure 7.13. *Continued*

Calculate the minimum level of effluent nitrate concentration achievable with the selected activated sludge flow scheme. Make all assumptions as necessary.

The characteristics of the industrial wastewater to be treated are given in Example 7.7 and in Table 7.3 and 7.4.

(1) Calculate $\hat{\mu}'_A$ and θ_{XM}:

$$\hat{\mu}'_A \text{ at } 20°C = 0.40 \ day^{-1}$$

$$S_O = 2.0 \; mg/l$$
$$K_{OA} = 0.4 \; mg/l$$

$$\hat{\mu}'_A = \hat{\mu}_A \frac{S_O}{K_{OA} + S_O} = 0.40 \times \frac{2.0}{0.4 + 2.0}$$

$$\hat{\mu}'_A = 0.33 \; days^{-1}$$

$$b_A = 0.10 \; day^{-1}$$

$$\theta_{XM} = \frac{1}{\hat{\mu}'_A - b_A} = \frac{1}{0.33 - 0.10} = 4.35 \; days$$

(2) Select θ_{XA}:

Select a safety factor, f_{SA} that would sustain autotrophs and secure an $S_{NH} < 10$ mg N/l.

$$f_{SA} = 1.2$$

$$\theta_{XAD} = f_{SA}\theta_{XM} = 1.2 \times 4.35 = 5.2 \; days$$

$$\text{Select } \theta_{XAD} = \theta_{XA} = 5.5 \; days$$

(3) Calculate S_{NH}:

$$S_{NH} = \frac{K_{NH}(1 + b_A\theta_{XA})}{\hat{\mu}'_A\theta_{XA} - (1 + b_A\theta_{XA})}$$

$$K_{NH} = 1.0 \; mg/l$$

$$S_{NH} = \frac{1.0 \times (1 + 0.1 \times 5.5)}{0.33 \times 5.5 - (1 + 0.1 \times 5.5)}$$

$$S_{NH} = \frac{1.55}{0.265} = 5.85 \; mg/l < 10 \; mg \; N/l$$

(4) Select V_D/V:

This is a high-strength wastewater with C_{TKN1} of 180 mg/l. Therefore a high V_D/V ratio of 0.4 is selected.

(5) Calculate θ_X and c:

$$\theta_X = \frac{\theta_{XA}}{1-\frac{V_D}{V}} = \frac{5.5}{1-0.4} = 9.2 \ days$$

Select $\theta_X = 10$ days used in the design for organic carbon removal.

Assume: $\eta = 0.8$

$$c = 1 - (1-\eta)\frac{V_D}{V}$$

$$c = 1 - (1-0.8) \times 0.4 = 0.92$$

(6) Calculate Y_{NH} and Y_{NHE}:

$$Y_{NH} = \frac{Y_H}{1+cb_H\theta_X}$$

$$Y_H = 0.64 \ g \ cell \ COD/g \ COD$$

$$b_H = 0.15 \ day^{-1}$$

$$Y_{NH} = \frac{0.64}{1+0.92 \times 0.15 \times 10} = 0.27 \ g \ cell \ COD/g \ COD$$

$$Y_{NHE} = Y_{NH}(1 + f_{EX}cb_H\theta_X)$$

$$f_{EX} = 0.2$$

$$Y_{NHE} = 0.27 \times (1 + 0.2 \times 0.92 \times 0.15 \times 10)$$

$$Y_{NHE} = 0.34 \ g \ COD/g \ COD$$

(7) Calculate N_X:

$$N_X = i_{XN}Y_{NHE}C_{S1}$$

$$C_{S1} = 1250 \ mg/l$$

$$i_{XN} = 0.08 \text{ g N/g cell COD}$$

$$N_X = 0.08 \times 0.34 \times 1250 = 34 \text{ mg N/l}$$

(8) Calculate N_{OX}:

$$N_{OX} = C_{N1} - i_{NXI}X_{I1} - i_{NSI}S_{I1} - N_X - S_{NH}$$

$C_{TKNI} = 180 \text{ mg N/l}$
$X_{I1} = 150 \text{ mg/l}$
$i_{NXI} = 0.03 \text{ g N/g COD}$
$S_{I1} = 100 \text{ mg/l}$
$i_{NSI} = 0.02 \text{ g N/g COD}$

$$N_{OX} = 180 - 0.03 \times 150 - 0.02 \times 100 - 34 - 5.85$$

$$N_{OX} = 133.7 \text{ mg N/l}$$

(9) Calculate N_{DP}:

$$N_{DP} = \frac{(1 - Y_{NHE})}{2.86}\left[S_{S1} + \frac{V_D}{V_R}(C_{S1} - S_{S1})\right]$$

$C_{SI} = 1250 \text{ mg/l}$
$S_{SI} = 225 \text{ mg/l}$

$$N_{DP} = \frac{(1 - 0.34)}{2.86} \times [225 + 0.40 \times (1250 - 225)]$$

$$N_{DP} = 0.22 \times (225 + 410)$$

$$N_{DP} = 147 \text{ mg N/l}$$

$$N_{DP} \geqslant N_{OX}$$

The wastewater is potentially suitable for pre-denitrification.

(10) Select R_T and Calculate S_{NO}:
Select the maximum allowable $R_T = 5.0$

Then,
$$S_{NO} = \frac{N_{OX}}{1 + R_T}$$

$$S_{NO} = \frac{133.7}{1 + 5} = 22.3 \text{ mg N/l}$$

Although not specially mentioned in the example, the effluent nitrate concentration, S_{NO} is likely to be higher than the applicable effluent limit. Since, the maximum achievable performance of the tested pre-denitrification system is limited by $R_T = 5$, other process options have to be tried for lower S_{NO} values, if necessary.

(11) Calculate N_D:

$$N_D = N_{OX} - S_{NO}$$

$$N_D = 133.7 - 22.3 = 111.4 \text{ mg N/l}$$

(12) Calculate P_X components:

$$P_{XH} = QY_{NH}C_{S1}$$

$$P_{XH} = 100 \times 0.27 \times 1.25 = 34 \text{ kg cell COD/day}$$

$$\begin{aligned}P_{XH} + P_{XP} &= P_{XH}(1 + f_{EX}cb_H\theta_X) \\ &= 34 \times (1 + 0.2 \times 0.92 \times 0.15 \times 10) \\ &= 43.4 \text{ kg COD/day}\end{aligned}$$

$$P_{XI} = Q\,X_{I1} = 100 \times 0.15 = 15 \text{ kg COD/day}$$

$$Y_{NA} = \frac{Y_A}{1 + b_A\theta_X} = \frac{0.24}{1 + 0.1 \times 5.5} = 0.15 \text{ g cell COD/g N}$$

$$\begin{aligned}P_{XA} &= Y_{NA}QN_{OX} \\ &= 0.15 \times 100 \times 0.133 \\ &= 2 \text{ kg COD/day}\end{aligned}$$

$$P_{XT} = P_{XH} + P_{XP} + P_{XA} + P_{XI}$$

$$P_{XT} = 43.4 + 2 + 15 = 60.4 \text{ kg COD/day}$$

$$\text{Assume } i_{SS,COD} = 0.9 \text{ kg TSS/kg COD}$$

$$P_{SS} = i_{SS,COD} \, P_{XT} = 0.9 \times 60.4 = 54.4 \text{ kg SS/day}$$

(13) Calculate M_{XT} and M_{SS}:

$$\begin{aligned} M_{XT} &= P_{XT} \theta_X \\ &= 60.4 \times 10 = 604 \text{ kg COD} \end{aligned}$$

$$M_{SS} = i_{SS,COD} M_{SS} = 0.9 \times 604 = 543.6 \text{ kg SS}$$

(14) Select $X_{SS} = 4000$ mg SS/l

(15) Calculate V_R

$$V_R = \frac{M_{SS}}{X_{SS}} = \frac{543.6}{4} = 136 \ m^3$$

(16) Calculate θ_h

$$\theta_h = \frac{V_R}{Q} = \frac{136}{100} = 1.36 \text{ days} = 32.6 \text{ hrs}$$

(17) Select SVI and Calculate X_{RSS}:

$$\text{SVI} = 120 \text{ ml/g}$$

$$X_{RSS} = \frac{10^6}{120} \cong 8300 \text{ mg SS/l}$$

(18) Calculate R_S and R_I:

$$R_S = \frac{1 - \dfrac{\theta_h}{\theta_X}}{\dfrac{X_{RSS}}{X_{SS}} - 1} = \frac{1 - \dfrac{1.36}{10}}{\dfrac{8300}{4000} - 1} = \frac{0.864}{1.075}$$

$$R_S = 0.8$$

$$R_I = R_T - R_S = 5.0 - 0.8 = 4.2$$

(19) Calculate OR:

$$OR_H = QC_{S1}[1 - Y_{NH}(1 + f_E cb_H \theta_X)]$$

$$f_E = f_{EX} + f_{ES} = 0.20 + 0.05 = 0.25$$

$$OR_H = 100 \times 1.25 \times [1 - 0.27 \times (1 + 0.25 \times 0.92 \times 0.15 \times 10)]$$

$$OR_H = 79.6 \text{ kg } O_2/\text{day}$$

$$OR_A = (4.57 - Y_{NA})QN_{OX}$$

$$OR_A = (4.57 - 0.15) \times 100 \times 0.133$$
$$= 58.8 \text{ kg } O_2/\text{day}$$

$$OR_D = 2.86 QN_D$$
$$= 2.86 \times 100 \times 0.111$$
$$= 31.7 \text{ kg } O_2/\text{day}$$

$$OR_T = OR_H + OR_A - OR_D$$

$$OR_T = 79.6 + 58.8 - 31.7 = 106.7 \text{ kg } O_2/\text{day}$$

(20) Check Alkalinity:

$$S_{ALK} = S_{ALK1} + \frac{1}{14}(S_{NO1} - S_{NO} + S_{NH} - S_{NH1})$$

$$\text{Assume } S_{NH1} = 60 \text{ mg/l}$$

$$S_{ALK1} = 460 \text{ mg l CaCO}_3 = \frac{460}{50} = 9.2 \text{ mmoles/l}$$

$$S_{ALK} = 9.2 + \frac{1}{14} \times (0 - 22.3 + 5.85 - 60)$$
$$= 9.2 - 5.5 = 3.7 \text{ mmoles/l}$$

$$S_{ALK} = 3.7 \times 50 = 185 \text{ mg/l CaCO}_3$$

$$S_{ALK} > 150 \text{ mg/l CaCO}_3$$

Sufficient alkalinity.

7.6 ENHANCED BIOLOGICAL PHOSPHORUS REMOVAL

Most industrial wastewaters do not require innovative treatment measures like EBPR to remove phosphorus. In fact, phosphorus deficiency for effective biological treatment is a more significant concern that should be checked for all industrial wastewaters as a prerequisite for appropriate design. This check should identify (i) a possible phosphorus deficiency, and (ii) the need for EBPR in addition to removal associated with microbial growth. It should be remembered that a relationship must be satisfied between the total phosphorus concentration, C_{PO1} and the total biodegradable COD, C_{S1} in the influent as a minimum nutritional requirement to ensure effective COD removal:

$$C_{PO1} > i_{XP} Y_H C_{S1} \tag{7.91}$$

This relationship indicates the need for external phosphorus supply. For a selected mode of operation the amount of phosphorus that would be potentially removed as incorporated into biomass can be calculated using expression (7.56). Since sludge wastage is the sole means of removing P in biological treatment systems, the amount of P removed is a function of the net sludge production and the P content of sludge. In entirely aerobic activated sludge systems, P content of the sludge can be assumed constant, as it is determined by nutrient requirements of microorganisms. The P removal in such systems is calculated from the net quantity of sludge wasted which in turn depends on the sludge age of the system.

It should also be noted that the soluble phosphorus concentration, S_{PO} is the significant parameter to be considered in meeting the effluent limitations because the particulate phosphorus, X_{PO}, mostly entrapped in the microbial flocs is removed from the liquid stream by biomass settling. In practice, it is also advisable to add a small coagulant dose of around 10–20 mg/l (usually iron salts) for improving the settling properties of the biomass and achieving better phosphorus removal.

However, in activated sludge systems designed for EBPR with an initial anaerobic stage, the P content of the sludge is far above the *regular* level (i.e. 0.02 g P/g TSS) dictated by the chemical composition of the biomass.

The extent of *enhanced* phosphorus accumulation in the biomass through EBPR depends mainly upon the relative magnitude of PAOs.

7.6.1 Factors affecting EBPR

Factors determining the selection of PAOs, and therefore affecting phosphorus removal performance, can be broadly classified into three groups: (i) wastewater characteristics; (ii) system parameters and (iii) environmental factors.

7.6.1.1 Wastewater characteristics

In EBPR systems, enhanced phosphorus uptake is closely related to storage and utilization of available organic matter. Therefore, the relative magnitude of the influent total phosphorus concentration, C_{PO1} with respect to biodegradable COD in the wastewater significantly affects the removal performance. The COD/P ratio of influent is often used as a useful index for estimating whether the process can be operated under phosphorus limited conditions. Phosphorus limitation implies that more than sufficient amount of organic carbon is available to remove all phosphorus present in the influent. Consequently, in a well designed phosphorus limited system, total P removal can be achieved. A carbon limited system however can only result in partial P removal where the effluent S_{PO} is determined by the difference between C_{PO1} and the P removal potential of the system. The threshold for phosphorus limited conditions is usually associated with a COD/P ratio of around 20 (Ekama and Marais, 1984b; Abu-ghararah and Randall, 1991; Janssen *et al.*, 2002).

It should be born in mind that the influent COD/P ratio has only an overall index value since it does not provide any indication about the biodegradable fraction of the total COD. In this framework, COD fractionation is a significant prerequisite for the evaluation of EBPR potential. In fact, the most decisive factor in the selection of PAOs is the S_{S1}/C_{S1} ratio. The VFA fraction of the readily biodegradable fraction, S_A is directly and preferentially used by PAOs. The remaining fraction, S_F also becomes available for PAOs after being fermented by facultative heterotrophs under anaerobic conditions. Relatively high anaerobic SRT will be required to generate sufficient VFA when the S_A fraction of the influent COD is low, provided that sufficient fermentable COD, S_F is available during the anaerobic period. The COD/TKN ratio can also drastically affect the EBPR performance of a nutrient removal SBR system as explained in the previous sections. If the COD/TKN ratio is not high enough to provide the desired nitrogen removal and EBPR for a mixed (anoxic/anaerobic) period even if it is increased to the acceptable highest limit, an anaerobic period, and hence EBPR, can only be obtained at the expense of nitrate removal.

7.6.1.2 System parameters

The sludge age and its anaerobic/aerobic fractions play an important role in the design of EBPR systems. The aerobic sludge age, θ_{XA} must be selected so as to allow selective growth of PAOs. A θ_{XA} value of 2 to 3 days can be sufficient enough, depending on temperature (Mamais and Jenkins, 1992). If nitrification is also envisaged, it will control the selection of the aerobic sludge age. If it is not required however, selection of an aerobic sludge age value short enough to prevent nitrification may be appropriate since produced nitrate will be recycled to the anaerobic (mixed) period and will upset the EBPR process. In practice, the composition of industrial wastewaters does not necessitate simultaneous nitrogen and phosphorus removal systems. If it is required, as in the case of domestic sewage/industrial effluent mixtures for example, care should be taken to adopt the minimum sludge age value that ensures the required nitrification, because longer aerobic (and anoxic) sludge ages would lead to a higher total sludge age with a lower sludge wastage and hence, less P removal. Moreover, excessive aeration causes depletion of stored substrate leading to a *carbon limited* system and secondary release of P due to decay reactions.

The anaerobic sludge age, θ_{XAN} required for EBPR is closely related to the influent COD composition. Since storage of VFAs is a very rapid process, θ_{XAN} can be as short as 0.5 day at 20°C, if the necessary amount of VFAs is present in the wastewater (Grady *et al.*, 1999). The required θ_{XAN} value may increase when hydrolysis of slowly biodegradable substrate is needed before fermentation due to the insufficient readily biodegradable COD in the influent. For simultaneous nitrogen and phosphorus removal systems, the anaerobic sludge age must be much longer, mainly due to utilization of readily biodegradable substrate under anoxic conditions.

7.6.1.3 Environmental factors

The growth of PAOs is affected by temperature. This effect is conveniently expressed in terms of an Arrhenius type of equation, the same way for heterotrophs and nitrifiers. Experimental observations suggest that PAOs are less sensitive to temperature changes as compared to nitrifiers. The proposed default value in ASM2 for the maximum specific growth rate at 20°C is 1.0/day both for PAOs and nitrifiers, which is reduced at 10°C to 0.35/day for nitrifiers and only to 0.67/day for PAOs (Henze *et al.*, 1995). This type of a temperature dependency means that the minimum aerobic sludge age for nitrifiers is higher than that of PAOs at temperatures below 20°C and thus controls the selection of the aerobic sludge age in simultaneous nitrogen and phosphorus removal systems. These values also indicate that at temperatures

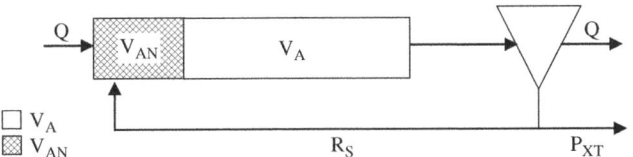

Figure 7.14. Schematic diagram of the conventional EBPR process

over 20°C it is difficult to prevent concomitant nitrification even when the objective is solely P removal.

7.6.2 System design for EBPR without nitrogen removal

A basic activated sludge system for EBPR involves a two stage reactor arranged in an anaerobic/aerobic sequence where the anaerobic volume fraction receives the wastewater to be treated (Figure 7.14). The biochemical mechanisms responsible for EBPR are too complex and a simple design approach based on system stoichiometry is not possible. Modeling is always a viable option for design and it is certainly quite beneficial for verification of the selected system. Different modeling alternatives are now available for this purpose with design methodologies based on these models (Marais *et al.*, 1983; Wentzel *et al.*, 1990; Henze *et al.*, 1995). System design still relies on engineering experience. As previously mentioned the aerobic fraction of the sludge age selected for design should prevent possible nitrification to the extent, but on the other hand, its anaerobic fraction should allow sufficient S_A production to achieve the desired degree of phosphorus removal. The literature generally recommends an anaerobic volume fraction of around 15% and an anaerobic hydraulic detention time in the range of 1.0–2.0 hrs (Randall *et al.*, 1982).

Occasionally, nitrification occurs by itself especially under summer conditions, regardless of the short sludge ages selected for the design and operation of EBPR systems. As a preventive measure during these periods, the designed

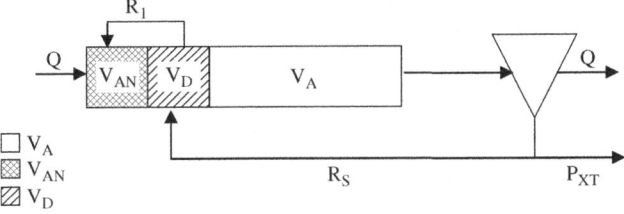

Figure 7.15. Nitrate removal with a post-anoxic reactor in EBPR

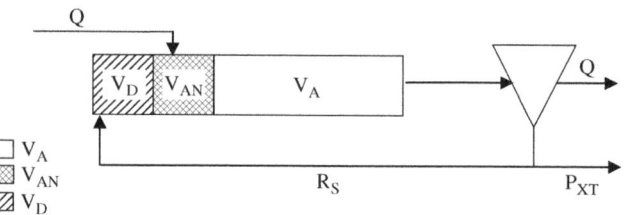

Figure 7.16. Nitrate removal with a pre-anoxic reactor in EBPR

system must have the operation flexibility to avoid direct mixing of the nitrate-containing sludge recycle stream with the wastewater. A number of alternatives may be envisaged for this purpose (i) sludge is recycled into a *post-anoxic reactor*, which removes nitrate in this stream and biomass is introduced to the anaerobic reactor by internal recycling from the post-anoxic volume as shown in Figure 7.15; (ii) sludge is recycled into a *pre-anoxic reactor*, which removes recycled nitrate by means of endogenous respiration. Influent is fed to the following anaerobic reactor (Figure 7.16).

7.6.3 System design for EBPR with nitrogen removal

The design of simultaneous nitrogen and phosphorus removal systems from industrial wastewaters, when required, should start with the same principles previously set forth for biological nitrogen removal. This requires a long sludge age necessary for nitrification and denitrification. The EBPR part should be integrated with the system in a way to maximize enhanced phosphorus removal despite the relatively low excess sludge production. The design should specifically emphasize (i) minimizing the amount of nitrate and dissolved oxygen introduced to the anaerobic volume, and (ii) securing optimal fermentation conditions for the influent readily biodegradable COD in order to maximize available S_A. In these systems, the maximum allowable non-aerated volume fraction should inevitably include the required anaerobic and anoxic phases. As a first approximation 35–40% of the total volume may be devoted to the anoxic volume and 10–15% to the anaerobic volume. Adjustments can be made later depending on the required effluent nitrate and phosphate concentrations.

When the COD/N ratio of the wastewater is sufficiently high, systems like a *five-stage modified Bardenpho process* as shown in Figure 7.17, may achieve complete nitrogen removal together with EBPR. When EBPR is the prime concern with less stringent effluent nitrogen requirements a dual anoxic system such as the *Johannesburg process* schematically given in Figure 7.18, may be

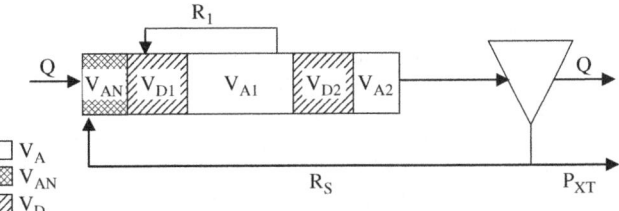

Figure 7.17. Five- stage – *modified Bardenpho* – process for biological N and P removal

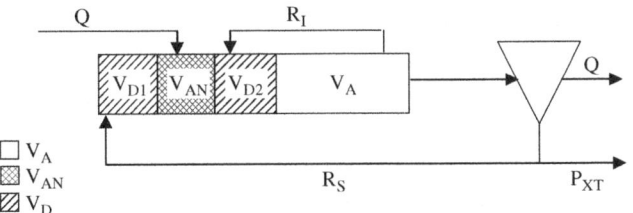

Figure 7.18. Johannesburg process for biological N and P removal

selected. In systems like the *Bardenpho process* the EBPR efficiency may be affected by intrusion and interference of nitrate. However activated sludge configurations like the *Johannesburg process* have provisions to avoid nitrate-containing sludge recycling into the anaerobic volume. In these systems, nitrogen and phosphorus removal mechanisms can be evaluated independently.

7.7 REFERENCES

Abu-ghararah, Z.H. and Randall, C.W. (1991) The effect of organic compounds on biological phosphorus removal. *Water Sci. Technol.* **23**, 585-594.

Archievala, S.J. (1981) *Wastewater Treatment and Disposal.* Marcel Dekker Inc., New York, N.Y.

Artan, N. and Orhon, D. (2005) *Mechanism and Design of Sequencing Batch Reactors for Nutrient Removal.* Scientific and Technical Report No.19, IWA Publishing, London, UK.

ATV, (1991) *Bemessung von einstufingen Belebungsanlagen ab 5000 Einwohnerwerten.* Arbeitsblatt A 131.

Benefield, L.D. and Randall, C.W. (1980) *Biological Process Design for Wastewater Treatment.* Englewood Cliffs, NJ: Prentice-Hall Inc.

Bohnke, B. (1989) Design of the nitrogen removal step in sewage treatment plant. *Korrespondenz Abwasser,* **36**(13), 65.

Christensen, M.H. and Harremoes, P. (1977) Biological Denitrification of Sewage: A Literature Review. *Prog. Wat. Tech.* **8**(4-5), 509.

Dold, P.L., Ekama, G.A. and Marais, G.v.R. (1980) A general model for the activated sludge process. *Prog. Wat. Tech.* **12**(6), 47.

Eckenfelder, W.W. Jr. and Grau, P. (1992) *Activated Sludge Process Design and Control: Theory and Practice.* Technomic Publishing Co. Inc. Lancaster, PA.

Eckhoff, D. and Jenkins, D. (1967) *Activated Sludge System Kinetics of the Steady and Transient States.* SERL Report No. 67-2. University of California, Berkeley.

Ekama, G.A. and Marais, G.v.R. (1984a) Nature of Municipal Wastewaters. In *Theory, Design and Operation of Nutrient Removal Activated Sludge Process*, Chapter 2, Water Research Commission, Pretoria, South Africa.

Ekama, G.A. and Marais, G.R. (1984b) *Biological excess phosphorus removal - design and operation of nutrient removal activated sludge process.* Water Research Commission, Pretoria S.A.

Grady, C.P.L. Jr. and Williams, D.E. (1975) Effects of influent substrate concentrations on the kinetics of natural microbial pollutions in continuous culture. *Water Res.* **9**, 171–180.

Grady, C.P.L., Daigger, G.T. and Lim, H.C. (1999) *Biological Wastewater Treatment.* 2nd edition, Marcal Dekker Inc., New York.

Groves, K.P., Daigger, G.T., Simpkin, T.J., Redmon, D.T. and Ewing, L. (1992) Evaluation of oxygen transfer efficiency and alpha factor on a variety of diffused aeration systems. *Water Environ. Res.* **64**, 691.

Gujer, W., Henze, M., Mino, T. and van Loosdrecht, M. (2000) *Activated Sludge Model No.3.* IWA Scientific and Technical Report No.9, (eds. Activated Sludge Models ASM1, ASM2, ASM2D and ASM3, Henze, M., Gujer, W., Mino, T., van Loosdrecht, M.), London, IWA.

Henze, M., Grady, C.P.L. Jr., Gujer, W., Marais, G.v.R. and Matsuo, T. (1987) *Activated Sludge Model No.1.* IAWPRC Scientific and Technical Report No.1, London: IAWPRC.

Henze, M., Gujer, W., Mino, T., Matsuo, T., Wentzel, M.C. and Marais, G.v.R. (1995) *Activated Sludge Model No.2.* IAWPRC Scientific and Technical Report No.2, London, IAWQ.

Janssen, P.M.J., Meinema, K. and van der Roest, H.F. (2002) *Biological Phosphorus Removal. Manuel for design and operation.* eds., IWA Publishing, Stowa.

Jenkins, D. and Garrison, W.E. (1968) Control of activated sludge by mean cell residence time. *J. Wat. Pollut. Control Fed.* **40**, 1905.

Mamais, D. and Jenkins, D. (1992) Effects of MCRT and temperature on enhanced biological phosphorus removal. *Water Sci. Technol.* **26**(5-6), 955-965.

Marais, G.v.R., Loewenthal, R.E. and Siebritz, I.P. (1983) Observations Supporting Phosphate Removal by Biological Excess Uptake – A Review. *Water Sci. Technol.* **15**, 15.

McKinney, R.E. (1962) Mathematics of complete-mixing activated sludge. *Proc. Am. Soc. Civil Eng.* **88**(SA3), 87-113.

Metcalf & Eddy Inc. (2003) *Wastewater Engineering, Treatment and Reuse.* New York: McGraw Hill.

Orhon, D. (1991) Design of pre-treatment. In *Pretreatment of Industrial Wastewaters*, (O. Tunay, D. Orhon, and A. Bederli, eds.), pp.131-173, ISO-SKATMK, Istanbul, (in Turkish).

Orhon, D. and Artan, N. (1994) *Modelling of Activated Sludge Systems.* Technomic Publishing Co. Inc., Lancaster, PA.

Orhon, D. (1998) Evaluation of industrial biological treatment design on the basis of process design. *Water Sci. Technol.* **38**(4-5), 1-8.

Orhon, D., Taslı, R. and Sozen, S. (1999) Experimental basis of activated sludge treatment for industrial wastewaters – The state of the art. *Water Sci. Technol.* **40**(1), 1-11.

Orhon, D., Insel, G. and Karahan, O. (2007) Respirometric assessment of biodegradation characteristics of the scientific pitfalls of wastewaters. *Water Sci. Technol.* **55**(10), 1-10.

Pearson, E.A. (1968) Kinetics of Biological Treatment. In *Advances in Water Quality Improvement*. Vol. 2, E. F. Gloyna and W. W. Eckenfelder, Jr., eds., University of Texas Press, Austin, TX.

Randall, C.W., Barnard, L.B. and Stensel, H.D. (1992) *Design and Retrofit of Wastewater Treatment Plants for Biological Nutrient Removal*. Technomic Publishing Inc., Lancaster, PA.

Sozen, S., Artan, N., Orhon, D. and Avcıoglu, E. (2002) Assessment of the denitrification potential for biological nutrient removal processes using OUR/NUR measurements, *Water Sci. Technol.* **46**(9), 237-246.

Spanjers, H., Vanrolleghem, P.A., Olsson, G. and Dold, P.L. (1998) *Respirometry in Control of the Activated Sludge Process: Principles*. IAWQ Scientific and Technical Report No. 7. London, UK.

U.S. EPA (1989) Fine Pore Aeration Systems. *U.S. EPA Technology Transfer Design Manual.* EPA/625/1-89/023.

U.S. EPA (1996) *Pretreatment Facility Inspection*. 3rd edition. U.S. EPA, California State University, Sacramento, USA.

WEF and ASCE (1991) *Design of Municipal Wastewater Treatments Plants*. Vol. 1. WEF Manual Practice No. 8., Brattleboro, VE: Book Press, Inc.

WEF (1994) *Pretreatment of Industrial Wastes*. Manual of Practice No. FD-3, Water Environment Federation, Virginia, USA.

Wentzel, M.C., Ekama, G.A., Dold, P.L. and Marais, G.v.R. (1990) Biological Excess Phoshorus Removal – Steady State Process Design. *Water S.A.*, **16**(1), 29-48.

WPCF (1983) *Nutrient Control*. Manual of Practice No. FD-7, Water Pollution Control Federation, Washington, DC, USA.

8
Sequencing batch reactor technology

8.1 INTRODUCTION

The *sequencing batch reactor* (SBR) is perhaps the most promising and viable activated sludge modification today for the removal of organic carbon and nutrients. In a relatively short period, it has become increasingly popular and now, it is well recognised as an effective biological treatment technology for different industrial wastewaters as well as domestic sewage (Artan *et al.*, 1996; Carucci *et al.*, 1999; Tilche *et al.*, 2001; Wilderer *et al.*, 2001; Murat *et al.*, 2002). Its recognition and growing success are largely due to significant system stability and a wide spectrum of operation alternatives offered in a very simple physical structure.

The SBR is basically a single tank that serves both as a biological reactor and settler in a temporal sequence; it inherently involves a cyclic operation,

© 2009 IWA Publishing. *Industrial Wastewater Treatment by Activated Sludge*, by Derin Orhon, Fatos Germirli Babuna and Ozlem Karahan. ISBN: 9781789065282. Published by IWA Publishing, London, UK.

each cycle incorporating the same pattern of successive selected phases. As contrasted to continuous flow systems, wastewater feeding is provided during the desired portions of the cycle, which then secures a *batch* wise, *fill-and-draw* type of an operation.

8.1.1 Historical development

The fill-and-draw scheme, constituting the basic principle of the SBR system was the basic mechanism that was used for the development of the original activated sludge. The pioneering experiments of Arden and Lockett (1914) supported by similar fill-and-draw experiments conducted within the same year at the University of Illinois (Mohlmann, 1917) found immediate recognition and the concept of *activated sludge* was soon utilized in practice both in England and in the US. A treatment plant designed as a fill-and-draw system was constructed for Daryhulme, Manchester (Orhon and Artan, 1994). In the US, a similar fill-and-draw plant was put into operation in 1915, in Milwaukee, Wisconsin. From 1914 to 1920, several full-scale fill-and-draw systems of different size were installed both in England and in the US (Wilderer *et al.*, 2001). After 1920, engineering efforts were then directed towards operating the process under continuous flow conditions. Successful experience with this new approach, together with the promotion of the diffused air process as a feasible means of air provision established the continuous flow principle as the major practical method for activated sludge operation.

After the choice for continuous flow processes, the interest for batch-fed systems was revived in the 1960s, with the development of new technology and equipment. The system in its newly devised scheme with a periodic influent feeding and periodic discharge was called the *sequencing batch reactor*. The new concept was supported by substantial research, which remained quite empirical in nature, mainly stressing the advantages of the SBR over the conventional continuous flow system: Dennis and Irvine (1979) reported that sludge settleability characteristics varied markedly with different fill/react ratios. Hoepker and Schroeder (1979) showed that the lower feed strength resulted in better effluent quality, and semi-batch treatment was more suitable in minimizing dispersed growth. Irvine *et al.* (1979) demonstrated the feasibility of nitrification-denitrification, given proper design and operation. Ketchum and Liao (1979) explored the potential of the SBR for phosphorus removal. The results of these studies consolidated the acceptance of the SBR in practice as a reliable process and as a viable and resourceful alternative for the biological treatment of domestic and industrial wastewaters.

8.1.2 Current experience

SBR is now extensively utilized both in full-scale application and as an experimental setup for fundamental and applied research. Its operational characteristics are well suited for investigating different microbial processes involved in biological wastewater treatment. With an SBR system, the functional relationships between the relevant parameters of a selected process may be observed in a much better way as compared to a continuous flow, completely mixed reactor, (CSTR). In the latter, each parameter can only be observed as a single value at a given steady-state operation; this value reflects the combined resulting effect of all the biochemical mechanisms involved. However, the SBR shows during each cycle the transient responses of all the observed parameters, such as COD, nitrogen forms, biochemical storage products, oxygen uptake rate, etc. Moreover, the SBR is also much better suited for this purpose as opposed to a simple batch reactor, mainly because it is operated at steady-state with identical responses in each cycle, reflecting this way the microbial culture history for selected operating conditions, a very important factor often overlooked in batch experiments.

In this context, SBR has successfully served and continues to serve as an experimental reactor for the appropriate interpretation and modelling of a number of fundamental issues dealing with different aspects of biological treatment (Imura *et al.,* 1993, Brenner, 2000; Karahan *et al.,* 2003; Yagci *et al.,* 2003).

The SBR has also been extensively utilized, both at laboratory and pilot-scale for the experimental assessment of treatment efficiencies both for domestic and industrial wastewaters. A comprehensive review of different SBR applications is reported by Mace and Mata-Alvarez (2002). In these applications, system flexibility of the cyclic operation offering additional system parameters such as cycle time, fill time, etc., mostly selected at random, have been elaborated as substantial competitive advantages of the SBR against the continuous flow configuration. Most of the studies were conducted with a pre-selected set of operating parameters and almost always reported satisfactory removal performance for the particular case of application.

Application of the SBR technology for the treatment of industrial wastewaters is now widely explored and tested: Satisfactory and reliable performance has been reported for winery wastewaters (Torrijos and Moletta, 1997), brewery wastewaters (Ling and Lo, 1999), food industry wastewaters (Raper and Green, 2001), dairy wastewaters (Mohseni and Bazari, 2000), slaughterhouse wastewaters (Belanger *et al.,* 1986), piggery wastewaters (Lee *et al.,* 1997), pulp and paper mill effluents (Franta and Wilderer, 1997) and tannery effluents (Carucci *et al.,* 1999). Similar to the experience with domestic sewage, very little evidence, if any, is available in the related literature to justify the specific operating parameters and design criteria selected to ensure the desired performance of the SBR system for the particular industrial wastewater. A remarkable exception to this statement is the comprehensive evaluation of the SBR treatment of tannery wastewater by Murat *et al.* (2002) involving appropriate wastewater characterization and COD fractionation of the wastewater for modelling, experimental assessment of model coefficients, mass balances for carbon and nitrogen removal, performance prediction of SBR operation based on process stoichiometry and model calibration of system performance. Effects of temperature were investigated in a following study (Murat *et al.*, 2004), which established the nitrogen balance of the SBR system treating tannery wastewater for a wide temperature range between 9 and 30° C and evaluated the experimental results by means of model calibration of COD, nitrate and ammonia nitrogen concentration profiles during cyclic operation.

The urge for simpler and more reliable treatment systems, the comfort of system flexibility and additional operating parameters together with extensive applied research highlighted SBR as a novel process with competitive advantages against the conventional continuous flow activated sludge. These factors have been the major ingredients for the increasing success of this process in full-scale application, both for domestic and industrial wastewaters. Today, full-scale application of the SBR process is wide spread with more than 1200 full-scale plants in North America; more than 700 plants in Japan, mainly as small installations in rural agricultural areas; around 150 plants in Germany with a large portion for industrial applications; more than 100 plants in Australia (Wilderer *et al.*, 2001). In Turkey, close to 1000 sewage treatment applications exist, mostly package and standardized plants for effluents from holiday resorts and small communities, in addition to some 50 plants for a variety of industrial wastewaters. More than 20 of these industrial applications were reported by Artan *et al.* (1996).

The application of the SBR process has not been limited with suspended growth systems. SBR alternatives using biofilm systems have also been developed with the generic name of *sequencing batch biofilm reactor* (SBBR) and found application in the treatment of domestic and industrial wastewaters (Wilderer, 1992; Pujol *et al.*, 1998).

8.1.3 Unified basis for modelling and design

While the SBR process offers a great flexibility and variety of operation, it should be considered as an activated sludge process in terms of basic principles related to modelling and design. This flexibility of operation, if well understood and interpreted in terms of governing biochemical processes, may provide additional advantages for appropriate design.

The performance of the SBR is now fully interpreted in terms of basic principles incorporated into recent activated sludge models. For such evaluations, the process offers the advantage of observing and modelling concentration transients for selected key parameters such as COD, N forms etc., It also provides the necessary flexibility of operation to transmit the outputs of kinetic evaluations into application, by appropriate adjustment of cycles and manipulation of aerated and non-aerated phases. Principles of a systematic approach to the design of SBR can be formulated similar to those for continuous flow systems. However, SBR has inherent constraints besides some advantages and requires different design considerations. While a simpler approach may be tolerated for organic carbon from domestic sewage, industrial effluents require a more elaborate wastewater characterization for design and the support of process modelling.

In recent years, principles of a systematic approach for the SBR design have been defined for using basic stoichiometry in a way that allows comparative evaluation with continuous flow activated sludge (Orhon and Artan, 1994; Artan *et al.*, 2001). These studies highlighted the fact that model simulation of SBR performance could provide useful and reliable information for a selected set of different operating conditions. However, interpretation of the simulation results for process design is only meaningful when support is provided in terms of relevant process stoichiometry and mass balance relationships for model components.

Until recently, the design and operation of the SBR plants relied largely upon empirical knowledge. Specific cyclic operation schemes were arbitrarily selected to suit the needs of a particular case, mostly as an engineering option and not because they represented the most suitable operation alternative from a process standpoint. It is now commonly agreed that the only way to provide a reliable basis for design and operation is to identify a rational mechanistic description of the process in terms of microbial kinetics and mass balance. A recent report defined a unified basis for a rational design approach for SBR systems, primarily based on relevant process stoichiometry (Artan and Orhon, 2005). In this report, specific emphasis has been placed upon the fact that such a unified design approach is also by nature the determining factor for the selection of the most appropriate cyclic operation scheme, the sequence of necessary phases and filling patterns for the particular application. This chapter outlines the stepwise design approach presented in this report for both organic carbon and nutrient removal, primarily focusing on the specific characteristics and features of industrial wastewaters.

8.2 PROCESS DESCRIPTION

The SBR process basically consists of a single tank, serving both as a biological reactor and settler: In SBR systems, biological processes and settling are carried out in a temporal sequence within the same reactor, whereas they occur simultaneously but in different tanks in continuous flow systems. Consequently a portion of the reactor operation is considered to be inactive in terms of the microbial processes involved.

The SBR process incorporates a variable-volume reactor. The total reactor volume, V_T, is composed of two independently controllable fractions, namely a stationary volume V_0, that basically holds the settled sludge and a volume V_F which is the volume of wastewater that is filled and discharged every cycle (Figure 8.1). Therefore, the reactor volume reaches its maximum level at the end of the influent filling period.

The process involves a cyclic operation at steady-state with intermittent feeding during selected portions or the entire duration of the cycle. The duration of a cycle and the feeding pattern within the cycle can be independently adjusted for a selected mode of operation.

In this context, the definition of the SBR system, as an activated sludge process involving variable-volume reactors with a cyclic operation and each cycle having a fixed pattern of several phases in succession, inevitably includes the following basic parameters inherently associated with system operation.

8.2.1 Cycle frequency (m)

Number of cycles per day, m, is an important parameter to be selected in SBR design and operation. It defines the total cycle time, T_C and the fill volume per cycle, V_F:

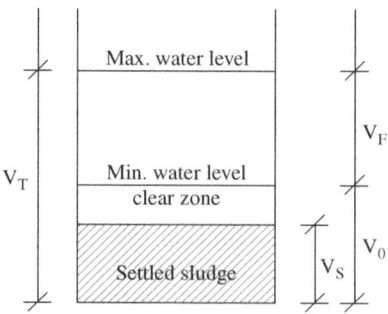

Figure 8.1. Schematic representation of SBR

$$T_C = \frac{1}{m} \quad (8.1)$$

$$V_F = \frac{Q}{m} \quad (8.2)$$

where Q is the daily volumetric flow rate of the wastewater to be treated in the SBR system.

8.2.2 Nominal hydraulic retention time (HRT)

A nominal hydraulic retention time, θ_h is usually defined for SBRs similar to continuous flow activated sludge systems as V_T/Q and can be expressed as a function of T_C using above equations:

$$\theta_h = \frac{V_T}{Q} = \frac{V_0 + V_F}{mV_F} = \left(1 + \frac{V_0}{V_F}\right)T_C \quad (8.3)$$

It should be noted that V_0/V_F ratio corresponds to the recycle ratio of the system. Similarly, θ_h may also be expressed in terms of the volumetric exchange ratio, VER, which is defined as the ratio of the fill volume to the total volume.

$$\theta_h = \frac{V_T}{mV_F} = \frac{T_C}{VER} \quad (8.4)$$

8.2.3 Duration of phases in a cycle

The SBR is a biological treatment process with an inherent cyclic operation, each cycle incorporating the same pattern of several successive phases. Traditionally, a cycle in the SBR process is conceived to consist of five phases: *Fill, react, settle, draw* and *idle* (Wilderer et al., 2001). In the *fill phase* (T_F), influent wastewater is fed into the SBR reactor, starting the relevant biochemical reactions through mixing/aeration as necessary, with the exception of static filling. In the *react phase* (T_R), influent filling stops and the SBR is operated as a batch reactor where mixing/aeration continues until reaching the desired level of completion for targeted biochemical conversions. Mixing/aeration are discontinued at the beginning of the *settle phase* (T_S), and the sludge is allowed to settle to the reactor bottom throughout the phase. The clear supernatant over the settled sludge layer is decanted and discharged during the *draw phase* (T_D). The *idle phase* (T_I) serves as a reserve period intended to increase the overall operational flexibility of the system. It can be used to extend the duration of one or more of the other phases when need arises.

In today's practice, the duration of wastewater feeding to the SBR may be varied from an almost instantaneous dump fill, mostly for creating substrate accumulation or sustaining selector conditions, to a continuous fill persisting throughout the entire process cycle including the settle and draw phases, making *react* period unidentifiable as a discrete process phase. In this context, wastewater filling, (T_F), provided during the desired portions of the cycle, should not be considered as a mechanistic component of the cycle, mainly because it does not necessarily precede a distinct react phase. Biological reactions obviously occur during wastewater filling, which only affects the structure of the relevant mass balance equations. Therefore, in lieu of fill and react phases, adoption of the concept of a *process phase* (T_P), covering the whole period adjusted to sustain the selected biochemical reactions in the reactor, seems more appropriate from a process standpoint. As the cycle includes, as mentioned before, an additional period ($T_S+T_D+T_I$) assumedly with no biological conversion for simplicity of mechanistic evaluation and modelling, the total cycle time, T_C, can be defined as:

$$T_C = T_P + T_{S+D+I} \qquad (8.5)$$

The duration of fill time, T_F, can range from a small fraction of total cycle time to total process time, or even to the total cycle time and fill time ratio, FTR, is defined as

$$FTR = T_F/T_C \qquad (8.6)$$

Fill time ratio affects reactor hydraulics seriously. The lower the fill time ratio the more pronounced becomes the substrate gradient within a cycle.

8.2.4 Duration of periods in a process phase

The process phase can be fully aerobic as in the systems for COD removal or it can sustain various environmental conditions adjusted by energy input. In nutrient removal SBR systems, the process phase T_P consists of aerated periods, (T_A), and mixed periods, (T_M), which can be anoxic or anaerobic depending on the presence of nitrate:

$$T_P = T_M + T_A = T_{AN} + T_{AOX} + T_A \qquad (8.7)$$

8.2.5 Number of tanks

The SBR process can be carried out in a single reactor or in multiple reactors in parallel. If equalization is not used, a continuous wastewater flow can be accommodated by providing multiple reactors; for n reactors, the equation below must be observed.

$$T_C = nT_F \tag{8.8}$$

Thus, n is inversely proportional to the FTR:

$$FTR = \frac{1}{n} \tag{8.9}$$

8.2.6 Sludge retention time (SRT)

The sludge retention time (SRT) for SBR system, θ_X, is defined as the mass of sludge contained in the reactor, M_{XT}, divided by the sludge wasted per day, P_{XT}, as for the continuous flow systems.

$$\theta_X = \frac{M_{XT}}{P_{XT}} \tag{8.10}$$

Sludge wasting from the mixed liquor (i.e. before settling) may provide a simple and direct means of controlling θ_X. In this case, θ_X may be defined, as follows, in terms of the total reactor volume, V_T, the volume of the sludge wasted from mixed liquor each cycle, V_W, and the cycle time, T_C. In practice however, excess sludge wasting takes place mostly after settle phase on a daily or even weekly basis.

$$\theta_X = \frac{V_T X_T}{m V_W X_T} = \frac{V_T}{V_W} T_C \tag{8.11}$$

8.2.7 System modelling

Interpretation and control of the efficiency of the SBR systems first require a proper kinetic description of the reactions taking place in the reactor. *Reaction kinetics* however cannot be directly evaluated and interpreted as process performance without any consideration of the *reactor hydraulics* defining the mixing characteristics of reactors where biological reactions are allowed to occur. In order to obtain accurate information about the fate of any component in the reactor, the conversion rate of the specific component should be processed as dictated by reactor hydraulics in terms of appropriate mass balances around the system.

On the basis of its hydraulic characteristics, SBR may be considered as a batch reactor during the portion of the cycle with no filling. This analogy however is not accurate and may at times be misleading. Actually, SBR is a *continuous* activated sludge system that reaches, if properly operated, a cyclic steady-state for a given set of operating conditions such as feed composition, structure of the cycle, etc. The term *batch* in the name of the process only implies intermittent feeding and discharge and nothing else. At steady state, the progression of each cycle is identical and the overall behaviour of the system in terms of concentration transients of significant parameters within the cycle remains the same from one cycle to the following one. The necessary period for a SBR system to reach steady-state is variable but at least two sludge ages.

As previously mentioned, wastewater feeding in SBR operation may take place during the selected fractions of the cycle but effluent withdrawal occurs periodically so that the system needs to be evaluated as a *variable volume* reactor. During the fill phase, the volume of the each tank operated in parallel gradually increases from an initial level, V_0, to the total reactor volume, V_T, defined as:

$$V_T = V_0 + Q_{in} T_F \qquad (8.12)$$

where, Q_{in} = the volumetric flow rate of the influent

The mass balance for any model component present in the influent, C_A, may be formulated as follows, taking the whole reactor volume as the control volume, with the assumption that an SBR always operates as a completely mixed reactor,

$$\frac{d[VC_A]}{dt} = Q_{in} C_{A\,in} + Vr_A \qquad (8.13)$$

where, V = reactor volume at a given time, t
$C_{A\,in}$ = influent concentration
r_A = observed conversion rate for CA

Taking the differential of the left-hand side and rearranging to include the definition of flow rate, $dV/dt = Q$, Equation (8.13) may be rewritten as:

$$\frac{dC_A}{dt} = \frac{Q_{in}}{V_{0i} + Q_{in}t}(C_{Ain} - C_A) + r_A \qquad (8.14)$$

where, C_A is the reactant concentration at a given time during the process phase.

During the *react* portion of the process phase (i.e. after fill phase) where the SBR functions as a completely mixed batch reactor having constant volume, mass balance equations are reduced to only the reaction terms:

$$\frac{dC_A}{dt} = r_A \qquad (8.15)$$

When distance is translated into the flow-through time, the above equation and the equation which describes the fate of a model component in a plug-flow reactor (PFR) are identical. They yield identical concentration profiles if they are integrated for the same concentration limits.

With the commonly accepted assumption that no biological conversion takes place during settling, draw and idle phases, the concentration of a soluble component at the end of the process phase remains the same till the beginning of the next cycle. When the system reaches a cyclic steady-state through continuous operation with constant influent conditions, the initial concentration of a soluble component, S_{ini}, becomes equal to the concentration of the same component at the end of the cycle i.e. the effluent concentration, S_e. The mass and concentration of a particulate component, however, is bound to change due to drawing of supernatant and sludge wasting according to the selected SRT to maintain cyclic steady-state operation.

8.3 ORGANIC CARBON REMOVAL

8.3.1 Basic principles

System design should be envisaged with the basic understanding that the SBR process is an activated sludge configuration and therefore it benefits from the same mechanistic evaluation related to basic stoichiometry and applicable mass balances already defined for the design of continuous flow systems. The adopted approach should however account for the specific features of the SBR incorporating a higher flexibility of operation and therefore, a higher level of options for selecting the optimum design strategy. In this context, a unified basis of a systematic design approach for different SBR alternatives can be formulated in a way quite similar to that currently applicable for continuous flow activated sludge, provided that it accurately accounts for specific characteristics and functions of the SBR. In general, a steady-state design model may be adopted since a long-term SBR operation under constant influent and operation conditions yields identical cycle profiles and effluent quality, which can be interpreted as *cyclic* steady-state conditions.

The design also requires relevant information for appropriate evaluation. For industrial effluents, this information is vital for a reliable design approach as it identifies the specific nature of the wastewaters generated from a wide spectrum of different industrial activities. As elaborated in detail in the preceding chapters, it involves (i) wastewater characterization and (ii) stoichiometric and kinetic data specifically generated for the wastewater stream to be treated. Reliable information on wastewater flow characteristics, with average and peak values, daily variations and seasonal or specific fluctuations are also a prerequisite for a proper design.

For industrial wastewaters, COD should be adopted as the main design parameter reflecting the organic carbon content, mainly because BOD_5, becomes meaningless and useless in most cases, due to slowly biodegradable, inhibitory and sometimes toxic components. COD may be used as a direct parameter to yield the stoichiometric equivalent of carbonaceous substrate, provided that its biodegradable fraction is determined. The fact that biodegradable COD yields the electron balance between substrate, biomass and the electron acceptor is a positive asset for appropriate design but requires accurate experimental identification of inert COD fractions. In this context, wastewater characterization involves the assessment of major fractions of the wastewater total COD, (C_{T1}), at least as biodegradable COD (C_{S1}), inert soluble COD (S_{I1}) and inert particulate COD (X_{I1}). If required, C_{S1} can be further fractionated into readily biodegradable COD, (S_{S1}), soluble biodegradable COD, (S_{H1}) and particulate slowly biodegradable COD, (X_{S1}). Extensive information is available in the literature on characterization with emphasis on COD fractionation and stoichiometric/kinetic data for a wide range of industrial wastewaters, as summarized in the preceding chapters.

Aside from relevant COD fractions, wastewater characterization should primarily cover all parameters prescribed in effluent discharge regulations for specific industries; it should also give indications for pre-treatment, if needed and supply the necessary information on the suspended solids content, nutrient balance (wastewater N and P contents; COD/N; COD/P ratios), pH, buffer capacity (alkalinity) etc.

Evaluation of organic carbon utilization involves a systematic analysis of the behaviour of heterotrophic microorganisms, preferentially using dissolved oxygen as the final electron acceptor. A rational design for organic carbon removal does not require elaborate model simulation as an integral part of the procedure. It should however adopt a model to derive the necessary mass balances for COD components. The simplest biokinetic model should include as major processes growth, hydrolysis and endogenous respiration. Slowly biodegradable COD is hydrolyzed to S_S prior to being used by heterotrophs for growth. Soluble and particulate inert COD, (S_P and X_P) are produced as part of endogenous respiration. Effluent COD contains mostly soluble inert COD in the influent or produced in the system. Biodegradable COD concentration in the effluent can be neglected in well designed systems. Under the assumption that all biodegradable COD is consumed, the kinetic parameters of the adopted model are not required except for the endogenous respiration rate coefficient, b_H, for heterotrophs.

Process stoichiometry is basically defined by means of the heterotrophic yield coefficient, Y_H, setting the necessary mass balance relationships for excess sludge generation, oxygen consumption, etc. On the basis of aerobic growth energetics, this coefficient assumes a constant value when expressed in terms of g cell COD/cell COD (e⁻ equivalent biomass/e⁻ equivalent substrate). Ekama and Marais (1984) reported a Y_H value of 0.66 g cell COD/g COD based on experiments to characterize domestic wastewaters. Henze *et al.* (1995) suggested a default Y_H value of 0.64 g cell COD/g COD for the Activated Sludge Model No.2, (ASM2). Experimental results indicate that a similar Y_H range is also applicable for industrial wastewaters. The design procedure also requires the inert fractions of biomass f_{EX} and f_{ES} coefficients defining the stoichiometry of microbial products generation through endogenous respiration.

The kinetic and stoichiometric data are quite wastewater specific and should be experimentally determined as required by the structure of the model adopted. While the process kinetics and stoichiometry are well documented for domestic sewage, experimental support is still needed for most industrial wastewaters.

8.3.2 Specific design parameters

Despite similar basic principles with continuous flow activated sludge, specific features of SBR require definition of the following set of new parameters for appropriate system design.

8.3.2.1 Effective sludge age

The sludge retention time, SRT, or the sludge age, θ_X is also a significant parameter for the SBR, setting the basis for system operation and major microbiological processes. It is directly related to the average net specific growth rate of the system by means of the following basic mass balance for heterotrophic biomass:

$$-P_{XH} + V_T X_H \frac{T_P}{T_C} (\mu_H - b_H) = 0 \qquad (8.16)$$

The mass balance reflects the specific character of the SBR where biomass production can only take place during the process time, T_P, since a portion of the cycle has to be devoted to settle and draw. This type of an operation leads to the definition of an effective sludge age, θ_{XE} for SBR systems:

$$\frac{1}{\theta_{XE}} = \mu_H - b_H \qquad (8.17)$$

These expressions show that the SBR needs to sustain a net specific growth rate during the process phase higher than that indicated by the overall sludge age,

270   Industrial Wastewater Treatment by Activated Sludge

θ_X, that is, by the net average specific growth rate defined as the mass of biomass produced per unit time, P_{XH}, divided by the biomass contained in the reactor using the mass balance for active biomass, X_H. In other words, the system generates during the process phase, T_P, the amount of biomass to be wasted after one cycle, T_C, to maintain the cyclic steady-state. In this context, the mass balance equation (8.16), combined with expression (8.10) defining θ_X, may also be arranged to yield:

$$\mu_H - b_H = \frac{1}{\theta_X}\frac{T_C}{T_P} = \frac{1}{\theta_{XE}} \tag{8.18}$$

Important features of SBR operation, such as excess sludge production and oxygen requirement as well as the effluent quality mainly depend on the selected effective SRT, θ_{XE}. The range of typical SRT values commonly adopted for continuous flow activated sludge configurations is also applicable for SBR. An important point to remember is that effective and total SRT are different in SBR while they are the same in continuous flow systems. An appropriate effective SRT is often selected based on experience in the case of COD removal from municipal wastewaters. However, for the biological treatment industrial wastewaters and leachates, the use of kinetic information from treatability studies is recommended.

8.3.2.2 The net yield

For SBR systems, the expression defining the net heterotrophic yield, Y_{NH} should be corrected for the effective sludge age, θ_{XE}:

$$Y_{NH} = \frac{Y_H}{1 + b_H \theta_{XE}} \tag{8.19}$$

Similarly, the net yield, Y_{NHE}, accounting both for heterotrophic growth and endogenous respiration can be calculated using the following expression:

$$Y_{NHE} = (1 + f_E b_H \theta_{XE})\frac{Y_H}{1 + b_H \theta_{XE}} \tag{8.20}$$

where: Y_H = heterotrophic yield, g COD(g COD)$^{-1}$
f_{EX} = particulate inert fraction of the biomass

8.3.2.3 Excess sludge production and reactor biomass

The excess sludge production rate, P_{XT} is another key design parameter and it should include all relevant particulate components such as active biomass, endogenous residues, and influent inert particulate matter of organic nature. Similar to continuous flow activated sludge systems, these components may be integrated in an expression defining P_{XT} in terms of the effective sludge age, θ_{XE}, with the assumption that all available biodegradable COD in the influent, C_{SI} is depleted:

$$P_{XT} = Q(Y_{NHE}C_{SI} + X_{II}) \qquad (8.21)$$

or,

$$P_{XT} = Q\frac{Y_H}{1 + b_H\theta_{XE}}(1 + f_{EX}b_H\theta_{XE}) + QX_{II} \qquad (8.22)$$

The total amount of biomass sustained in the reactor, M_{XT}, is also a significant parameter in SBR design. It can be directly derived from the basic definition of the sludge age, θ_X, based on the calculated value of P_{XT}:

$$M_{XT} = V_T X_T = P_{XT}\theta_X = P_{XT}\theta_{XE}\frac{T_C}{T_C - T_{S+D}} \qquad (8.23)$$

The above equation shows that M_{XT} cannot be arbitrarily selected but it is a function of the selected effective sludge age and wastewater characteristics. It also indicates that the process stoichiometry can only determine M_{XT} and not its components V_T and X_T separately, suggesting free selection of one of the variables (volume or biomass concentration) within reasonable limits, based on practical experience.

In the adopted design procedure it is helpful to express both P_{XT} and M_{XT} in terms of total suspended solids (TSS) parameter. For this purpose the coefficient, $i_{SS,COD}$ may be used to convert COD into TSS. ASM3 suggests a default value of 0.9 g TSS/g COD for this coefficient (Gujer et al., 2000). This way P_{SS} (kg TSS/day) and M_{SS} (kg TSS) may be computed as follows:

$$P_{SS} = i_{SS,COD}P_{XT} \qquad (8.24)$$

and,

$$M_{SS} = i_{SS,COD}M_{XT} \qquad (8.25)$$

8.3.2.4 Reactor volume

In SBR design, sizing the reactor usually involves selection of T_C, and calculation of the corresponding hydraulic retention time, HRT, instead of selecting a suitable MLSS concentration. Achievable settling characteristics of activated sludge expected to occur during the settle phase, as well as mixing and oxygen transfer considerations should justify appropriate T_C selection. The first constraint is the initial volume, V_0 which should be large enough to hold the settled biomass with a safety factor, SF:

$$V_0 = SF\, V_S = SF\frac{M_{XT}}{X_R} = SF\frac{M_{SS}}{X_{RSS}} \qquad (8.26)$$

where, X_{RSS} = the settled biomass concentration (mg TSS/l)

It is obvious that smaller V_S and hence V_0 levels may be obtained for higher settled sludge concentrations, X_{RSS}. The maximum achievable value of X_{RSS} depends upon the settling characteristics of sludge and the duration of settling phase, T_S; it can be estimated from a selected value for the sludge volume index, SVI (ml/g), defining the volume (in ml) occupied by a gram of sludge after 30 minutes of quiescent settling. From this definition,

$$X_{RSS}(g/m^3) = \frac{10^6}{SVI} \qquad (8.27)$$

and using the value of M_{XT} in Equation (8.23), V_0 may be re-expressed as:

$$V_0 = SF\, P_{SS}\, \theta_X\, SVI(10^{-3}) \qquad (8.28)$$

Selection of T_C also sets the fill volume, V_F, as defined by expressions (8.1) and (8.2). Then, the total reactor volume can be calculated as shown below:

$$V_T = V_0 + V_F = SF\, P_{SS}\, \theta_X\, SVI(10^{-3}) + QT_C \qquad (8.29)$$

It should be noted that inherent properties of SBR operation only allow internal recycling and the V_0/V_F ratio reflects the same function of the recycle ratio in continuous flow activated sludge systems. Using the basic definitions of V_0 and V_F, this ratio can be calculated directly as a function of T_C or m:

$$\frac{V_0}{V_F} = SF\, \frac{P_{SS}}{Q}\, \theta_X\, SVI(10^{-3})m \qquad (8.30)$$

Incorporating the effective sludge age, θ_{XE}, the above expression can be rearranged as follows:

$$\frac{V_0}{V_F} = SF\, \frac{P_{SS}}{Q}\, SVI(10^{-3})\, \frac{\theta_{XE}}{(T_P/T_C)}\, \frac{24}{T_C} \qquad (8.31)$$

This expression indicates that the selection of T_C and its effective fraction, T_P jointly determine the recycle ratio, V_0/V_F, for a given set of wastewater and sludge characteristics and effective SRT. The hydraulic retention time, θ_h may then be calculated from equation (8.3). Fill volume requirement increases with increasing T_C; however, higher T_C does not mean always higher HRT. Because, V_0/V_F is inversely proportional with T_C; decreasing T_C also results in decreasing the effective fraction, T_P/T_C, since the time to be devoted to settling and draw is almost constant for all selected cycle times. Therefore, HRT may increase with decreasing T_C, due to a higher V_0/V_F value as a result of lower T_C and T_P/T_C fraction as can be seen from equation (8.31). Thus, for each set of operation condition dictated by wastewater strength and effective SRT, a cycle time, T_C that will minimize the total volume requirements can be determined.

With the simplifying assumption that the amount of biomass at the end of the react phase remains the same after solids settling, the MLSS concentration in the reactor after fill, X_T, can be calculated as a function of recycle ratio in terms of TSS:

$$X_{SS} = \frac{V_0/V_F}{1 + V_0/V_F} \frac{X_{RSS}}{SF} \qquad (8.32)$$

Optimum T_C values usually yield recycle ratios of higher than 1.0 depending on wastewater strength and the effective sludge age used. For industrial wastewater treatment, T_C values which will yield high recycle ratios up to an acceptable limit (for example, $V_0/V_F < 4.0 - 5.0$) should be selected to minimize HRT.

8.2.3.5 Design parameters for aeration

Aeration is the most important part of SBR operation for organic carbon removal. Appropriate design for aeration is much more complex as compared to continuous flow activated sludge, mainly because (i) SBR is basically a batch operation within each cycle, and (ii) influent feeding is intermittent. Consequently, significant transients of oxygen demand are likely to occur within cyclic operation.

The oxygen requirement at steady state is basically determined by the effective sludge age, θ_{XE}, for a given industrial wastewater. The heterotrophic oxygen requirement, OR_H can be calculated as:

$$OR_H = QC_{SI}[1 - Y_{NH}(1 + f_E b_H \theta_{XE})] \qquad (8.33)$$

The OR_H value calculated from the expression above should be considered for the most critical wastewater temperature. This requirement must be satisfied with the fraction of the oxygen supply that is available to the heterotrophic biomass. As the oxygen transfer efficiency, SOTE, is commonly defined under standard conditions, OR_H should first be converted into standard oxygen requirement, SOR. The required air supply to the reactor may then be calculated as a function of SOR and SOTE. Details of this evaluation can be found elsewhere (Grady *et al.*, 1999; Orhon and Artan, 1994).

The specific characteristics of SBR operation dictate that the daily amount of oxygen demand must be supplied to the reactor during aerated periods of the cycle.

Accordingly, an oxygen requirement rate, AOR, must be defined for the effective portions of the daily operation:

$$AOR = \frac{OR_H}{mT_A} = \frac{OR_H}{24(T_A/T_C)} \qquad (8.34)$$

For a given type of aeration device, the total power input required for oxygen transfer will be determined by this quantity of oxygen which is essentially independent of the reactor volume. Generally, for economic reasons, the same aeration equipment is used to satisfy the mixing requirements necessary to keep the MLSS in suspension. A minimum volumetric power input is required for mixing. Thus, the volumetric oxygen transfer rate at the maximum water level, r_{OT} (g O_2/m^3.h) must be considered in design. This parameter can be estimated from the following basic mass balance, neglecting the difference between influent and effluent DO concentrations:

$$V_T r_{OT} = OR_H/mT_A \qquad (8.35)$$

The volumetric oxygen transfer rate, r_{OT} may be redefined in terms of C_{SI} and θ_h, combining equating expressions (8.34) and (8.35):

$$r_{OT} = \frac{QC_{SI}[1 - Y_{NH}(1 + f_E b_H \theta_{XE})]}{V_T m T_A} \qquad (8.36)$$

or,

$$r_{OT} = \frac{[1 - Y_{NH}(1 + f_E b_H \theta_{XE})]C_{SI}}{\theta_h (T_A/T_C)} \qquad (8.37)$$

Constraints associated with r_{OT}, should also be observed for the selection of T_C, and hence the reactor volume, in addition to considerations outlined in the previous section. It is evident from the above equation that higher r_{OT} levels are required for shorter T_C values due to shorter aerated time fractions (i.e. T_A/T_C) for the same θ_h. This may somewhat offset the benefit acquired in the total reactor volume requirement by decreasing T_C. The advantage of operating with a smaller reactor volume will have to be evaluated and compared, especially for weaker wastewaters, with the additional expense in aeration equipment due to longer fractional time devoted to settling and draw and consequently less time available for aeration.

Volumetric power input should satisfy two conditions: It must be (i) lower than a critical upper value to avoid disruption of biological flocs and deterioration of settling properties, and (ii) a high enough level to sustain the necessary mixing conditions in the reactor. Suggested values for the upper limit range from 40–80 g O_2/m^3.hr depending on the type of aeration equipment.

This limit must be observed for the initial reactor volume (i.e. V_0) if aeration starts together with filling as usually practised for carbon removal systems. Rewriting equation (8.37) for V_0, the volumetric oxygen transfer rate at the minimum water level, $r_{OT@0}$, will be:

$$r_{OT@0} = \frac{[1 - Y_{NH}(1 + f_E b_H \theta_{XE})]C_{S1}}{T_A(V_0/V_F)} \tag{8.38}$$

Inserting the value of V_0/V_F in equation (8.31), $r_{OT@0}$ may be redefined as a function of the effective SRT, independently from the selected T_C:

$$r_{OT@0} = \frac{[1 - Y_{NH}(1 + f_E b_H \theta_{XE})]QC_{S1}}{SF \, P_{SS} \, \theta_{XE} \, SVI(10^{-3})} \tag{8.39}$$

Recommended air input to satisfy mixing requirements for diffused air systems is in the range of 20–30 m³ air/min per 1000 m³ of reactor volume, and 15–30 W/m³ of reactor volume for vertical shaft mechanical aerators; this corresponds to 15–30 g O_2/m³.hr with the assumption that mechanical aerators transfer approximately 1.0 g O_2/W.hr to the mixed liquor in activated sludge systems. If the calculated r_{OT} is lower than this value, the mixing requirements will control the selection of the aeration equipment, a situation often encountered in systems operated at high SRT values. In SBR systems, r_{OT} can also be increased by selecting a lower T_C if required to obtain a lower θ_h. Consequently, T_C should be selected so as to optimize the requirements for both reactor volume and aeration capacity until mixing requirements control the sizing of the aeration equipment.

8.3.3 Process design

The basic design data that needs to be compiled remains the same regardless of the selected activated sludge configuration. It involves wastewater flow; total COD, C_{T1} and its biodegradable and non biodegradable fractions; other wastewater characteristics such as TSS, VSS, TKN, TP, pH, alkalinity, critical design temperature, specific pollutants/inhibitors, if necessary; structure of the adopted model; relevant kinetic and stoichiometric coefficients; applicable discharge limitations.

The design procedure has a unified basis as it involves a similar stepwise evaluation approach already described for continuous flow activated configurations. The suggested design algorithm for SBR systems is schematically given in Figure 8.2. As shown in the figure, it involves four basic steps that need to be checked by means of specific expressions and relationships defining significant design parameters and functions of SBR: (i) Selection of the effective sludge

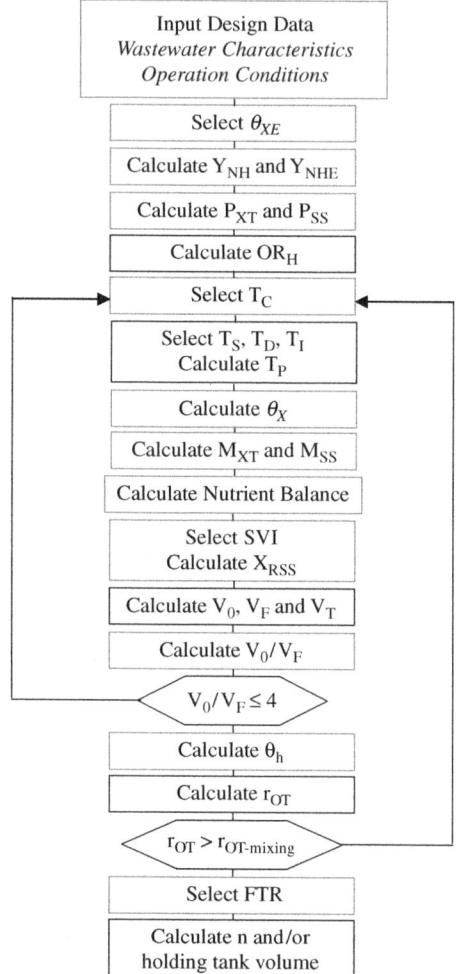

Figure 8.2. The recommended design algorithm of organic carbon removal for SBR

age, θ_X; (ii) selection of the cycle time, T_C; (iii) calculation of the recycle ratio, V_0/V_F, and (vi) calculation of the hydraulic retention time, θ_h. The recommended design procedure for organic carbon (COD) removal from the typical industrial wastewater, already evaluated in Example 7.6 in the preceding chapter, is now illustrated for the SBR system in Example 8.1.

Example 8.1 SBR design for organic carbon removal

Design a SBR system for organic carbon removal, for an industrial wastewater with characteristics given below and in Tables 7.4 and 7.5.

Wastewater flow, $Q = 100 \, m^3/day$
Total Influent COD, $C_{TI} = 1500 \, m^3/day$
Total influent nitrogen, $C_{NI} = 85 \, mg/l$
Total influent phosphorus, $C_{POI} = 10 \, mg/l$

Make all assumptions as necessary.

(1) Select θ_{XE}:

A design value of 8 days is selected for θ_{XE} similar to continuous flow activated sludge system for the same industrial wastewater.

(2) Calculate Y_{NH} and Y_{NHE}:

$$Y_{NH} = \frac{Y_H}{1 + b_H \theta_{XE}}$$

$$Y_H = 0.64 \, g \, cell \, COD/g \, COD$$
$$b_H = 0.15 \, day^{-1}$$

$$Y_{NH} = \frac{0.64}{1 + 0.15 \times 8} = 0.29 \, g \, cell \, COD/g \, COD$$
$$Y_{NHE} = Y_{NH}(1 + f_{EX} b_H \theta_{XE})$$
$$f_{EX} = 0.20$$

$$Y_{NHE} = 0.29 \times (1 + 0.20 \times 0.15 \times 8)$$
$$Y_{NHE} = 0.36 \, g \, COD/g \, COD$$

(3) Calculate P_{XT} and P_{SS}:

$$P_{XT} = Q(Y_{NHE} C_{S1} + X_{I1})$$

$$C_{S1} = 1250 \, mg/l = 1.25 \, kg/m^3$$
$$X_{I1} = 150 \, mg/l = 0.15 \, kg/m^3$$
$$Q = 100 \, m^3/day$$

$$P_{XT} = 100 \times (0.36 \times 1.25 + 0.15)$$
$$= 60 \, kg \, COD/day$$

$$P_{SS} = i_{SS,COD}P_{XT} = 0.9 \times 60 = 54 \text{ kg SS/day}$$

(4) Calculate OR_H:

$$OR_H = QC_{S1}[1 - Y_{NH}(1 + f_E b_H \theta_{XE})]$$

$$f_E = f_{EX} + f_{ES} = 0.20 + 0.05 = 0.25$$

$$OR_H = 100 \times 1.25 \times [1 - 0.29(1 + 0.25 \times 0.15 \times 8)]$$

$$OR_H = 77.9 \text{ kg O}_2/\text{day}$$

(5) Select T_C:

A T_C value should be selected to optimize reactor volume and aeration requirements. Selection should start from an acceptable low value.

As a first approximation a T_C value of 6 hours is selected.

(6) Select T_S, T_D, and T_I and calculate T_P:

A total 2 hours is selected for $T_S + T_D + T_I$

$$T_P = T_C - (T_S + T_D + T_I) = 6 - 2 = 4 \text{ hrs}$$

(7) Calculate θ_X:

$$\theta_X = \theta_{XE} \frac{T_C}{T_P} = 8 \times \frac{6}{4} = 12 \text{ days}$$

(8) Calculate M_{XT} and M_{SS}:

$$M_{XT} = P_{XT}\theta_X = 60 \times 12 = 720 \text{ kg COD}$$

$$M_{SS} = i_{SS,COD}M_{XT} = 0.9 \times 720$$

$$= 648 \text{ kg TSS}$$

(9) Calculate The Nutrient Balance:

The nutrient balance of the wastewater is the same as in Example 7.6.

The industrial wastewater is deficient in phosphorus.

(10) Select SVI and Calculate X_{RSS}:

Select an SVI of 120 ml/g for the industrial wastewater.

$$X_{RSS} = \frac{10^6}{SVI} = \frac{10^6}{120} = 8300 \text{ mg SS/l} = 8.3 \text{ kg TSS/m}^3$$

(11) Calculate V_O, V_F and V_T:

$$V_0 = SF \frac{M_{SS}}{X_{RSS}}$$

Select a SF value of 1.3 to avoid escape of biomass with the effluent.

$$V_0 = 1.3 \times \frac{648}{8.3} = 101.5 \text{ m}^3 \cong 100 \text{ m}^3$$

$$V_F = QT_C = 100 \times \frac{6}{24} = 25 \text{ m}^3$$

$$V_T = 100 + 25 = 125 \text{ m}^3$$

(12) Calculate V_O / V_F:

$$\frac{V_0}{V_F} = \frac{100}{25} = 4.0$$

V_O/V_F can also be directly calculated using equation (8.30):

$$\frac{V_0}{V_F} = SF \frac{P_{SS}}{Q} \theta_X \frac{SVI}{10^3} m$$

$$= 1.3 \times \frac{54}{100} \times 12 \times \frac{120}{10^3} \times 4 \cong 4$$

(13) Calculate θ_h:

$$\theta_h = \left(1 + \frac{V_0}{V_F}\right) T_C = \frac{V_T}{Q}$$

$$\theta_h = (1 + 4) \times \frac{6}{24} = 1.25 \text{ days} = 30 \text{ hrs}$$

(14) Calculate r_{OT}:

From equation (8.35):

$$r_{OT} = \frac{OR_H}{V_T m T_A}$$

$$r_{OT} = \frac{77.9}{125 \times 4 \times 4} = 0.039 \text{ kg/m}^3.\text{hr} = 39 \text{ g/m}^3.\text{hr}$$

r_{OT} is below it is upper limit range set for avoiding deterioration of setting properties.

(15) Calculate FTR:

The industrial wastewater to be treated is strong in terms of its COD content. Therefore a T_F period of 2 hrs is selected to tamper high substrate and oxygen demand gradients. Then,

$$FTR = \frac{T_F}{T_C} = \frac{2}{6} = 0.33$$

(16) Calculate n:

The number of parallel tanks, n:

$$n = \frac{1}{FTR} = \frac{T_C}{T_F} = \frac{6}{2} = 3 \text{ parallel tanks}$$

The number of parallel tanks can be reduced when designed together with a holding tank.

8.4 BIOLOGICAL NITROGEN REMOVAL

As previously mentioned, SBR has been promoted as a novel process of biological treatment. The novelty is however limited to the operation features of the system. The fundamentals of the biochemical processes involved are inevitably the same as applied to traditional, continuous flow activated sludge. This is often overlooked in some studies which attempt to manipulate SBR in an empirical way without the basic understanding of the process, mostly resulting as trial and error exercises simply testing a sequence of parameter combinations. Recently, significant research effort has been devoted to the mechanistic understanding of SBR, mainly for nutrient removal from wastewaters with different characteristics (Andreottola *et al.*, 1997; Tasli *et al.*, 2001; Artan *et al.*, 2001; 2002; 2003); these studies emphasized basic stoichiometry, process kinetics and simulation of applicable models for the assessment of system behaviour and performance. Wilderer *et al.* (2001) provided a comprehensive treatise of SBR systems. Artan and Orhon (2005) offered a detailed analysis of SBR design for nutrient removal.

The appropriate approach for the understanding of nitrogen removal in SBR is to formulate fundamentals of microbial processes in terms of meaningful parameters applicable for the selected mode of operation. This is best done by setting the necessary process analogy between the continuous flow activated sludge and SBR.

The operation flexibility of SBR is particularly suitable for nitrogen removal. Provision of a non-aerated phase which sustains anoxic conditions, T_{AOX}, establishes the sequence of anoxic/aerobic phases necessary for a *pre-denitrification* system. The system receives the wastewater feeding, T_F, within the anoxic phase for deriving full benefit from the denitrification potential of the influent. Anoxic conditions are obtained through nitrate generated in the aerobic phase of the previous cycle. In all pre-denitrification systems recycle ratio limits the level of denitrification and hence, the level of achievable nitrogen removal. If this level is not sufficient, a post-denitrification phase may be introduced to overcome this limitation, providing a *dual anoxic phase* operation. In this mode of operation wastewater feeding may be split between the anoxic phases, if necessary. The SBR may also be operated with more frequent shifts in aeration (*intermittent aeration*) with a multiple sequence of anoxic/aerobic phases, as required. In this case, it may not be practical to restrict wastewater filling to only anoxic phases and filling may continue during the entire process phase. SBR is flexible enough for other process options as well. However, it should be born in mind that these options are not to be selected at random but they should be evaluated starting from the simplest alternative for their suitability to the wastewater studied.

8.4.1 Nitrogen balance

Nitrogen removal performance in SBR systems, as in all activated sludge configurations, depends upon the balance between the three key parameters, namely the nitrification capacity, N_{OX}, the denitrification potential, N_{DP}, and the available nitrate or the magnitude of oxidized nitrogen supplied to the mixed periods, N_A. All three parameters are conveniently defined in mass per unit volume of wastewater, allowing simple manipulation in mass balance. The nitrogen mass balance previously formulated in equation (7.66) for continuous activated sludge systems for defining N_{OX} also applies to SBR. The denitrification potential, N_{DP} indicates the concentration of nitrate nitrogen that may be potentially removed, provided that enough nitrate nitrogen is supplied to the non-aerated period. Using the analogy between continuous flow systems, equation (7.80) may be adapted by replacing the term T_{AOX}/T_P by the anoxic volume fraction, to define N_{DP} for the SBR.

$$N_{DP} = \frac{(1 - Y_{NH})}{2.86}\left[S_{S1} + \eta \frac{T_{AOX}}{T_P}(C_{S1} - S_{S1})\right] \quad (8.40)$$

where, η = overall correction factor for reduced heterotrophic activity under anoxic conditions

The following mass balance expressions reflecting the NO_3^--N equivalents of the principal internal organic carbon sources available for the heterotrophic growth may also be used for the same purpose (Artan and Orhon, 2005). They reflect different N_{DP} fractions associated with the readily biodegradable substrate, N_{SS}, the slowly biodegradable substrate, N_{XS}, and with endogenous respiration, N_{ER}. These expressions assume that all biodegradable COD is totally consumed for the selected effective sludge age:

$$N_{SS} = (1 - Y_H)\frac{S_{S1}}{2.86} \quad (8.41)$$

$$N_{XS} = (1 - Y_H)\frac{X_{S1}}{2.86} \quad (8.42)$$

$$N_{ER} = (1 - f_E)b_H\theta_{XE}\frac{Y_H}{1 + b_H\theta_{XE}}\frac{(S_{S1} + X_{S1})}{2.86} \quad (8.43)$$

The available fractions of these N_{DP} components may be combined in the same way to yield the overall N_{DP} of the SBR system.

$$N_{DP} = \eta \frac{T_{AOX}}{T_P}(N_{SS} + N_{XS} + N_{ER}) \quad (8.44)$$

For the SBR operation in a pre-denitrification mode, the above expression may be arranged as follows:

$$N_{DP} = N_{SS} + \eta \frac{T_{AOX}}{T_P}(N_{XS} + N_{ER}) \quad (8.45)$$

Oxidized nitrogen is supplied to the first anoxic period by recycle and to the following anoxic periods from the preceding aerobic period. In an SBR operation with a single anoxic/aerobic sequence (pre-denitrification), the recycle ratio, V_O/V_F, is the key parameter that sets the magnitude of nitrate nitrogen that will be introduced into the anoxic period:

$$N_A = \frac{V_O}{V_F}S_{NO} \quad (8.46)$$

where, S_{NO} = the effluent nitrate nitrogen concentration

However, in the case of a multiple sequence of mixing/aeration phases within a cycle, recycle supplies nitrate only to the first anoxic period while the others utilize the nitrogen oxidized in the preceding aerobic periods. Hence, in these SBR systems, the total available nitrate is not determined solely by the V_O/V_F ratio. So, both N_{DP} and N_A depend on selected process options i.e. aeration and filling strategies employed.

SBR technology 283

The denitrification efficiency of a process is limited by either the denitrification potential, N_{DP}, or the available oxidized nitrogen, N_A. In the case of N_{DP} limitation ($N_{DP} < N_A$), the effluent nitrate concentration will be determined by the difference between N_{OX} and N_{DP}:

$$S_{NO} = N_{OX} - N_{DP} \qquad (8.47)$$

In the case of available nitrate limitation ($N_A < N_{DP}$), the effluent nitrate concentration will be determined by the difference between N_{OX} and N_A:

$$S_{NO} = N_{OX} - N_A \qquad (8.48)$$

Introducing the definition of N_A for *pre-denitrification* systems in equation (8.46) into the above expression, the effluent nitrate concentration and hence the denitrification efficiency can be defined as a function of V_0/V_F:

$$S_{NO} = \frac{N_{OX}}{1 + V_0/V_F} \qquad (8.49)$$

and

$$E = \frac{N_A}{N_{OX}} = \frac{V_0/V_F}{1 + V_0/V_F} \qquad (8.50)$$

8.4.2 Selection of process options

Wastewater characteristics and effluent requirements are the most important factors affecting process selection. Regulations are normally set in terms of the allowable total nitrogen concentration, $C_{N,E}$ in the effluent. Based on this effluent limit, a design effluent nitrate nitrogen concentration, $S_{NO,D}$ that would safely provide compliance with regulatory restrictions may be calculated, considering the remaining ammonia concentration, the inert fraction of the soluble organic nitrogen and the nitrogen fraction of the biomass escaping in the effluent:

$$S_{NO,D} = C_{N,E} - S_{NH,E} - i_{N\,SI}S_{I1} - i_{N\,VSS}X_{VSS,E} \qquad (8.51)$$

Assessment of an appropriate value for $S_{NO,D}$ determines $N_{OX} - S_{NO,D}$, the fraction of the oxidized nitrate concentration that needs to be removed through denitrification. The condition below should be satisfied as an indication that the desired process efficiency is potentially achievable.

$$N_{DP} \geq N_{OX} - S_{NO,D} \qquad (8.52)$$

If not, the selected anoxic fraction should be increased to the acceptable high limit to provide additional denitrification. In the case when N_{DP} is not limiting,

8.4.2.1 Pre-denitrification

A typical SBR operation in a pre-denitrification mode is schematically illustrated in Figure 8.3. In this mode of operation, recycling is the only instrument for the S_{NO} supply into the anoxic zone. The required available nitrate, N_A is a function of N_{OX} and the V_0/V_F ratio as indicated by the following mass balance:

$$N_{OX} - S_{NO,D} = N_A = \frac{V_0/V_F}{1 + V_0/V_F} N_{OX} \qquad (8.53)$$

Pre-denitrification should be evaluated as the priority option because it offers the simplest SBR operation for nitrogen removal and yields the highest denitrification potential for the same anoxic fraction. It should be considered with the practical recycling constraint which brings an upper limit of $V_0/V_F = 4$, generally accepted for SBR operation. It is obvious that designs proposing higher V_0/V_F ratios would involve an excessively high reactor volume running the risk of not being economically competitive. With this constraint, the expected nitrate removal efficiency may be at best slightly over 80% as dictated by the mass balance equation (8.53). For higher removal rates, other process options should be selected. Pre-denitrification design also involves an appropriate balance between three key parameters, V_0/V_F ratio, cycle time, T_C and the resulting hydraulic retention time, θ_h. This balance will be further elaborated in the following design section.

8.4.2.2 Step feeding with dual anoxic phases

As mentioned before, this process configuration includes a second anoxic phase and offers an improved option when the effluent nitrate limitation cannot be met with simple pre-denitrification. In this context, an SBR operation with dual mixed filling basically requires two sets of mixed and aerobic periods, T_{A1}/T_{A2} and T_{AOX1}/T_{AOX2} within the same cycle as schematically shown in Figure 8.4.

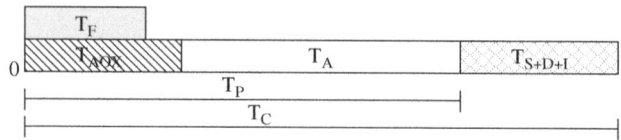

Figure 8.3. SBR operation in a pre-denitrification mode

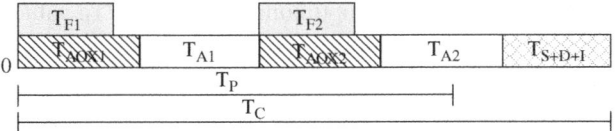

Figure 8.4. SBR operation with dual anoxic phases and step feeding

A β fraction of the total fill volume ($V_{F2} = \beta V_F$), is diverted to the second anoxic period to create the necessary denitrification potential:

$$N_{DP2} = \beta N_{SS} + \eta \frac{T_{AOX2}}{T_E}(\beta N_{XS} + N_{ER}) \tag{8.54}$$

Similarly, denitrification potential of the first anoxic reactor, N_{DP1} can be expressed as:

$$N_{DP1} = (1-\beta)N_{SS} + \eta\left(\frac{T_{AOX}}{T_E} - \frac{T_{AOX2}}{T_E}\right)[(1-\beta)N_{XS} + N_{ER}] \tag{8.55}$$

The nitrate nitrogen generated in the first aerobic period, N_{A1} is supplied to the second anoxic period and this way, additional nitrate removal can be achieved by post denitrification. Assuming that the ammonia load introduced into the first stage is entirely oxidized in this stage, N_{A1} will be proportional to (1-β). In order to achieve complete ammonia oxidation in each stage, fractionation of total aerobic period should be done in proportion to the ammonia load introduced into each anoxic period since it will be converted into nitrate at the following aerobic period. In this case N_{A1} would be (1- β) fraction of the overall N_{OX}:

$$N_{A1} = (1-\beta)N_{OX} \tag{8.56}$$

Denitrification potential of the second anoxic period, N_{DP2} should be sufficient to remove all nitrate introduced from the first stage (N_{A1}). This way, the oxidized nitrate generated in the second stage, βN_{OX}, will partly recycle to the first anoxic period with sludge, in proportion with V_0/V_F and it will provide the available nitrate, N_{A2} for denitrification in the first stage:

$$N_{A2} = \frac{V_0/V_F}{1 + V_0/V_F}\beta N_{OX} \tag{8.57}$$

The remaining portion of βN_{OX} will leave the system as effluent nitrate nitrogen:

$$S_{NO} = \frac{\beta N_{OX}}{1 + V_0/V_F} \tag{8.58}$$

For a given total anoxic fraction (T_{AOX}/T_E), optimum β and T_{AOX2}/T_E values to obtain minimum effluent nitrate concentration can be calculated using the following two key nitrogen balances to be satisfied around each sub-cycle:

$$N_{DP2} = N_{A1} \tag{8.59}$$

$$N_{DP1} = N_{A2} \tag{8.60}$$

8.4.2.3 Intermittent aeration

SBR operation with multiple stages - *intermittent aeration* - can supply more nitrate for denitrification for a limited recycle ratio and effluent nitrate will probably be determined by denitrification potential. In this type of operation, filling only during anoxic periods to increase the denitrification potential for a given total anoxic fraction may not be practical. Although a part of the denitrification potential is wasted, extended filling throughout the entire process time may still be preferable especially when the S_S fraction of wastewater is relatively low but COD/TKN ratio is high (Figure 8.5). Nitrate may not be completely depleted in each sub-cycle due to the frequent shift from mixed to aerated periods. As a result, effluent is determined by the difference between total nitrification capacity, N_{OX} and denitrification potential, N_{DP} of the system.

If the denitrification potential of each anoxic period, N_{DPi}, is sufficient to remove all nitrate introduced from the previous stage (N_{Ai}), oxidized nitrate in the last aerobic period will partly leave the system as effluent nitrate in proportion with V_0/V_F. The effluent S_{NO} concentration can be calculated again from equation (8.53); in this case, however, the β value will be determined by the number of sub-cycles provided that they are of equal duration.

8.4.3 SBR design for pre-denitrification

The suggested procedure for SBR design based on pre-denitrification is summarized in Figure 8.6. The starting step in the proposed design procedure is the selection of the aerobic sludge age, θ_{XA} that will ensure nitrification. This parameter should be selected as indicated in the preceding chapters, with full

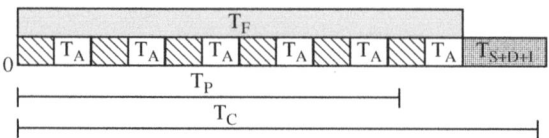

Figure 8.5. SBR operation in an intermittent aeration mode with continuous filling

SBR technology

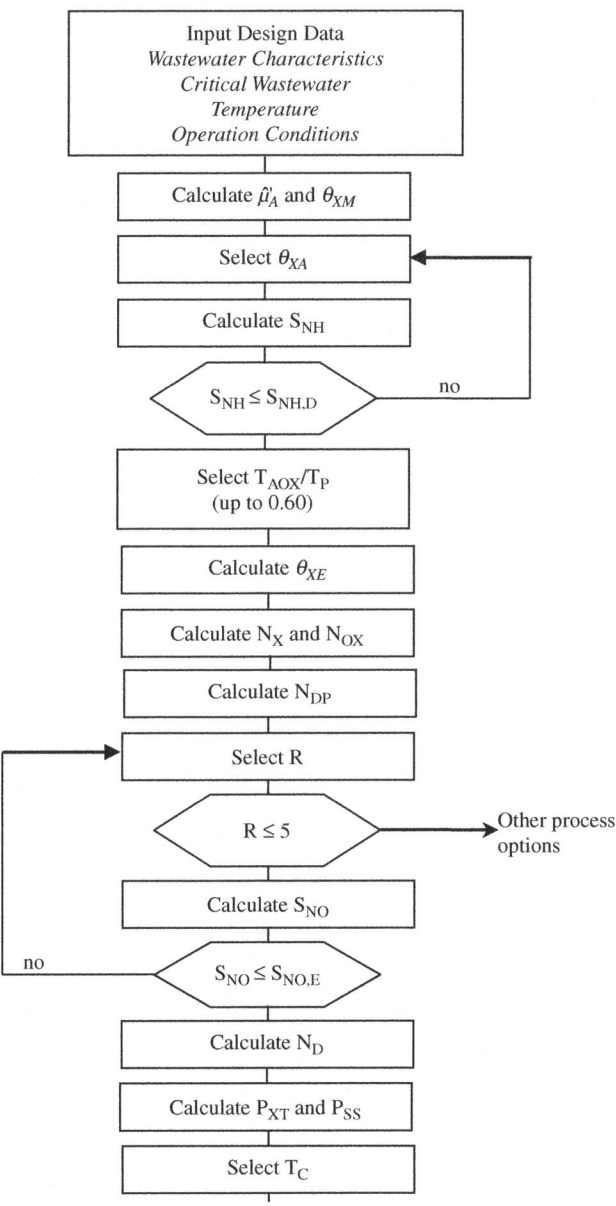

Figure 8.6. The recommended design algorithm of nutrient removal for SBR

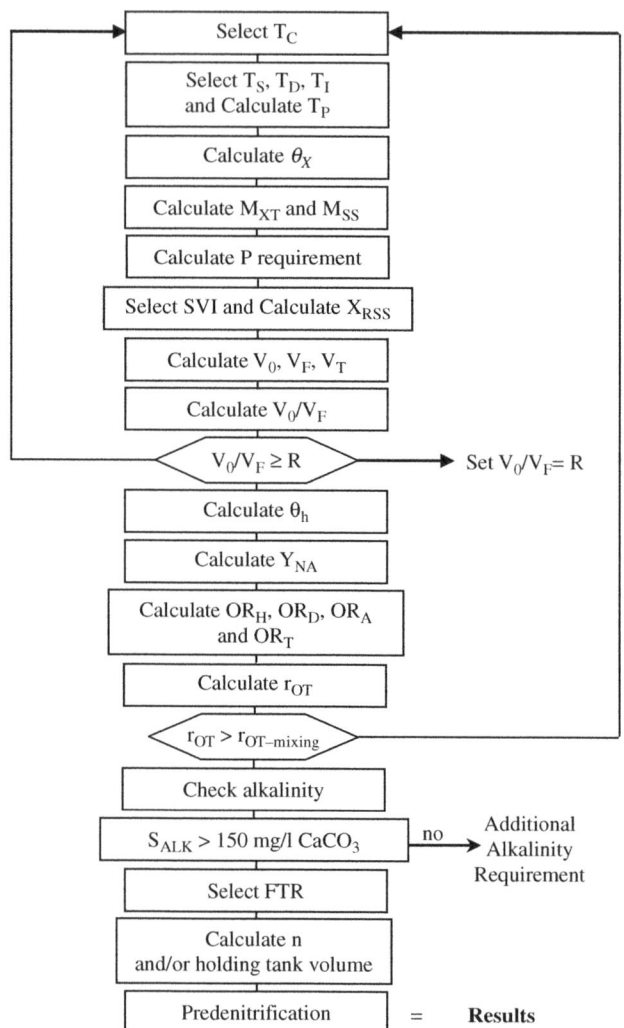

Figure 8.6. *Continued*

account of the inhibitory effects associated with industrial wastewaters. The θ_{XA} selection should be confirmed in terms of expected effluent ammonia concentration. The next step involves selection of an appropriate T_{AOX}/T_P value in view of the current practice where a constraint of $T_{AOX}/T_P < 0.6$ is generally accepted as a safety measure to ensure satisfactory settling properties of biomass. Major design parameters, such as N_X, N_{OX}, N_{NP} etc., can be calculated based on adopted design values for θ_{XA} and at this stage, carbon limitation is checked using expression (8.52), which ensures that the selected T_{AOX}/T_P value is acceptable in terms of the N_{DP} provided.

$$N_{DP} > N_{OX} - S_{NO,D} \tag{8.61}$$

This check, if not satisfied by successive iterations increasing T_{AOX}/T_P up to the maximum limit, indicates organic carbon deficiency in the wastewater for the required level of nitrogen removal. In this case other process options with external carbon addition should be envisaged. This part of the design ends by computing the required recycle ratio, R for pre-denitrification. A calculated R ratio higher than the acceptable limit (R > 5.0) also indicates the need for considering other process options.

The second part of the design involves assessing the basic parameters of SBR operation, namely the cycle time, T_C, the V_0/V_F ratio and the nominal hydraulic retention time, θ_h. Equation (8.3) gives the basic relationship between these parameters.

$$\theta_h = \left(1 + \frac{V_O}{V_F}\right) T_C \tag{8.62}$$

Selection of T_C determines V_F fraction in each cycle. V_0 is determined using equation (8.28) as the minimum required volume that is necessary to hold the settled biomass. The calculated V_0/V_F ratio should be equal or higher than the recycle ratio R required for the desired level of denitrification. Two remedial options may be considered if $V_0/V_F < R$: As a first option, the value of V_0/V_F may be increased by selecting a smaller value for T_C. For wastewaters with a low COD/N level, the required recycle ratio may not be achieved even if it is selected as short as the lowest acceptable level ($T_C = 4$ hrs, for example). The second option is to increase V_0 so that the resulting $V_0/V_F > R$. It should be noted that the V_0/V_F ratio calculated by means of equation (8.30) only relates to settling properties of the biomass in the reactor and approximates in a way, the sludge recycle in continuous flow systems. The additional internal recycle may be provided with a higher V_0, if needed. This would inevitably lead to a high θ_h value and a larger reactor volume. If this solution is not feasible, then a *step feeding* type of an operation alternative may be investigated as an appropriate solution. The following example illustrated the recommended procedure for the design of SBR operated in a pre-denitrification mode for nitrogen removal from industrial wastewaters. The same industrial wastewater characteristics previously used in Example 7.7 for denitrification in continuous flow activated sludge are adopted for the purpose of comparing the results of the design.

Example 8.2 SBR design for pre-denitrification

Design a SBR system operated in a pre-denitrification mode for an industrial wastewater with characteristics selected for example 7.7 and summarized below. The system is required to denitrify under summer conditions considering that the applicable design temperature is set as 20 °C.

Calculate the minimum level of effluent nitrate concentration achievable with the selected SBR system. Make all assumptions as necessary.

Wastewater flow, $Q = 100 \, m^3/day$
Total influent biodegradable COD, $C_{S1} = 1250 \, mg/l$
Total influent nitrogen, $C_{TKN1} = 180 \, mg/l$
Influent alkalinity, $S_{ALK1} = 460 \, mg/l \, CaCO_3$
Dissolved oxygen concentration, $S_O = 2.0 \, mg/l$
$Y_A = 0.24$ g cell COD/g N
$\hat{\mu}_A$ at 20 °C = 0.40 day^{-1}
$K_{NH} = 1.0 \, mg/l$
$K_{OA} = 0.4 \, mg/l$
$b_A = 0.10 \, day^{-1}$

Additional characteristics of the industrial wastewater to be treated are given in Table 7.4 and 7.5.

(1) Calculate $\hat{\mu}_A'$ and θ_{XM}:

$$\hat{\mu}_A \text{ at } 20°C = 0.40 \, day^{-1}$$
$$S_O = 2.0 \, mg/l$$
$$K_{OA} = 0.4 \, mg/l$$

$$\hat{\mu}_A' = \hat{\mu}_A \frac{S_O}{K_{OA} + S_O} = 0.40 \times \frac{2.0}{0.4 + 2.0}$$

$$\hat{\mu}_A' = 0.33 \, day^{-1}$$

$$b_A = 0.10 \, day^{-1}$$

$$\theta_{XM} = \frac{1}{\hat{\mu}_A' - b_A} = \frac{1}{0.33 - 0.10} = 4.35 \, days$$

SBR technology

(2) Select θ_{XA}:

Select a safety factor, f_{SA} that would sustain autotrophs and secure an $S_{NH} < 10$ mg N/l.

$$f_{SA} = 1.2$$

$$\theta_{XA} = f_{SA}\theta_{XM} = 1.2 \times 4.35 = 5.2 \, days$$

Select $\theta_{XA} = 5.5$ days

(3) Calculate S_{NH}:

$$S_{NH} = \frac{K_{NH}(1 + b_A\theta_{XA})}{\hat{\mu}'_A\theta_{XA} - (1 + b_A\theta_{XA})}$$

$$K_{NH} = 1.0 \, mg/l$$

$$S_{NH} = \frac{1.0 \times (1 + 0.1 \times 5.5)}{0.33 \times 5.5 - (1 + 0.1 \times 5.5)}$$

$$S_{NH} = 5.85 \, mg/l < 10 \, mg \, N/l$$

(4) Select T_{AOX}/T_P:

Similar to continuous flow activated sludge system, a high T_{AOX}/T_P ratio of 0.4 is selected.

(5) Calculate θ_{XE}:

$$\theta_{XE} = \frac{\theta_{XA}}{1 - \frac{T_{AOX}}{T_P}} = \frac{5.5}{1 - 0.4} \cong 9 \, days$$

(6) Calculate Y_{NH} and Y_{NHE}:

$$Y_{NH} = \frac{Y_H}{1 + b_H\theta_{XE}}$$

$$Y_H = 0.64 \, g \, cell \, COD/g \, COD$$

$$b_H = 0.15 \, day^{-1}$$

$$Y_{NH} = \frac{0.64}{1 + 0.15 \times 9} = 0.27 \, g \, cell \, COD/g \, COD$$

$$Y_{NHE} = Y_{NH}(1 + f_{EX}b_H\theta_{XE})$$

$$f_{EX} = 0.20$$

$$Y_{NHE} = 0.27 \times (1 + 0.20 \times 0.15 \times 9)$$

$$Y_{NHE} = 0.34 \text{ g COD/g COD}$$

(7) Calculate N_X and N_{OX}:

$$N_X = i_{XN} Y_{NHE} C_{S1}$$

$$C_{S1} = 1250 \text{ mg/l}$$

$$i_{NX} = 0.08 \text{ g N/g cell COD}$$

$$N_X = 0.08 \times 0.34 \times 1250 = 34 \text{ mg N/l}$$

$$N_{OX} = C_{N1} - i_{NXI} X_{I1} - i_{NSI} S_{I1} - N_X - S_{NH}$$

$$C_{TKN1} = 180 \text{ mg N/l}$$
$$X_{I1} = 150 \text{ mg/l}$$
$$i_{NXI} = 0.03$$
$$S_{I1} = 100 \text{ mg/l}$$
$$i_{NSI} = 0.02$$

$$N_{OX} = 180 - 0.03 \times 150 - 0.02 \times 100 - 34 - 5.85$$

$$N_{OX} = 133.65 \text{ mg N/l}$$

(8) Calculate N_{DP}:

$$N_{DP} = \frac{(1 - Y_{NHE})}{2.86} \left[S_{S1} + \eta \frac{T_{AOX}}{T_P} (C_{S1} - S_{S1}) \right]$$

$$S_{S1} = 225 \text{ mg/l}$$
$$C_{S1} - S_{S1} = 1050 \text{ mg/l}$$
$$\eta = 0.9$$

$$N_{DP} = \frac{(1 - 0.34)}{2.86} \times [225 + 0.9 \times 0.40 \times 1050]$$

$$N_{DP} = 0.23 \times (225 + 378) = 139 \text{ mg N/l}$$

$$N_{DP} \geq N_{OX}$$

The industrial wastewater is potentially suitable for pre-denitrification.

(9) *Select R_T and Calculate S_{NO}:*

Since the wastewater has a high C_{TKN} level of 180 mg N/l, the maximum allowable R = 5 is selected.

$$S_{NO} = \frac{N_{OX}}{1 + R_T} \frac{133.65}{1 + 5} \cong 22 \text{ mg N/l}$$

If the required $S_{NO,E}$ level is lower than 22 mg N/l, other process options other than pre-denitrification alone (i.e. dual anoxic SBR system, etc.,) should be selected.

(10) *Calculate N_D:*

$$N_D = N_{OX} - S_{NO} = 133.65 - 22 \cong 111 \text{ mg N/l}$$

(11) *Calculate P_{XT} and P_{SS}:*

The input of autotrophic biomass is neglected.

$$\begin{aligned} P_{XT} &= Q \left(Y_{NHE} C_{SI} + X_{II}\right) \\ &= 100 \times (0.34 \times 1.25 + 0.15) \\ &= 57.5 \text{ kg COD/day} \end{aligned}$$

$$P_{SS} = i_{SS,COD} P_{XT} = 0.9 \times 57.5 = 51.8 \text{ kg TSS/day}$$

(12) *Select T_C:*

As a first approximation a T_C value of 6 hours is selected.

(13) *Select T_S, T_D, T_I and calculate T_P:*

Select $T_S + T_D + T_I = 2$ hrs

$$T_P = T_C - (T_S + T_D + T_I) = 6 - 2 = 4 \text{ hrs}$$

(14) *Calculate θ_X:*

$$\theta_X = \theta_{XE} \frac{T_C}{T_P} = 9 \times \frac{6}{4} = 13.5 \text{ days}$$

(15) Calculate M_{XT} and M_{SS}:

$$M_{XT} = P_{XT}\theta_X = 57.5 \times 13.5 = 776 \, \text{kg COD}$$

$$M_{SS} = P_{SS}\theta_X = 51.8 \times 13.5 = 700 \, \text{kg TSS}$$

(16) The Nutrient Balance:

As in Example 8.1.

(17) Select SVI and Calculate X_{RSS}:

SVI = 120 ml/g industrial wastewater.

$$X_{RSS} = \frac{10^6}{SVI} = \frac{10^6}{120} = 8300 \, \text{mg TSS/l} = 8.3 \, \text{kg TSS/m}^3$$

(18) Calculate V_0, V_F and V_T:

$$V_0 = SF \frac{M_{SS}}{X_{RSS}}$$

$$SF = 1.2$$

$$V_0 = 1.2 \times \frac{700}{8.3} \cong 100 \, \text{m}^3$$

$$V_F = QT_C = 100 \times \frac{6}{24} = 25 \, \text{m}^3$$

$$V_T = V_0 + V_F = 100 + 25 = 125 \, \text{m}^3$$

(19) Calculate V_0/V_F:

$$\frac{V_0}{V_F} = \frac{100}{25} = 4.0$$

$$\frac{V_0}{V_F} < R$$

The selected T_C is maintained and V_0/V_F ratio is set as 5.0.

$$\frac{V_0}{V_F} = 5.0 \qquad V_0 = 5 \times 25 = 125 \, \text{m}^3$$

$$V_T = 125 + 25 = 150 \, \text{m}^3$$

SBR technology

(20) Calculate θ_h:

$$\theta_h = \frac{V_T}{Q} = \frac{150}{100} = 1.5 \text{ days} = 36 \text{ hrs}$$

(21) Calculate Y_{NA}:

$$Y_{NA} = \frac{Y_A}{1 + b_A \theta_{XA}}$$

$$Y_A = 0.24 \, g \, cell \, COD/g \, N$$
$$b_A = 0.10 \, day^{-1}$$

$$Y_{NA} = \frac{0.24}{1 + 0.1 \times 5.5} = 0.15 \, g \, cell \, COD/g \, N$$

(22) Calculate OR components and OR_T:

$$OR_H = QC_{SI}[1 - Y_{NH}(1 + f_E b_H \theta_{XE})]$$

$$f_E = f_{EX} + f_{ES} = 0.20 + 0.05 = 0.25$$

$$OR_H = 100 \times 1.25 \times [1 - 0.27 \times (1 + 0.25 \times 0.15 \times 9)]$$

$$OR_H \cong 80 \, kg \, O_2/day$$

$$OR_A = (4.57 - Y_{NA})QN_{OX}$$

$$OR_A = (4.57 - 0.15) \times 100 \times 0.133$$
$$= 58.8 \, kg \, O_2/day$$

$$OR_D = 2.86 Q N_D$$
$$= 2.86 \times 100 \times 0.111 = 31.7 \, kg \, O_2/day$$

$$OR_T = OR_H + OR_A - OR_D$$

$$OR_T = 80 + 58.8 - 31.7 \cong 107 \, kg \, O_2/day$$

(23) Calculate r_{OT}:

$$r_{OT} = \frac{OR_T}{V_T m T_A} = \frac{OR_T}{V_T m 0.6 T_P}$$

$$r_{OT} = \frac{107}{150 \times 4 \times 0.6 \times 4} = 0.074 \, \text{kg/m}^3.\text{hr}$$

$$r_{OT} = 74 \, \text{g/m}^3.\text{hr}$$

r_{OT} is close to the upper limit defined for satisfactory settling. In case of adverse effects T_C may be increased to 8 hrs.

(24) Check alkalinity:

As in Example 7.8.

(25) Calculate FTR and n:

As in Example 8.1.

8.5 ENHANCED BIOLOGICAL PHOSPHORUS REMOVAL

Enhanced Biological Phosphorus Removal (EBPR), is now commonly associated with the metabolic activities of *phosphorus accumulating organisms*, (PAOs), which have a competitive advantage only when the activated sludge system is operated in an anaerobic/aerobic sequence. Obviously, the biochemical reactions identified for EBPR in continuous flow activated sludge systems remain the same and the mechanistic models developed on the basis of these reactions are equally applicable to SBRs.

The characteristics of SBR operation are particularly suitable for complex processes like EBPR. It allows observation of transient behaviour of major parameters during different phases within a cycle while being operated at steady-state. For a selected mode of operation concentration profiles of major process components like S_A, S_{PHA}, S_{PO}, etc., may be followed and remedial action may be readily implemented, when necessary. However, the single tank structure of the SBR may also be a drawback as sludge and internal recycle cannot be differentiated and independently manipulated as in continuous flow activated sludge. Consequently, nitrification and resulting nitrate intrusion becomes a much more significant concern for the EBPR in the SBR systems.

As explained in detail in the preceding chapters, several factors such as *wastewater characteristics, the sludge age, temperature,* etc. affect the selection of PAOs and hence the overall P content of the sludge required for the desired level of EBPR. Among these factors, the magnitude of the available organic carbon that triggers and sustains the EBPR mechanism is particularly important. It should be noted that the COD/P ratio commonly evaluated for this purpose is only a crude index as it does not indicate what fraction of the influent COD would be readily available for PHA storage in the anoxic phase. It is well known that only volatile fatty acids – mainly acetate – can be converted into internal storage products. In this context, phosphorus limited conditions with excess available organic carbon can best be assessed using the S_A fraction of the biodegradable COD. A recent study evaluated the SBR performance at steady state for EBPR using acetate as the sole organic carbon source (Yagci et al., 2007). The SBR performance was experimentally tested for different initial S_A/S_{PO} (COD/P) ratios in the range of 6.7 – 20 mg COD/mg P. The results obtained are outlined in Table 8.1. Among the four different experimental runs displayed in the table, only run 1 conducted with a S_A/S_{PO} ratio of 20 provided complete P removal, leading to the conclusion that the threshold S_A/S_{PO} level for phosphorus limitation is equal to or slightly below 20 g COD/g P. In practice, it is quite unusual for the wastewaters to satisfy this ratio by the available S_A in the influent, so that COD fractionation of the wastewater and the relative magnitude of different biodegradable COD fractions become quite significant for effective EBPR. Therefore, the anoxic phase of the SBR should be so designed to allow fermentation of the readily biodegradable COD fraction, S_F aside from S_A, and generation of additional S_F through hydrolysis of slowly biodegradable COD.

8.5.1 EBPR without nitrogen removal

Industrial wastewaters do not usually require treatment measures that would remove both nitrogen and phosphorus. Therefore, only for a few particular

Table 8.1. Characteristics of SBR performance for EBPR using acetate as the sole carbon source (Yagci *et al.*, 2007)

Run	VFA/P (inf.)	$P_{eff.}$ mg P/L	P_{rel} mg P/L	P_{upt} mg P/L	P_{rel}/VFA$_{ut}$ mg P /mg COD	P_{upt}/P_{rel} g P /g P	Removal efficiency
Ac-1	20.0	0.00	93.9	113.7	0.42	1.21	100%
Ac-2	11.4	2.99	105.6	132.4	0.46	1.25	91%
Ac-3	8.9	3.88	109.7	141.8	0.48	1.29	91%
Ac-4	6.7	8.35	112.1	147.1	0.49	1.31	86%

industrial effluents an SBR system may be required to perform EBPR together with COD removal. The design of anaerobic/aerobic SBR for only phosphorus and carbon removal relies (i) on the selection of an appropriate aerobic sludge, θ_{XA}, for the growth of PAOs; (ii) an appropriate anaerobic fraction for sustaining EBPR and (iii) prevention and/or minimization of nitrate intrusion into the anaerobic period by sludge recycle, should nitrification occur.

The aerobic sludge age should be selected, if possible, lower than minimum aerobic sludge age for nitrifiers. Otherwise, nitrification may result in the presence of nitrate in the reactor during the initial phase and may upset anaerobic conditions required for EBPR. In this case, higher T_C values resulting in lower recycle ratios (V_0/V_F) should be selected for minimizing a pre-anoxic period. Measures like mixing during idle phase could also be envisaged. Deriving accurate analytical expressions for the required anaerobic fraction is almost impossible due to the complexity of processes and the excessive number of components involved. Pilot plant studies or simulation programs implementing appropriate models should be utilized, when necessary.

8.5.2 Simultaneous nitrogen and phosphorus removal

In SBR systems simultaneous nitrogen and phosphorus removal can be achieved by adding an anaerobic phase to the process sequence as schematically shown in Figure 8.7. SBR operation does not allow elimination of the nitrate recycle to the first mixed period when nitrogen removal is desired. However, nitrate recycle can be minimized using lower V_0/V_F ratios allowing to establish anaerobic conditions after a short anoxic phase in the first mixed period. Nevertheless, it may be difficult to obtain high degrees of nitrogen removal along with EBPR unless the wastewater composition is particularly suitable.

After determining θ_{XA}, θ_{XE} and the required recycle ratio for the desired nitrogen removal that would be provided by means of pre-denitrification, system design requires selection of an anaerobic fraction of process time, T_{AN}/T_P, according to the desired phosphorus removal. It is difficult to give an analytical expression for the selection of T_{AN}/T_P. T_{AN} can be increased until the fraction of the total non-aerated fraction, T_M/T_P stays equal or below 0.6. When the

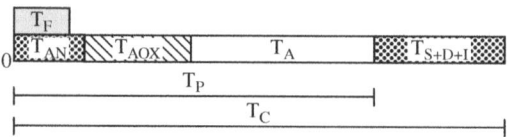

Figure 8.7. SBR operation for simultaneous nitrogen and phosphorus removal

required T_M/T_P ratio is higher than this threshold level, desired N and P removal cannot be accomplished without external carbon supplement.

8.6 REFERENCES

Andreottola, G., Bortone, G. and Tilche, A. (1997) Experimental validation of a simulation and design model for nitrogen removal in sequencing batch reactors. *Water Sci. Technol.* **35**(1), 113–120.

Ardern, E. and Lockett, W.T. (1914) Experiments on the oxidation of sewage without the aid of filters. *J. Soc. Chem. Ind.* **33**, 523.

Artan, N., Akkaya, M. and Artan, S.R. (1996) Experiences with the SBR treatment of industrial wastewaters. *Proceedings, 1st IAWQ Specialized Conference on SBR Technology*. 18–20 March, Munich, Germany.

Artan, N., Wilderer, P.A., Orhon, D., Morgenroth, E. and Ozgur, N. (2001) The mechanism and design of sequencing batch reactor systems for nutrient removal – the state of the art. *Water Sci. Technol.* **43**(3), 53–60.

Artan, N., Wilderer, P.A., Orhon, D., Tasli, R. and Morgenroth, E. (2002) Model evaluation and optimization of nutrient removal potential for sequencing batch reactors. *Water S.A.* **28**(4), 423–432.

Artan, N., Ozgur Yagci, N., Artan, S. and Orhon, D. (2003) Design of SBRs for biological nitrogen removal from high strength wastewaters. *J. Environ. Sci. Heal. A.* **38**(10), 2125–2134.

Artan, N. and Orhon, D. (2005) *Mechanism and Design of Sequencing Batch Reactors for Nutrient Removal*. IWA Scientific and Technical Report No.19. London, UK.

Belanger, D., Bergevin, P., Lapierre, G. and Zaloum, R. (1986) Conception control and efficiency of a sequencing batch reactor for the purification of wastewater from a slaughterhouse. *Sci. Tech. Eau,* **19**, 142–156.

Brenner, A. (2000) Modeling of N and P transformation in an SBR treating municipal wastewater. *Water Sci. Technol.* **42**(1–2), 55–63.

Carucci, A., Chiavola, A., Majone, M. and Rolle, E. (1999) Treatment of tannery wastewater in a sequencing batch reactor, *Water Sci. Technol.* **40**(1), 253–259.

Dennis, R.W. and Irvine, R.L. (1979) Effect of fill:react ratio on sequencing batch biological reactors. *J. Wat. Pollut. Control Fed.* **51**(2), 255–263.

Ekama, G.A. and Marais, G.v.R. (1984) Biological nitrogen removal. In *Theory, Design and Operation of Nutrient Removal Activated Processes.* Water Research Commission.

Franta, J. and Wilderer, P.A. (1997) Biological treatment of papermill wastewater by sequencing batch reactor technology to reduce residual organics. *Water Sci. Technol.* **35**(1), 129–136.

Grady, C.P.L., Daigger, G.T. and Lim, H.C. (1999) *Biological Wastewater Treatment*. 2nd edition, Marcal Dekker Inc., New York.

Gujer, W., Henze, M., Mino, T. and van Loosdrecht, M. (2000) *Activated Sludge Model No.3*. IWA Scientific and Technical Report No.9, (eds. Activated Sludge Models ASM1, ASM2, ASM2D and ASM3, Henze, M., Gujer, W., Mino, T., van Loosdrecht, M.), London, IWA.

Henze, M., Gujer W., Mino, T., Matsuo, T., Wentzel, M.C. and Marais, G.v.R. (1995) *Activated Sludge Model. No.2.* IAWQ Scientific and Technical Report No.3, London, UK.

Hoepker, E.C. and Schroeder, E.D. (1979) The effect of loading rate on batch activated sludge effluent quality. *J. Wat. Pollut. Control Fed.* **51**(2), 264–273.

Imura, M., Suzuki, E., Kitao, T. and Iwai, S. (1993) Advanced treatment of domestic wastewater using sequencing batch reactor activated sludge. *Water Sci. Technol.* **28**(10), 267–274.

Irvine, R.L., Miller, G. and Bhamrah, A.S. (1979) Sequencing batch treatment of wastewaters in rural areas. *J. Wat. Pollut. Control Fed.* **51**(2), 244–254.

Karahan-Gul, O., van Loosdrecht, M.C.M. and Orhon, D. (2003) Modification of Activated Sludge Model no. 3 considering direct growth on primary substrate. *Water Sci. Technol.* **47**(11), 219–225.

Ketchum, L.H. and Liao, P.C. (1979) Tertiary chemical treatment for phosphorus reduction using sequencing batch reactors. *J. Wat. Pollut. Control Fed.* **51**(2), 298–304.

Lee, S.I., Park, J.H., Ko, K.B. and Koopman, B. (1997) Effect of fermented swine wastes on biological nutrient removal in sequencing batch reactor. *Water Res.* **31**(7), 1807–1812.

Ling, L. and Lo, K.V. (1999) Brewery wastewater treatment using suspended and attached growth sequencing batch reactors. *J. Environ. Sci. Heal. A,* **34**(2), 341–355.

Mace, S. and Mata-Alvarez, J. (2002) Utilization of SBR technology for wastewater treatment: an overview. *Ind. Eng. Chem.* **41**, 5539–5553.

Mohlmann, F.W. (1917) The activated sludge method of sewage treatment. *Water survey series, No.14, University of Illinois Bulletin,* **15**, 75.

Mohseni, B.A. and Bazari, H. (2000) Biological treatment of milk factory wastewater by sequencing batch reactor. *Proceedings, 2nd International Symposium on Sequencing Batch Reactor Technology,* Volume 2, 153–156, IWA Publishing Co, London.

Murat, S., Ates Genceli, E., Tasli, R., Artan, N. and Orhon, D. (2002) Sequencing batch reactor treatment of tannery wastewater for carbon and nitrogen removal. *Water Sci. Technol.* **46**(9), 219–227.

Murat, S., Insel, G., Artan, N. and Orhon, D. (2004) Effect of temperature on the nitrogen removal performance of a sequencing batch reactor treating tannery wastewater. *Water Sci. Technol.* **48**(11), 319–326.

Orhon, D. and Artan, N. (1994) *Modelling of Activated Sludge Systems.* Technomic Publishing Co. Inc., Lancaster.

Pujol, R., Lemmel, H. and Gousailles, M. (1998) A keypoint of nitrification in an upflow biofiltration reactor. *Water Sci. Technol.* **38**(3), 43–49.

Raper, W.G.C. and Green, J.M. (2001) Simple process for nutrient removal from food processing effluents. *Water Sci. Technol.* **43**(3), 123–130.

Tasli, R., Artan, N. and Orhon, D. (2001) Retrofitting SBR systems to nutrient removal in sensitive tourist areas. *Water Sci. Technol.* **44**(1), 121–128.

Tilche, A., Bortone, G., Malaspina, F., Piccinini, S. and Stante, L. (2001) Biological nutrient removal in a full-scale SBR treating piggery wastewater: results and modelling. *Water Sci. Technol,* **43**(3), 363–371.

Torrijos, M. and Moletta, R. (1997) Winery wastewater depollution by sequencing batch reactor. *Water Sci. Technol.* **35**(1), 249–257.

Wilderer, P.A. (1992) Sequencing batch biofilm reactor technology. In *Harnessing biotechnology for the 21st century* (ed. M.R. Ladisch and A. Bose). pp. 475–479, American Chemical Society, Washington, D.C.

Wilderer, P.A., Irvine, R.L. and Goronszy, M.C. (2001) *Sequencing Batch Reactor Technology*. Scientific and Technical Report No. 10, IWA Publishing, London UK.

Yagci, N., Artan, N., Ubay Cokgor, E., Randall, C.W. and Orhon, D. (2003) Metabolic model for acetate uptake by a mixed culture of phosphate-and glycogen-accumulating organisms under anaerobic conditions. *Biotechnol. Bioeng.* **84**(3), 359–373.

Yagci, N., Ubay Cokgor, E., Artan, N., Randall, C. and Orhon, D. (2007) The Effect of Substrate on the Composition of Polyhydroxyalkanoates in Enhanced Biological Phosphorus Removal. *J. Chem. Technol. Biot.* **82**(3), 295–303.

9
Management of textile wastewaters

9.1 GENERAL ASPECTS

The textile industry representing one of the largest and most wide spread production sectors in the world, has a significant role in the economies of many countries. Different sizes of manufacturing premises from small traditional enterprises performing on a commission basis to large integrated facilities; variety of raw materials (cotton, synthetics, wool, silk, linen, blends etc.); various types of equipment (jig, beck, package, thermosol, skein, pad-roll, jet, pad-steam etc.) operating in either batch, semi-continuous or continuous manner; an array of dyes from direct, disperse, acid to sulphur, basic, reactive; countless auxiliary chemicals and many products (carpet, knit fabric, yarn, hosiery, woven fabric etc.) are involved in this sector. Reflecting this complexity, many variable emissions of pollutants requiring a case by case evaluation, are generated by the textile mills.

© 2009 IWA Publishing. *Industrial Wastewater Treatment by Activated Sludge*, by Derin Orhon, Fatos Germirli Babuna and Ozlem Karahan. ISBN: 9781789065282. Published by IWA Publishing, London, UK.

Before going further into the details of wastewater management in the textile industry, it is beneficial to briefly overview the current understanding of industrial management. It is a well known fact that while planning the future investments, the manufacturers must attribute special importance to the environmental issues. Otherwise, through an initial strategy that focuses only on production without accounting for environmental aspects, mitigation of the negative environmental impacts at later stages can be difficult and costly. Besides as previously emphasized in Chapter 6 the industrial premises must adopt a stage-wise approach that includes in-plant control measures prior to defining an end-of-pipe treatment. The reflections of this approach that covers both production and end-of-pipe treatment processes in a stage-wise manner can be traced back in the legislative framework. For example in the European Council Integrated Pollution Prevention and Control (IPPC) Directive it is stated that: The facilities must adopt not only treatment but also production processes in accordance with the ones described in relevant *Reference Documents on Best Available Techniques* (BREF) (European Council Directive 96/61/EC; European Commission, 2003). The US Pollution Prevention Act underlines the importance of devoting attention to in-plant control in a similar manner (US Pollution Prevention Act of 1990). Moreover the new understanding of industrial management dictates an integrated way of looking at the environmental medium as a whole instead of concentrating on a single media. By doing so pollutant emissions to air, water and land are handled in an interrelated and comprehensive manner (EPA, 1997; European Council Directive 96/61/EC). In conclusion, the prevention of multimedia pollution via in-plant control measures followed by end-of-pipe treatment alternatives must be adopted as the only roadmap to reach a sound industrial management.

Poor housekeeping practices, extensive use of synthetic auxiliaries (some of which can even be classified as xenobiotics), missing the opportunities of valuable by-product recovery, inefficient operations leading to the loss of expensive chemicals, unnecessary water use are the common pinpoints of the environmental problems arising from textile finishing industries. All the enumerated malpractices that in turn might represent an additional economic burden to the textile manufacturer, originate from the lack of implementing in-plant control measures. Fortunately, the possible in-plant control alternatives applicable to textile wet processing are well defined. In this context the first part of this chapter is devoted to a brief outline covering in-plant control measures, end-of-pipe treatment options and polluting sources together with their relevant characterization for textile mills. In the last part, a case study on water conservation and wastewater recovery and reuse is presented.

9.1.1 In-plant control measures applicable to textile mills

In-plant control measures applicable to textile mills can be categorized under five headings: Water conservation; wastewater reclamation and reuse; material reclamation; substitution of chemicals; and process modifications.

9.1.1.1 Water conservation

With the help of eliminating the unnecessary water consumption throughout the textile mills considerable savings can be obtained. Implementation of good housekeeping practices in an industry is the most easily applicable in-plant control measure: Hoses left running, broken and missing valves, malfunctioning toilets etc. can be avoided by a simple check. The unnecessary washing of the goods can be eliminated by reducing the dirt, grease and rust in the production areas (Smith, 1988). Preventing the accidental loss of process chemical baths and avoiding the preparation of larger baths than required together with limiting the excessive water use in rinsing operations can be regarded as worthwhile measures of water conservation (Van Veldhuisen, 1991). The washing or rinsing operations are very important for conserving the process water. There are some reported cases where the use of hot water instead of cold water has halved the water consumption (UNEP IE, 1994).

The application of counter-current washing or rinsing is a common way of reducing the process water use. This cheap, easy to implement operation basically involves the reuse of the least contaminated wastewater from the final wash for the next-to-last wash until the water reaches the first wash stage before being discharged (Smith 1986; 1988). The counter-current washing is applicable to the rinsings that take place after processes such as continuous dyeing, desizing, scouring and bleaching (UNEP IE, 1994).

Moreover, it is possible to conserve water with the help of process modifications. Dyeing employing reactive dyes that are known to have relatively low degrees of fixation on the fiber, can be optimized by elevating the process temperature. The outcome of such a process modification may conserve water and chemicals that in turn results in reducing the pollution load of the wastewater (Van Veldhuisen, 1991).

9.1.1.2 Wastewater reclamation and reuse

The most important issue in industrial wastewater reuse is the evaluation of the economic impact of these applications. A reduced amount of wastewater will be directed towards the end-of-pipe treatment with wastewater reclamation and reuse. However, a careful feasibility study must be performed by checking beforehand:

(i) the savings obtained from lowering the freshwater input.
(ii) the additional costs (both investment and operational) related to the installation and operation of new pipelines, pumping facilities etc.
(iii) investment and operational costs arising from the possible pre-treatment requirement of the reclaimed wastewater before reuse.
(iv) possible elevation of the end-of-pipe treatment costs due to dealing with a more concentrated effluent.

A case assessment covering these issues in a detailed manner is presented in Part 9.2 of this chapter.

The water reuse measures applicable to textile mills can be grouped into two (Van Veldhuisen, 1991):

(a) reuse of uncontaminated non-contact cooling water from condensers, heat exchangers, yarn dryers, pressure dyeing machines, air compressors etc. in processes requiring hot water.
(b) reuse of process wastewater from one operation in another unrelated operation. The most common examples are reuse of wash water from bleaching operations in caustic washing and scouring; reuse of mercerizing wash water for scouring, bleaching operations and for wetting the fabric; reuse of scouring rinses for desizing; reuse of final rinsing waters in bleaching for bath make up or primary rinsing baths.

A wide range of treatment schemes from activated carbon applications, ozonation to electrolysis and nanofiltration can be used for recovered wastewaters that require prior treatment (Sewekow, 1995).

9.1.1.3 Material reclamation

The main target of recovering the valuable substances in an industry is to lower the processing costs. Pollutant loads are also reduced due to this in-plant control measure. Among others dyebath reuse, size and caustic recoveries are the most commonly implemented material reclamation practices in the textile mills.

The exhaust dyebath typically contains some or all of the following ingredients: Water, dyestuffs, buffer systems for pH control, salt, dyeing auxiliaries (retarders, levelling agents, lubricants, carriers, softeners, surfactants etc.). During conventional dyeing only the dyestuff and a few additives are fixed on the good, so most of the other chemicals remain in the dyebath and consequently are discharged with the effluent (Smith, 1986). Fixation percentages of different dyes are as follows; direct dyes: 70–95 %, vat dyes: 80–95 %, sulphur dyes: 60–70 %, reactive dyes: 50–95 %, disperse dyes: 80–92 %, acid dyes: 80–93 %, cationic (basic) dyes: 97–98 % (Sewekow, 1995). Since reactive, sulphur and vat dyes undergo chemical processes their spent dyebath reuse is not well documented (Smith 1986). On the other hand acid, basic, direct and disperse dyes do not change their chemical properties during the dyeing process. Therefore the reuse of dyeing operations, listed below, can be easily managed (Smith, 1986; 1988):

(i) acid dyes applied on nylon and wool.
(ii) basic dyes applied on acrylic and certain copolymers.
(iii) direct dyes on cotton.
(iv) disperse dyes on synthetic polymers.

In order to reuse the dyebaths first of all the spent dyebath has to be analysed in terms of its chemical content. According to the results of this analysis, make up amounts of dyes and chemicals have to be added. Afterwards the bath can be reused for another dyeing.

Caustic soda can be recovered from wastewaters generated by mercerization process using membrane technologies or evaporation (Sewekow, 1995; UNEP IE, 1994; Bredereck, 1978). 98 % of the used caustic can be reclaimed while reducing the chemical costs by 80 % (The World Bank, 1997; Smith, 1988).

Synthetic sizing agents such as polyvinyl alcohol and poly(meth)acrylates can be recovered successfully by ultrafiltration. By doing so apart from obtaining a saving on the chemical costs, a poorly degradable wastewater constituent is minimized (Sewekow, 1995).

9.1.1.4 Substitution of chemicals

Due to the emerging new concepts such as xenobiotics, substitution of chemicals lately draws special attention. The aim of this in-plant control is to replace the process chemicals that either have high toxicities and/or low biodegradabilities by the environmentally friendly ones. Rather than comparing the pollutant strength of the chemicals measured in terms of conventional parameters; the levels of toxicity and recalcitrance imparted by them gained importance. Within the context of minimizing the pollution at source a lot of effort is given now to develop and design safer, benign chemicals (Rieger *et al.*, 2002). As mentioned earlier, countless numbers of auxiliary chemicals are used in the textile mills. All of these chemicals are not assessed for their characteristics. Besides it is hard to identify the possible physico-chemical mechanisms and impacts that arise from using a couple of them together under operating conditions. Therefore, a case for wise evaluation is required. In this context some concise recommendations for chemical substitutions applicable to textile mills are presented below.

For the oxidation of vat dyes, rather than using dichromate that contributes chromium content of the effluent, periodate or peroxide oxidizers can be applied (The World Bank, 1997; Smith, 1988).

The aquatic toxicities and biodegradabilities of surfactants vary through a wide range. Among other surfactants nonylphenol ethoxylate (AP) is known to have 25 % biodegradability, while linear alcohol ethoxylate (LAE) is completely biodegradable. Untreated LAE has a higher toxicity of $LC_{50} = 2$ mg/l compared to AP with a LC_{50} of 13 mg/l. On the other hand after treatment AP is stated to have substantial toxicity whereas LAE does not exert any. Thus, ethoxylated octyl or nonyl phenol (AP) must be substituted with LAE (The World Bank, 1997; Van Veldhuisen, 1991; Smith, 1988).

In case of dyeing cellulosic fibers to green and blue instead of using direct or reactive types of dyes that contain metals; metal free vat dyes can be applied (Smith, 1988).

Hypochlorite used in the bleaching of cotton can be substituted with hydrogen peroxide as considerably reduced AOX levels in the discharged wastewaters are obtained (Sewekow, 1995).

The usage of dye carriers containing chlorinated aromatics must be avoided (The World Bank, 1997).

When starch is applied in sizing; for desizing operations instead of using enzymes, oxidation by hydrogen peroxide is recommended (Smith, 1988).

9.1.1.5 Process modifications

The process modifications that either minimize or eliminate the pollutant discharges are almost completely site-specific. Items given below constitute a partial list of candidate process/equipment modifications (UNEP IE, 1994):

(a) solvent aided scouring and bleaching processes.
(b) hot mercerization instead of a conventional cold one often, enabling the elimination of separate scourings treatment.
(c) combined disperse and direct/reactive dyeing for fabric blends containing low percentages of cellulosics.
(d) use of padding method instead of exhaust method for dyeing, whenever possible.
(e) direct finishing of pigment printed goods and direct carbonising of disperse printed goods without intermediate washing.

Another example of this in-plant control practice is the application of pad-batch dyeing to woven and knit fabrics. Pad-batch dyeing is the most promising process modification as it eliminates the need for some speciality chemicals thereby reducing the cost and pollution at source (Smith, 1986; 1988; UNEP IE, 1994; The World Bank, 1997).

9.1.2 End-of-pipe treatment options applicable to textile mills

The complex nature of the textile industry that arises from the variety of the raw materials used, processes/operations/techniques employed, chemicals applied, products obtained and in-plant measures practiced; has its reflections on the quality and the quantity of the effluent generated. Moreover, due to the differences in the handled orders especially for facilities performing on a commission basis, the pollutant concentrations and loads vary widely among the industry. Therefore an end-of-pipe treatment configuration composed of a sequence of technologies that has an ability to adapt itself to the mentioned variations is a prerequisite. In this respect, a stable type of treatment train ensuring the discharge limits, yet a flexible one modifiable to ever changing conditions must be prescribed for textile effluents.

Another important issue in defining the appropriate treatment schemes for textile effluents is to identify the possible partial pre-treatment requirements of some segregated wastewater streams. In other words, the segregation of certain process discharges having either a highly toxic and/or recalcitrant nature from the general effluent and passing these discharges through a special type of partial pre-treatment; can ease the treatability of the general discharge. The discharges originating from the baths where specialty finishing agents such as carriers are added can be considered as good examples of the effluent streams that require partial treatment. However, at this point it is necessary to mention about the need for a careful evaluation on the outputs obtained from such partial pre-treatment especially in terms of toxicity. After partial chemical oxidation there exists a possibility that more toxic by-products will be obtained than the parent segregated effluents. As literature provides scarce information on by-product toxicity (Hao *et al.*, 2000), a case wise appraisal is necessary.

Screening followed by neutralization and equalization are the commonly applied primary treatment units prescribed for the effluents generated from the wet textile mills. The most commonly implemented biological treatment unit is activated sludge. Trickling filters and lagoons are rarely used biological treatment alternatives.

Some treatment schemes involve chemical treatment applied either before or after biological treatment. Although a common application, coagulation flocculation alone proves unreliable in meeting the discharge limitations. Oxidation by the use of ozone and UV, membrane technologies and adsorption processes are among the technologies applied on segregated process streams.

9.1.3 Polluting sources and characteristics

In the textile industry, different wastewater streams originating from various processes are combined to produce an end-of-pipe discharge, the quality of which depends on complex and interrelated array of factors from the equipment used and the level of in-plant control measures implemented to the types of auxiliary chemicals added, the procedures applied etc. Therefore, it is clearly apparent that every textile mill follows a unique way of manufacturing and the possibility of procuring data from facilities adopting a similar way of production is almost null. In this respect comparing the overall discharges generated by different textile mills can only be considered as rough preliminary assessments. Instead it is much more beneficial to present data on separate processes by indicating the details of the production. However it is stated in *Reference Document on Best Available Techniques (BREF) for the Textiles Industry* that the available data on wastewaters generated from specific textile processes is very poor (European Commission, 2003). In order to at least partially fill this information gap process based data rather than the overall data involving the whole textile effluent is given in this chapter. In this context data on 28 different cases engaging various processes are presented. Specifications related to the investigated cases are tabulated in Table 9.1. Appraisal of the conventional characterization data shown in Table 9.2 emphasizes the case specific nature of this sector. Due to the previously mentioned complex nature of the sector, different wastewater characteristics and pollution loads are obtained even for cases with similar process specifications. Case B2 and B3, both applying *optical brightening, kiering, peroxide bleaching* and *reactive dyeing* operations on cotton knit fabric, constitute a good example of this issue. According to the data summarized in Table 9.1, one can expect a similar wastewater characterization from these cases. On the contrary the data tabulated in Table 9.2 clearly shows that although exactly the same amount of wastewater is generated from both of the cases, a COD load and concentration that are twice as much as Case B2 are associated with Case B3. Usage of highly polluting auxiliary chemicals might be the reason for such a finding. Among other cases a distinctively concentrated effluent is discharged in Case G3 that involves a special type of finishing where an organic solvent called *white spirit* is applied (Kabdasli *et al.*, 2000; Orhon *et al.*, 2003). Apart from Case G3 the COD values of the presented cases vary in

Table 9.1. Specifications related to the investigated processes

Case	Type of material	Process specification	Applied dye or main specific agent
A1	Cotton denim fabric* Orhon et al., (2001a)	Desizing; stone bleaching	Desizing enzymes; pumice stone
A2	Cotton denim fabric* Orhon et al., (2001a)	Desizing; stone bleaching	Desizing enzymes; pumice stone
A3	Cotton denim fabric* Germirli Babuna et al., (1998a)	Desizing; stone bleaching	Desizing enzymes; pumice stone
B1	Cotton knit fabric Germirli Babuna et al., (1999)	Optical brightening; peroxide bleaching; dyeing	Optical brightener; H_2O_2; reactive dye
B2	Cotton knit fabric Dogruel, (2000)	Optical brightening; kiering; peroxide bleaching; dyeing	Optical brightener; H_2O_2; reactive dye
B3	Cotton knit fabric Orhon et al., (2001b)	Optical brightening; kiering; peroxide bleaching; dyeing	Optical brightener; H_2O_2; reactive dye
B4	Cotton knit fabric Orhon et al., (2000a)	Optical brightening; kiering; peroxide bleaching; dyeing	Optical brightener; H_2O_2; reactive dye
B5	Cotton knit fabric Orhon et al., (2000a)	Optical brightening; peroxide bleaching	Optical brightener; H_2O_2
B6	Cotton knit fabric Germirli Babuna et al., (1998b)	Peroxide bleaching; dyeing	H_2O_2; reactive dye
B7	Cotton knit fabric Orhon et al., (2000a)	Peroxide bleaching; dyeing	H_2O_2; reactive dye
B8	Cotton knit fabric Germirli Babuna et al., (1998a)	Peroxide bleaching; dyeing	H_2O_2; reactive dye
B9	Cotton knit fabric Orhon et al., (2000a)	Kiering, dyeing	Soda; reactive dye
C1	Cotton/PES knit fabric Germirli Babuna et al., (1998a)	Peroxide bleaching; dyeing	H_2O_2; reactive dye
C2	Cotton/PES knit fabric Orhon et al., (2000a)	Bleaching; dyeing	H_2O_2; reactive dye
D	PES knit fabric Germirli Babuna et al., (1998a)	Dyeing	Disperse dye
E	Wool/PES knit fabric Orhon et al., (2000b)	Dyeing	Metal complex, disperse dyes
F	Viscose rayon knit fabric Orhon et al., (2000a)	Dyeing	Reactive dye

Table 9.1. *Continued*

Case	Type of material	Process specification	Applied dye or main specific agent
G1	Silk**+Cotton*** woven fabric (Orhon et al., 1996)	Bleaching; desizing; kiering; dyeing; printing	H_2O_2; reactive, acid, pigment, disperse dyes; urea
G2	Cotton knit fabric (Kabdasli et al., 2000)	Rotation printing	Copolymer; pigment dye
G3	Cotton knit fabric (Kabdasli et al., 2000)	Tube and item printing	Ethyleneurea; white sprite; pigment dye
H	Wool woven fabric (Orhon et al., 2000b)	Dyeing	Metal complex, disperse dyes
J1	Wool yarn (Orhon et al., 2000b)	Dyeing	Chromium dye
J2	Wool yarn (Orhon et al., 2000b)	Dyeing	Metal complex, disperse dyes
K	Acrylic fiber and yarn (Germirli Babuna et al., 1999)	Dyeing	Basic dye
L	Wool, PES, linen woven fabric & yarn (Uner et al., 2006)	Dyeing	Cationic, reactive, disperse, metal complex, chromium dyes
M	Wool, PES woven fabric (Dulkadiroglu et al., 2002)	Finishing	Softeners; crease proofing agents
N1	Acrylic fiber carpet (Yildiz et al., 2007)	Dyeing; finishing	Cationic softeners; cationic basic dye
N2	Polyamide fiber carpet (Yildiz et al., 2008)	Dyeing; finishing; latex coating	Anionic softeners; anionic acidic dye

* previously dyed jeans; ** 80% of the production; *** 20% of the production

a wide range from 365 mg/l to 4740 mg/l. Generally soluble COD accounts for over 70 % of the total COD. Unlike the other wastewaters, the effluents obtained from cases A1, A2 and A3 have very high TSS contents of over 9700 mg/l. All of these three cases involve the application of a special type of finishing named *stone bleaching* on cotton denim fabric. Dissimilar to that of the other cases mentioned, VSS/TSS ratios of only 1 % are obtained due to the usage of inorganic pumice stone. Apart from Case G1 where urea is applied for printing; nitrogen and phosphorus deficient effluents are generated from all the cases. The pH values of the wastewaters vary from alkaline to acidic. Variable unit water

Textile wastewaters 313

Table 9.2. Conventional wastewater characterization of the investigated cases

Case	Total COD (mg/l)	Soluble COD (mg/l)	TSS (mg/l)	VSS (mg/l)	TKN (mg/l)	NH$_4$-N (mg/l)	Total P (mg/l)	pH	Q (m^3/ton fabric)	COD load (kg COD/ ton fabric)
A1 (Orhon et al., 2001a)	1910	1570	10400	124	31	9.4	18.5	8.0	70	130
A2 (Orhon et al., 2001a)	1940	1650	11200	100	32	1.0	35	8.9	70	136
A3 (Germirli Babuna et al., 1998a)	2400	1700	9700	70	35	5.6	34	9.3	68.4	155
B1 (Germirli Babuna et al., 1999)	2300	1900	135	80	14	ND	4.5	10.1	ND	ND
B2 (Dogruel, 2000)	955	675	105	85	ND	ND	ND	9.6	75	72
B3 (Orhon et al. 2001b)	1980	1210	170	130	25	21	27	10.2	75	148
B4 (Orhon et al., 2000a)	1180	890	100	90	14	ND	13	10.3	75	84.5
B5 (Orhon et al., 2000a)	4740	ND	70	60	45	ND	ND	ND	40	190
B6 (Germirli Babuna et al., 1998b)	2100	1558	700	ND	62	ND	13.6	10.5	ND	ND
B7 (Orhon et al., 2000a)	672	ND	48	34	ND	ND	ND	ND	99	67
B8 (Germirli Babuna et al., 1998a)	1470	1165	490	160	110	0.5	4	10.9	80	118
B9 (Orhon et al., 2000a)	828	ND	65	32	22	ND	10	ND	91	75
C1 (Germirli Babuna et al., 1998a)	2400	1690	370	180	20	0.2	7	10.2	80	192
C2 (Orhon et al., 2000a)	2070	ND	85	54	34	ND	29	ND	95	197
D (Germirli Babuna et al., 1998a)	1985	1485	213	22	27	1.7	9	5.8	20	40
E (Orhon et al., 2000b)	1445	1320	<10	ND	73	50	ND	7	151	236
F (Orhon et al., 2000a)	728	ND	29	28	16	ND	32	ND	113	82

Table 9.2. Continued

Case	Total COD (mg/l)	Soluble COD (mg/l)	TSS (mg/l)	VSS (mg/l)	TKN (mg/l)	NH_4-N (mg/l)	Total P (mg/l)	pH	Q (m^3/ton fabric)	COD load (kg COD/ ton fabric)
G1 (Orhon et al., 1996)	1070	620	105	90	110	62	2	8.2	ND	ND
G2 (Kabdasli et al., 2000)	785	ND	125	ND	30	20	ND	7.4	ND	ND
G3 (Kabdasli et al., 2000)	49170	ND	9500	ND	1765	368	ND	8.5	ND	ND
H (Orhon et al., 2000b)	650	ND	30	ND	ND	ND	ND	5.7	231	150
J1 (Orhon et al., 2000b)	1080	ND	3500	ND	ND	ND	ND	4.1	24	26
J2 (Orhon et al., 2000b)	365	ND	1450	ND	ND	ND	ND	6.2	38	14
K (Germirli Babuna et al., 1999)	1900	1590	90	43	72	ND	4.2	4.5	ND	ND
L (Uner et al., 2006)	700	610	22	22	16	4	ND	6.9	ND	ND
M (Dulkadiroglu et al., 2002)	690	455	85	80	20	8	0.8	7.1	81	56
N1 (Yildiz et al., 2007)	775	495	25	ND	12	7	0.1	5.3	26	30
N2 (Yildiz et al., 2008)	1890	1850	30	ND	25	2	41	7.3	67.5	127
Organized Industrial District* (Orhon et al., 1999)	932	580	225	130	54	ND	7.9	8.2	ND	ND
Domestic sewage (Orhon et al., 1997)	410	140	210	145	43	32	7.2	7.4	ND	ND

* predominantly textile; ND: not determined

consumption values (from 20 to 230 m^3/ton fabric) and unit COD loads (from 14 to 236 kg COD/ton fabric) are observed.

The COD fractions of the textile wastewaters generated from the investigated cases are presented in Table 9.3. Accordingly 4–35 % of the total COD is assessed to be inert in nature and the particulate inert COD fraction is observed to be either negligible or relatively low. Kinetic and stoichiometric coefficients of the textile effluents are given in Table 9.4 and 9.5. As evident from the latter table both single and dual hydrolysis models can be compatible for the textile effluents originating from different processes. A relatively low $\hat{\mu}_H$ value is obtained for Case N1 (where acrylic fiber carpet dyeing and finishing operations are performed), possibly due to the inhibitory effect of auxiliary chemicals added. Comparatively low hydrolysis rate coefficients, k_{hS}, are associated with Cases A1, A3, K and N1 involving denim processing and acrylic dyeing effluents. A higher hydrolysis rate coefficient is obtained for Case A2 having very similar process specifications as Cases A1 and A3, all dealing with denim fabric finishing operations. The tabulated data on COD fractionation and kinetic and stoichiometric coefficients strongly supports the need for separate evaluation of the cases.

The results of the physico-chemical treatment applied on raw textile effluents indicate limited COD removal efficiencies coupled with higher colour removals (Table 9.6). As given in Table 9.7, the effect of these physico-chemical treatments on the COD fractions of the effluents is also questioned. The figures obtained support the *sui generis* nature of the cases requiring separate appraisal.

Table 9.8 summarizes the characterization of untreated reusable fraction of effluent streams together with untreated (raw) remaining effluents obtained after in-plant control measures. The application of wastewater reuse minimizes the freshwater demand by 52, 22 and 34 % for cases B2, B4 and M respectively. As expected from such an application remaining wastewaters with increased pollutant concentrations are obtained. However, the biodegradability of remaining wastewaters is of importance in evaluating the feasibility of a reuse application. In this respect COD fractionation of the remaining effluents are presented in Table 9.9. Reuse application results in getting a higher soluble inert COD for Case B2. Such an output can create difficulties in fulfilling the effluent limitations. On the other hand the segregation of the reusable portion of the wastewater streams do not have any appreciable effect on the percent COD fractionation of the cases B4 and M, although stronger wastewaters are obtained in both cases.

The data presented in this chapter for textile finishing mills can be used as a guiding tool; nevertheless due to the distinctively diverse nature of the textile processes applied in every facility, a case by case evaluation is required to reach a sound management strategy for the studied premise.

Table 9.3. COD fractionation of the textile wastewaters

Case	C_{T1} (mg/l)	S_{T1} (mg/l)	S_{H1} (mg/l)	S_{S1} (mg/l)	X_{S1} (mg/l)	X_{I1} (mg/l)	S_{I1} (mg/l)	S_{I1}/C_{T1} (%)	C_{S1}/C_{T1} (%)
A1*** (Orhon et al., 2001a)	1910	1570	1005	325	340	N	240	13	87
A2*** (Orhon et al., 2001a)	1940	1650	1140	410	290	N	100	5	95
A3*** (Germirli Babuna et al., 1998a)	2400	1700	1270	330	700	N	100	4	96
B1 (Germirli Babuna et al., 1999)	2300	1900	1310	420	365	35	170	7	91
B2 Dogruel, (2000)	955	675	245	110	ND	ND	320	34	
B3 (Orhon et al., 2001b)	1980	1210	734	187	708	62	289	15	82
B4 (Orhon et al., 2000a)	1180	890	525	118	227	63	247	21	74
B6 (Germirli Babuna et al., 1998b)	2100	1558	ND	ND	517	25	317	15	
B8 (Germirli Babuna et al., 1998a)	1470	1165	575	330	288	17	260	18	81
C1 (Germirli Babuna et al., 1998a)	2400	1690	1275	165	598	112	250	10	85
D (Germirli Babuna et al., 1998a)	1985	1485	770	300	390	110	415	21	74
E (Orhon et al., 2000b)	1445	1320	833	340	ND	ND	147	10	
M (Dulkadiroglu et al., 2002)	687	455	203	220	160	72	32	5	85
N1 (Yildiz et al., 2007)	775	465	417	25	230	50	55	7	87
N2 (Yildiz et al., 2008)	1890	1660	40	12	1485*	ND	125	7	
Organized Industrial District** (Orhon et al., 1999)	932	580	411	139	352	N	20	2	97
Domestic sewage (Orhon et al.,1997)	450	155	97	40	250	45	18	4	86

* soluble; ** predominantly textile; *** after passing through a 6 hr of gravity settling; ND: not determined; N: negligible

Table 9.4. Kinetic and stoichiometric coefficients for different textile wastewaters

Case	Y_H g cell COD/g COD	$\hat{\mu}_H$ day^{-1}	K_S mg COD/l	b_H day^{-1}	Y_{SP}	f_{ES}
A1* (Orhon et al. 2001a)	0.68	3.2	20	0.14	0.044	0.065
A2* (Orhon et al. 2001a)	0.68	4.9	20	0.14	0.045	0.066
A3* (Germirli Babuna et al. 1998a)	0.68	3.6	15	0.14	0.045	0.066
B1 (Germirli Babuna et al. 1999)	0.62	3.2	13	0.19		0.063
B8 (Germirli Babuna et al. 1998a)	0.69	4.1	5	0.18	0.08	0.12
C1 (Germirli Babuna et al. 1998a)	0.69	5.3	5	0.14	0.06	0.09
D (Germirli Babuna et al. 1998a)	0.69	5.3	25	0.12	0.07	0.10
K (Germirli Babuna et al. 1999)**	0.60	3.9	10	0.17		0.089
N1 (Yildiz et al. 2007)	0.66	1.65	1.0	0.1		
N2 (Yildiz et al. 2008)	0.66	6	1.0	0.1		

* after passing through a 6 hr of gravity settling; ** effluent of the partial oxidation with H_2O_2

Table 9.5. Kinetics of hydrolysis for textile wastewaters

Case	k_{hS}	k_{hX}	K_{XS}	K_{XX}
	(day^{-1})		[g cell COD/(g COD)]	
A1*** (Orhon et al. 2001a)	1*			0.16*
A2*** (Orhon et al. 2001a)	2.5*			0.16*
A3*** (Germirli Babuna et al. 1998a)	0.8	0.5	0.05	0.15
B1 (Germirli Babuna et al. 1999)	0.8*			0.7*
B8 (Germirli Babuna et al. 1998a)	3	1	0.05	0.5
C1 (Germirli Babuna et al. 1998a)	3	1	0.05	0.2
D (Germirli Babuna et al. 1998a)	3.8*			0.65*
K (Germirli Babuna et al. 1999)**	1.6*			0.7*
N1 (Yildiz et al. 2007)	1.40	0.36		0.02
N2 (Yildiz et al. 2008)	3.50	0.72		0.04

* compatible with single hydrolysis; ** effluent of the partial oxidation with H_2O_2; *** after passing through a 6 hr of gravity settling

Table 9.6. Physico-chemical treatability of textile wastewaters

Case	Physico-chemical method	Agent type	Optimum dose (mg/l)	Initial COD (mg/l)	COD removal %	Colour removal %
B4 Orhon et al.,(2000a)	Precipitation	Sodium bentonite	2000	1180	47	69
B4 Orhon et al.,(2000a)	Oxidation	Ozone	43*	1180	9	69
B4 Orhon et al., (2000a)	Oxidation	Ozone	62*	1180	13	74
B4 Orhon et al., (2000a)	pH adjustment+ Oxidation	H_2SO_4+Ozone	14*	1180	18	36
B2 Dogruel, (2000)	Oxidation	Ozone	130*	955	11	83
B2 Dogruel, (2000)	Oxidation	Ozone	235*	955	19	92
B2 Dogruel, (2000)	Oxidation	Ozone	465*	955	21	94
B2 Dogruel, (2000)	Oxidation	Ozone	1385*	955	32	94
K Germirli Babuna et al., (1999)	Oxidation	H_2O_2	1.0**	1900	63	ND
L Uner et al., (2006)	Precipitation	Sodium bentonite	1000	700	58	93
L Uner et al., (2006)	Precipitation	Alum	50	700	62	89
L Uner et al., (2006)	Precipitation	$FeSO_4$	500	700	61	97

* utilized ozone; ** H_2O_2/COD with 500 mgl^{-1} Fe^{3+} and a day of reaction time

Table 9.7. COD fractionation of chemically pre-treated textile wastewaters

Case Pre-treatment	C_{T1} (mg/l)	S_{T1} (mg/l)	S_{H1} (mg/l)	COD Components					S_{I1}/C_{T1} (%)	C_{S1}/C_{T1} (%)
				S_{S1} (mg/l)	X_{S1} (mg/l)	X_{I1} (mg/l)	S_{I1} (mg/l)		
Case B4, with 2000 mg/l sodium bentonite Orhon et al. (2000a)	630	375	163	70	192	63	142	23	67
Case B2, with 130 mg/l utilized ozone Dogruel, (2000)	850	580	240	40	ND	ND	300	35	ND
Case B2, with 220 mg/l utilized ozone Dogruel, (2000)	775	545	200	50	ND	ND	295	38	ND
Case B2, with 1385 mg/l utilized ozone Dogruel, (2000)	650	480	165	45	ND	ND	270	42	ND
Case K, with H_2O_2*, Germirli Babuna, et al., (1999)	710	700	596	86	N	N	28	4	96

ND: not determined; N: negligible; *H_2O_2/COD with 500 mg/l Fe^{3+} and a day of reaction time

Table 9.8. Characterization of segregated streams for reuse application

Parameter	Case B2		Case B4		Case M	
	Raw reusable wastewater	Raw remaining wastewater	Raw reusable wastewater	Raw remaining wastewater	Raw reusable wastewater	Raw remaining wastewater
Total COD (mg/l)	315	1220	350	1475	180	1460
Soluble COD (mg/l)	190	850	200	1215	120	970
Colour (Pt-Co)	30	770	25	990	20	440
TSS (mg/l)	60	125	80	115	15	190
VSS (mg/l)	60	95	80	94	ND	180
Cl^- (mg/l)	275	2530	320	5210	ND	ND
TDS (g/l)	1.18	ND	1.1	12.5	340	640
pH	7.4	9.7	5.2	10.6	7.1	6.2
Flowrate (%)	52	48	22	78	34	66
Reference	Dogruel, (2000)		Orhon et al., (2000a)		Dulkadiroglu et al., (2002)	

ND: not determined

Table 9.9. COD fractionation of remaining textile wastewaters

Case	C_{T1} (mg/l)	S_{T1} (mg/l)	S_{H1} (mg/l)	COD Components S_{S1} (mg/l)	X_{S1} (mg/l)	X_{I1} (mg/l)	S_{I1} (mg/l)	S_{I1}/C_{T1} (%)	C_{S1}/C_{T1} (%)
Case B2, Untreated remaining wastewater Dogruel, (2000)	1220	850	410	65	ND	ND	365	30	ND
Case B4, Untreated remaining wastewater Orhon et al., (2000a)	1475	1215	776	132	182	78	307	21	74
Case M, Untreated remaining wastewater Dulkadiroglu et al., (2002)	1460	970	418	485	341	149	67	5	85

ND: not determined

9.2 A CASE ON WATER CONSERVATION AND WASTEWATER RECOVERY AND REUSE

Effectuation of a sound in-plant control strategy for textile industries may be adopted for the main purpose of accomplishing momentous reductions in water use, raw material and energy consumption, wastewater production and in some cases even wastewater load (UNEP IE, 1994). The amount of wastewater generated can be lowered as a natural consequence of water conservation. Application of appropriate wastewater recovery and reuse practices may further decrease the quantity of the effluents. Wastewater minimization by water conservation and/or wastewater reuse is the most readily acquirable in-plant control strategy and yet it is often overlooked. As mentioned earlier one of the major features of the textile industry is the high water consumption. Malpractices such as the usage of inefficient washing equipment, poor housekeeping applications, feeding freshwater at all operations requiring water and the application of longer washing cycles than required, can be considered as the contributors of generating excessive amounts of wastewater in the textile industry (UNEP IE, 1994). Significant reductions however can be achieved simply by identifying and preventing the unnecessary water consumption. On the other hand, a part of the wastewater originating from one operation may be of sufficient quality to be reused in a second operation, directly or after appropriate treatment.

While applying water conservation measures one of the main concerns is the possible adverse effect of such an activity on wastewater quality. Segregating the relatively less polluted wastewater fraction for reuse (Orhon *et al.*, 2000b; 2001b) and preventing unnecessary water consumption are observed to generate a stronger wastewater (Dulkadiroglu *et al.*, 2002; Erdogan *et al.*, 2004) likely to require a higher level of treatment before discharge. Thus, the feasibility of the mentioned in-plant control applications must be evaluated by comparing the savings obtained on fresh water demand versus elevated end-of-pipe wastewater treatment costs together with cost of treating reusable streams, where applicable (Dulkadiroglu *et al.*, 2002). The following case shows the methodology of implementing water conservation along with wastewater reclamation and reuse in a textile finishing mill.

9.2.1 Characteristics of the plant operation

The wool finishing plant investigated handles previously *dyed wool, wool-lycra, wool-polyester* and *wool-polyester-lycra* blends fabric finishing operations (Dulkadiroglu *et al.*, 2002). The plant operates nine different processes all in batch-wise modes. Acetic acid together with different types of detergents, crease-proofing agents and softeners are added to either fill and draw or shower baths in order to obtain the required finishing on fabrics. As can be seen from the production data summarized in Table 9.10, three types of previously dyed fabric namely; *A type, B type* and *C type*; are subjected to finishing operations in the plant. When a dyeing process is applied to fabric, such fabrics are defined as A type of fabrics. The fabrics manufactured from dyed yarns are named as B type of fabrics. Lastly, C type fabrics are the fabrics produced from dyed tops or dyed fibers.

Five different processes namely, B type-100% wool and 96% wool + 4% lycra fabric; C type-100% wool and 96% wool + 4% lycra fabric; A type-50% wool + 50% polyester and 48% wool + 48% polyester + 4% lycra fabric; B type-50% wool + 50% polyester and 48% wool + 48% polyester + 4% lycra fabric; and C type-50% wool + 50% polyester and 48% wool + 48% polyester + 4% lycra fabric finishing processes are selected for the investigation in terms of identifying their recoverable wastewater streams and water conservation practices. The selected processes constitute approximately 82% of the total production. A type (100% wool and 96% wool + 4% lycra) fabric finishing operations are not included in the survey as the used wide washing equipment is not suitable for wastewater recovery and water conservation applications. Besides the processes located under '*others*' item in Table 9.10 are not covered

Table 9.10. Production data

Process description	Process number	Processed product (kg/day)	Type of equipment	Production (%)
100% Wool and 96% Wool + 4% Lycra				
A Type dyed fabric finishing	I	907	Wide washing	13.6
→B Type dyed fabric finishing	II	483	Turbo	7.2
→C Type dyed fabric finishing	III	1256	Turbo	18.8
50% wool + 50% PES and 48% Wool + 48% PES + 4% Lycra				
→A Type dyed fabric finishing	IV	1135	Spiral	17.0
→B Type dyed fabric finishing	V	1263	Turbo	18.9
→C Type dyed fabric finishing	VI	1338	Turbo	20.1
Others				
A Type dyed fabric finishing	VII	52	Various	0.8
B Type dyed fabric finishing	VIII	196	Various	2.9
C Type dyed fabric finishing	IX	39	Various	0.6

→Investigated processes

as due to their relatively lower production percentages they are not applied even on a weekly basis. (Please note that the investigated processes are the ones applied on a day to day basis.)

The total daily water consumption of the investigated processes is 444.4 m^3 for an average of 5475 kg of fabric processed. These values yield a unit water consumption rate of 81 m^3 of water usage per ton of fabric that can be considered as a typical level obtained for similar textile finishing mills (Germirli Babuna, et al., 1998a; Orhon, et al., 2000a). However since the investigated plant only applies finishing operations on previously dyed raw materials and does not implement any kind of dyeing process as usually performed by comparable textile mills, a much lower unit water consumption value must be expected from the facility.

9.2.2 Adopted methodology for in-plant control

9.2.2.1 Description of the processes and evaluation of the segregated effluent characteristics

The first step in identifying the unnecessary water consuming points and reusable wastewater streams throughout an industrial process is to investigate the production schemes in detail. Such an effort involves every step of the production using water and/or generating wastewater. The obtained results must then be evaluated in a way to assess the unnecessary water consumption points and to identify reusable wastewater streams throughout the operations.

The schematic flowcharts of investigated five processes are illustrated in Figures 9.1, 9.2, 9.3, 9.4 and 9.5 (Dogruel et al., 2002). As can be seen from the figures, the wastewaters either originate from fill and draw rinsings or shower rinsings. A sample from each fill and draw rinsing spent dyebath was collected. On the other hand the sampling frequencies indicated in the mentioned figures were used for obtaining the samples from shower rinsings. In this respect according to the data given in Figure 9.1 for first sampling, one sample from 1 Fill and Draw Rinsing spent bath (Sample A.2.1); one sample from 2. Fill and Draw Rinsing spent bath (Sample A.2.2), one sample from 10 Minutes Shower Rinsing (Sample A.2.7 obtained after 10 minutes), one sample from 3. Fill and Draw Rinsing spent bath (Sample A.2.8), three samples from 30 Minutes Shower Rinsing (Sample A.2.10 obtained after 10 minutes, Sample A.2.12 after 20 minutes and Sample A.2.14 after 30 minutes), one sample from Softening spent bath (Sample A.2.17) were collected.

The next step is to characterize the segregated streams in terms of polluting parameters. When considered within the light of (i) the reuse criteria given in

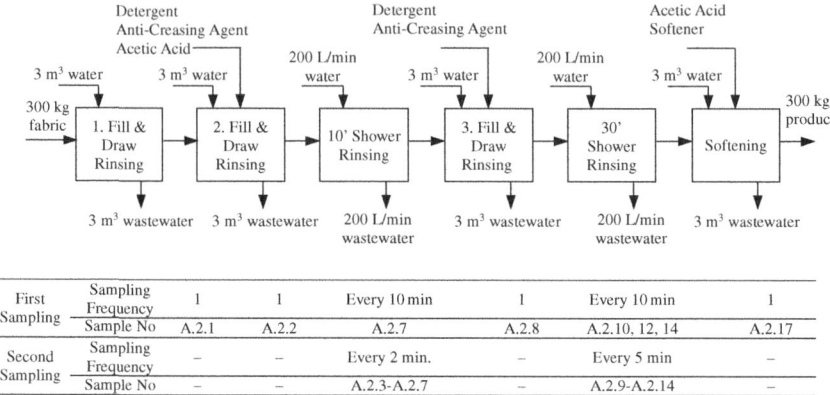

Figure 9.1. Schematic flowchart of Process II

the literature, and (ii) specific demands of the manufacturer on product quality; this quality analysis eases the route to reach the optimum results.

The characterization of segregated wastewater streams are tabulated in Tables 9.11, 9.12, 9.13, 9.14 and 9.15 (Dogruel et al., 2002). Data corresponding to shower rinsings are highlighted in gray. Due to the fact that all the collected wastewater samples contain fibers coming from the processed fabric, considerably different total COD values are obtained for replicate analysis (the figures given in tables represent the mean values). In this respect, it is concluded that soluble COD values are more dependable than total COD for a sound evaluation. Therefore, further assessments are based on soluble COD

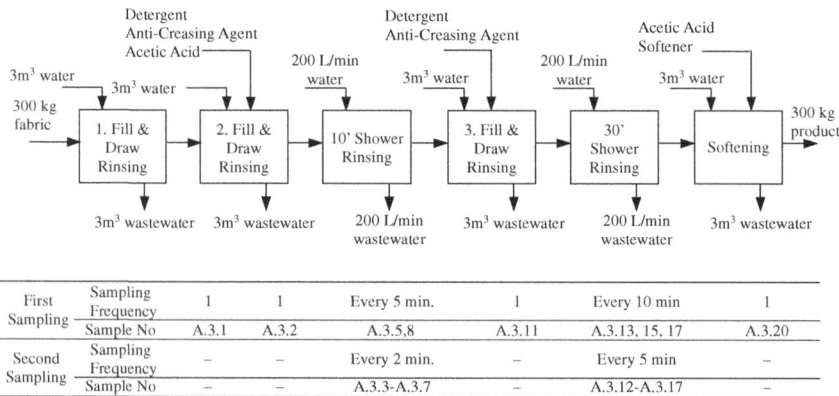

Figure 9.2. Schematic flowchart of Process III

Textile wastewaters 327

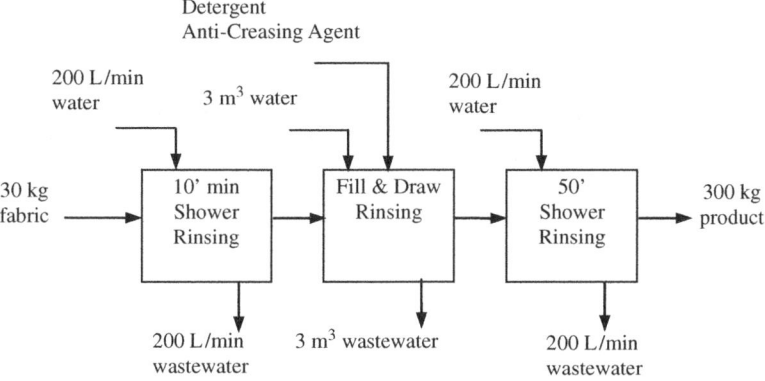

Sampling Frequency	Every 5 min.	1	Every 10 min.
Sample No	B.1.1-B.1.2	B.1.3	B.1.4-B.1.9

Figure 9.3. Schematic flowchart of Process IV

values. As the values tabulated in Tables 9.11, 9.12 and 9.15 indicate the necessity of an additional characterization study, shower rinsings originating from these processes are further investigated in terms of soluble COD by collecting other samples (i.e. Since Sample A.2.7, collected at the end of 10 Minutes Shower Rinsing of Process II, has a soluble COD level of <10 mg/l at first sampling, a second sampling program is conducted by taking a sample in every two minutes from the same rinsing step, namely Samples A.2.3–A.2.7).

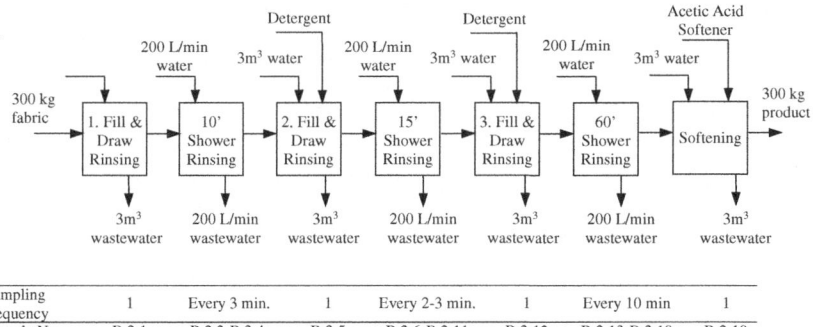

Sampling Frequency	1	Every 3 min.	1	Every 2-3 min.	1	Every 10 min	1
Sample No	B.2.1	B.2.2-B.2.4	B.2.5	B.2.6-B.2.11	B.2.12	B.2.13-B.2.18	B.2.19

Figure 9.4. Schematic flowchart of Process V

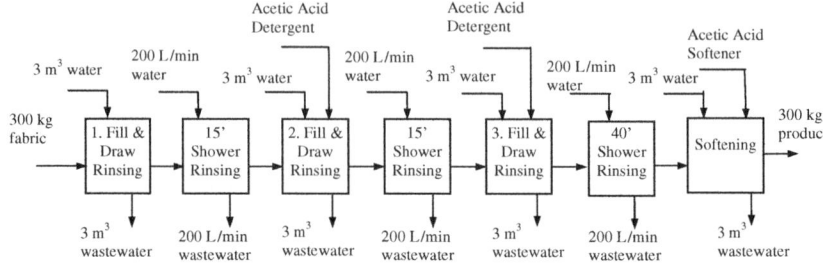

Figure 9.5. Schematic flowchart of Process VI

The results are shown in Tables 9.16, 9.17 and 9.18. The differences observed in soluble COD values detected on two samples originating from the same process can be attributed to the usage of various auxiliaries added in different runs.

It must be kept in mind that the most inevitable part of the adopted methodology involves the evaluation of the envisaged water conservation and reuse measures in terms of the effects on product quality. The possible negative effects detected can easily be avoided at this stage.

Table 9.11. Wastewater characterization for Process II (First sampling)

Sample no	Sampling frequency (minutes)	pH	Conductivity (µS/cm)	Total COD(mg/l)	Soluble COD (mg/l)
A.2.1		7.2	592	1390	375
A.2.2		5.0	649	2830	1620
A.2.7	10	7.5	487	155	<10
A.2.8		7.3	493	1325	895
A.2.10	10	7.7	494	40	30
A.2.12	20	7.8	494	30	25
A.2.14	30	7.8	496	20	20
A.2.17		5.0	649	695	565

Data highlighted in gray: shower rinsings

Table 9.12. Wastewater characterization for Process III (First sampling)

Sample no	Sampling frequency (minutes)	pH	Conductivity (μS/cm)	Total COD (mg/l)	Soluble COD (mg/l)
A.3.1		7.5	485	290	200
A.3.2		7.1	491	1000	675
A.3.5	5	7.6	466	220	140
A.3.8	10	7.5	454	125	50
A.3.11		7.4	513	1900	1565
A.3.13	10	7.7	490	320	290
A.3.15	20	7.8	484	80	45
A.3.17	30	7.7	486	40	30
A.3.20		7.6	526	160	150

Data highlighted in gray: shower rinsings

The below given practical issues can be recommended as the outcomes of the applied methodology:

(i) *for water conservation:* In pre-evaluations the unnecessarily elongated shower rinsings are indicated as the most rewarding operations within all the processes for conserving water. Therefore it is suggested to stop the shower rinsings at a point where the soluble COD of the segregated wastewaters reach maximum 50 mg/l. However, due to the monitored deterioration in product quality under the mentioned conditions, the application period of shower rinsings is recommended not to be shortened below 5 minutes even though soluble COD values of the wastewaters obtained within this period can go under 50 mg/l;

(ii) *for wastewater reuse:* It is recommended to reuse the effluent streams having soluble COD values lower than 650 mg/l. This set soluble COD

Table 9.13. Wastewater characterization for Process IV (First sampling)

Sample no	Sampling frequency (minutes)	pH	Conductivity (μS/cm)	Total COD (mg/l)	Soluble COD (mg/l)
B.1.1	5	7.7	687	790	550
B.1.2	10	7.8	623	170	65
B.1.3		8.1	549	2500	1800
B.1.4	5	8.1	579	700	550
B.1.5	15	8.0	599	215	125
B.1.6	25	8.2	605	80	50
B.1.7	35	8.2	604	45	25
B.1.8	45	8.2	613	45	15
B.1.9	50	8.2	610	45	<10

Data highlighted in gray: shower rinsings

Table 9.14. Wastewater characterization for Process V (First sampling)

Sample no	Sampling frequency (minutes)	pH	Conductivity (µS/cm)	Total COD (mg/l)	Soluble COD (mg/l)
B.2.1		6.23	440	950	305
B.2.2	3	6.47	410	760	220
B.2.3	6	6.59	400	260	75
B.2.4	9	6.54	400	250	70
B.2.5		5.85	540	2460	1430
B.2.6	2	6.62	400	500	350
B.2.7	4	6.47	400	440	190
B.2.8	6	6.65	380	280	150
B.2.9	9	6.37	390	260	110
B.2.10	12	6.50	390	150	110
B.2.11	15	6.55	390	140	75
B.2.12		6.33	410	940	700
B.2.13	10	6.79	400	110	65
B.2.14	20	6.91	400	75	55
B.2.15	30	9.96	400	50	40
B.2.16	40	6.96	390	50	35
B.2.17	50	6.98	390	50	30
B.2.18	60	7.14	400	30	20
B.2.19		4.91	625	1020	700

Data highlighted in gray: shower rinsings

Table 9.15. Wastewater characterization for Process VI (First sampling)

Sample no	Sampling frequency (minutes)	pH	Conductivity (µS/cm)	Total COD (mg/l)	Soluble COD (mg/l)
B.3.1		7.7	555	2960	1820
B.3.4	5	7.8	507	915	650
B.3.7	10	7.6	487	115	75
B.3.9	15	7.6	487	55	35
B.3.10		5.6	676	2380	2110
B.3.13	5	7.4	507	540	380
B.3.16	10	7.7	481	60	90
B.3.18	15	7.7	478	125	50
B.3.19		7.1	517	1120	400
B.3.21	10	7.8	479	60	60
B.3.23	20	7.8	490	50	30
B.3.25	30	7.7	486	40	25
B.3.27	40	7.8	483	30	20
B.3.30		5.7	675	800	530

Data highlighted in gray: shower rinsings

Table 9.16. Soluble COD values of shower rinsings for Process II (Second sampling)

Sample no	Sampling frequency (minutes)	Shower rinsing no	Soluble COD (mg/l)
A.2.3–A.2.7	2, 4, 6, 8, 10	1	<40
A.2.9	5	2	80
A.2.10	10	2	70
A.2.11–A.2.14	15, 20, 25, 30	2	<35

Table 9.17. Soluble COD values of shower rinsings for Process III (Second sampling)

Sample no	Sampling frequency (minutes)	Shower rinsing no	Soluble COD (mg/l)
A.3.3	2	1	215
A.3.4	4	1	70
A.3.6–A.3.7	6, 9	1	<40
A.3.12–A.3.17	5, 10, 15, 20, 25, 30	2	<40

figure seems high at a first glance; nevertheless a final reusable wastewater stream with a soluble COD value of around 200 mg/l is obtained because of the dilutions. The quality of the segregated effluent streams generated from 1st Fill and Draw Rinsings cannot be predicted soundly as these wastewaters might contain contaminants originating from the previous dyeing operations. A similar uncertainty is also valid for the effluent quality of Fill and Draw Rinsings with different brands of

Table 9.18. Soluble COD values of shower rinsings for Process VI (Second sampling)

Sample no	Sampling frequency (minutes)	Shower rinsing no	Soluble COD (mg/l)
B.3.2	2	1	350
B.3.3	4	1	160
B.3.5	6	1	100
B.3.6	9	1	60
B.3.8–B.3.9	12, 15	1	<50
B.3.11	2	2	190
B.3.12	4	2	170
B.3.14	6	2	85
B.3.15	9	2	65
B.3.17	12	2	45
B.3.20	5	3	75
B.3.21–B.3.27	10, 15, 20, 25, 30, 35, 40	3	<50

auxiliary additions. Therefore discharges of 1st Fill and Draw Rinsings and Fill and Draw Rinsings in which auxiliary chemicals are added must not be summed up with reusable wastewaters.

For example let us evaluate the data tabulated in Table 9.14 and Figure 9.4 for Process V (B type 50% wool + 50% polyester and 48% wool + 48% polyester + 4% lycra dyed fabric finishing operation) within the context of the proposed in-plant control measures. According to the outlined characterization figures in Table 9.14, 60 minutes shower rinsing must be stopped after the first 20 minutes as after this point soluble COD values lower than 50 mg/l are obtained. By doing so, a water conservation of 6 m^3 per 300 kg finished fabric is obtained. All the wastewater streams originating from 10 and 15 minutes Shower Rinsings together with the rest of 60 minutes Shower Rinsing (first 20 minutes) can be added to the reusable fraction. As a result 9 m^3 out of total 29 m^3 per 300 kg finished fabric, can be directed towards reusable wastewaters collection tank. The results of this evaluation for Process V are tabulated in Table 9.19.

9.2.2.2 Technical basis of the feasibility analysis

Basic data. As previously mentioned the water consumption of the investigated processes with an average of 5475 kg of fabric processed per day is 444.4 m^3/day in total. These figures indicate a unit water consumption rate of 81 m^3 of water per ton of fabric processed. At a first glance this unit water consumption value can be considered as a typical level obtained for similar textile finishing mills (Germirli Babuna, *et al.*, 1998a; Orhon, *et al.*, 2000a). However, a detailed evaluation guiding to an insight of the case shows clearly that the investigated plant only applies finishing operations on previously dyed raw materials and the processes do

Table 9.19. In-plant control applications for Process V

Type of operation	Wastewater generation	Reusable streams	Water conservation streams
	(m^3/300 kg fabric)		
1st Fill&draw rinsing	3	–	–
10 min shower rinsing	2	2	–
2nd Fill&draw rinsing	3	–	–
15 min shower rinsing	3	3	–
3rd Fill&draw rinsing	3	–	–
60 min shower rinsing	12	4 (first 20 min)	6 (from 20 to 60 min)
Softening	3	–	–
Total	**29**	**9**	**6**

not involve any kind of dyeing procedure. Therefore, a comparison based on the other textile mills performing dyeing operations can be misleading. A much lower unit water consumption figure must be expected from the investigated facility.

According to the tabulated values in Table 9.20:

(i) water conservation measures: A 150 m^3/day reduction in water consumption is possible, decreasing the water usage to 294 m^3/day and the unit water consumption rate to 54 m^3/ton of fabric.

(ii) water conservation together with reuse measures: The proposed in-plant control measures also allow for the reuse of 100 m^3/day of the wastewater stream bringing down the water consumption to 194 m^3/day, which ultimately corresponds to a 35 m^3/ton of fabric.

Thus approximately 60 % reduction in water consumption (which can be directly transferred to the amount of generated wastewater) can be achieved with the proposed in-plant control measures. However a check in terms of the technical and economical feasibility is required beforehand.

Wastewater quality. An increase in the wastewater strength is expected when water conservation practices are applied in an industrial premise. Moreover by segregating the reusable wastewater streams together with conserving water a much stronger end-of-pipe effluent likely to have more complex nature towards biotreatment will be generated (Orhon *et al,* 2001b). In this respect the flowrates and characterization of

(i) the raw wastewater before in-plant control application (W*astewater A)*;
(ii) the wastewater after water conservation *(Wastewater B)*
(iii) the remaining wastewater after water conservation and segregation of reusable streams *(Wastewater C)*

are given in Table 9.21 with relevant effluent limitations for discharge to sewer and to receiving waters (Erdogan *et al.,* 2004).

Before any in-plant control application the textile mill has a highly colored wastewater, with an average COD content of 687 mg/l and a TSS content of only 85 mg/l. It is obvious that Wastewater A is somewhat weak in character due to excessive water use.

A stronger effluent (Wastewater B) is obtained after implementing water conservation measures. Application of both water conservation and wastewater reuse practices ends up in an effluent (Wastewater C) that is even stronger in character as expected. In order to have the right perception about this effluent a further effort for displaying the COD fractions is required.

Table 9.20. Results of the proposed in-plant control applications

Process	Processed product (kg fabric/day)	Water usage Wastewater generation (m³/300 kg fabric)	Water usage Wastewater generation (m³/day)	Water conservation (m³/day)	Water conservation (%)	Reusable streams (m³/day)	Reusable streams (%)
100% wool and 96% wool + 4% Lycra							
B Type dyed fabric finishing	483	22	35.4	11.3	32	4.8	14
C Type dyed fabric finishing	1256	23	96.3	33.5	35	12.6	13
50% Wool + 50% PES and 48% Wool + 48% PES + 4% Lycra							
A Type dyed fabric finishing	1135	15	56.8	26.5	47	18.9	33
B Type dyed fabric finishing	1263	29	122.1	25.3	21	37.9	31
C Type dyed fabric finishing	1338	30	133.8	53.5	40	26.8	20
TOTAL	5475	–	444.4	150.1	34	101	23

Table 9.21. Flowrates and characterization of wastewaters generated from the investigated mill and relevant discharge standards

Parameter	Wastewaters			Discharge standards	
	A	B	C	Receiving water	Sewer
Flowrate (m^3/day)	444	294	194	–	–
Total COD (mg/l)	687	1038	1460	300	800
Soluble COD (mg/l)	455	687	970	–	–
TSS (mg/l)	85	128	190	100	350
VSS (mg/l)	80	121	180	–	–
TDS (mg/l)	380	574	640	–	–
TKN (mg/l)	20	29	30	–	–
николаев NH_4-N (mg/l)	8	12	18	–	–
TP (mg/l)	0.8	1.0	1.2	–	10
Conductivity (μS/cm)	620	635	655	–	–
Alkalinity (mg $CaCO_3$/l)	108	106	106	–	–
Colour (Pt-Co)	220	332	440	–	–
pH	7.1	6.9	6.2	6–9	6–10

The wastewater characterization conducted on the two samples collected from the reusable streams equalization tank is tabulated in Table 9.22. The table also represents two sets of reuse criteria suggested in the literature for reuse in the processes (Dulkadiroglu et al., 2002). The reusable wastewater stream is apparently much weaker in character with a COD of around 230 mg/l, a total dissolved solids (TDS) of 340 mg/l and very low in colour.

Table 9.22. Reusable wastewater characterization and reuse criteria

Parameter	Raw reusable wastewater		Reuse criteria	
	Sample I	Sample II	Li and Zhao (1999)	Hoehn (1998)
Flowrate (m^3/day)	101	101	–	–
Total COD (mg/l)	180	235	0–160	<50
Soluble COD (mg/l)	120	175	–	–
TSS (mg/l)	15	15	0–50	<500
TDS (mg/l)	340	345	100–1000	–
Total hardness (mg $CaCO_3$/l)	0	0	0–100	90
Chloride (mg/l)	<100	<100	100–300	<150
Total chromium (mg/l)	<0.5	<0.5	–	0.1
Iron (mg/l)	<1	<1	0–0.3	0.1
Manganese (mg/l)	<0.3	<0.3	<0.05	0.05
Conductivity (μS/cm)	550	625	800–2200	–
Alkalinity (mg $CaCO_3$/l)	135	120	50–200	–
Colour (Pt-Co)	20	30	–	–
pH	7.10	7.19	6.5–8.0	6.5–7.5

9.2.2.3 Appropriate treatment alternatives

Appropriate treatment alternatives applicable to reusable wastewater portion and end-of-pipe effluents are evaluated separately below.

TSS, TDS, chloride, total hardness, alkalinity and conductivity levels of the untreated reusable wastewater stream are below the limits indicated by the reuse criteria (Table 9.22). On the other hand total chromium, manganese and iron contents of the reusable wastewater are within the required limits prescribed by the manufacturer. Therefore the reusable stream requires only polishing in terms of COD and colour removal without adversely affecting the TDS content. Due to its relatively high cost and extra care need, membrane treatment is not desired by the manufacturers. Ozonation and chemical treatment are among the most commonly applied treatment methods to reusable portion of the effluents (Sewekow, 1995). The results of ozonation experiments performed on Sample II (Table 9.23), indicated that even with extreme ozone dosages, the soluble COD reduction efficiency can only be improved to 40%, giving an effluent soluble COD

Table 9.23. Results of ozonation tests applied to reusable wastewater stream

Parameter		Ozone feeding time			
	–	5 minutes	10 minutes	15 minutes	30 minutes
Ozone flux (mg/min)	–	58	58	58	58
Utilized ozone (mg)	–	25	150	405	1255
Soluble COD (mg/l)	175	155	140	130	105
Soluble COD removal (%)	–	11	20	26	40
Conductivity (µMhos/cm)	625	640	620	620	600
Alkalinity (mg CaCO$_3$/l)	120	110	90	60	60
Colour (Pt-Co unit)	30	10	0	0	0
Colour removal (%)	–	67	100	100	100
pH	7.19	7.68	7.52	7.38	7.31

Table 9.24. Results of chemical treatability applied to reusable wastewater stream

Parameter	Alum			Sodium bentonite		
Dosage (mg/l)	50	75	100	500	1000	1500
Total COD (mg/l)	55	65	30	75	85	55
Total COD removal (%)	69	64	83	58	53	69
Conductivity (µS/cm)	600	600	600	580	660	900
Alkalinity (mgCaCO$_3$/l)	50	90	65	155	180	240
Colour (Pt-Co unit)	0	0	0	0	0	0
TDS (mg/l)	405	405	400	350	455	615
SVI (ml/g)	140	105	105	30	25	15
pH	6.03	6.26	6.26	7.63	7.07	7.59

concentration of above 100 mg/l (Dulkadiroglu et al., 2002). Consequently the application of ozone to reusable streams is regarded as a useless effort to be avoided.

Table 9.24 presents briefly the results of chemical treatability tests conducted with different doses of alum and bentonite on Sample II (Dulkadiroglu et al., 2002). Usually the textile manufacturers do not prefer the usage of iron salts as coagulants while treating the reusable portion of the wastewaters. The reuse of

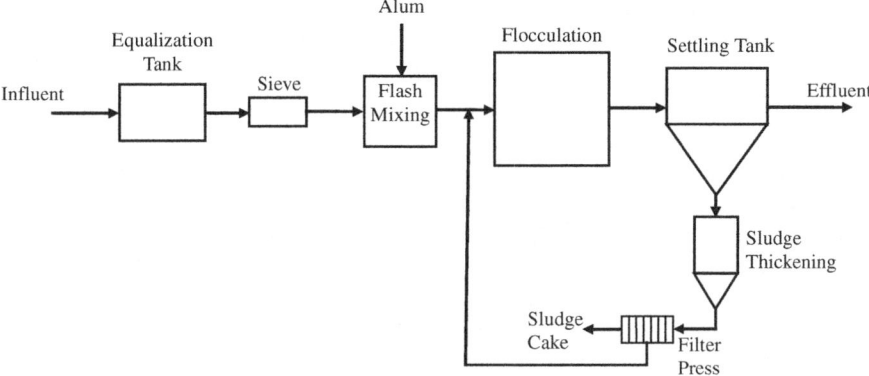

Figure 9.6. Schematic diagram of chemical treatment of reusable effluents using alum

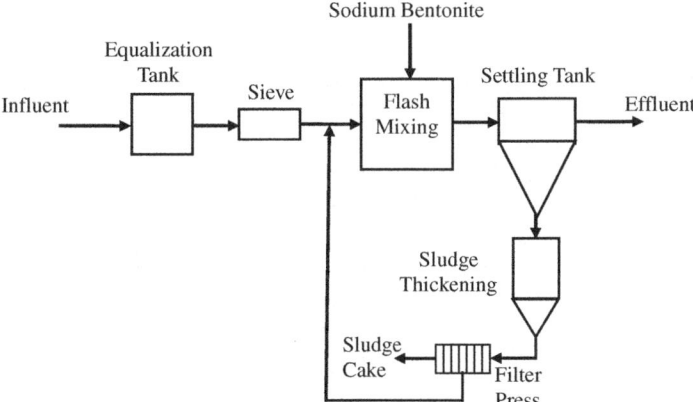

Figure 9.7. Schematic diagram of chemical treatment of reusable effluents using bentonite

treated effluents with iron salts in the process may cause staining problems in the final textile good. Chemical treatment with either alum or bentonite is proved to be efficient in providing the required total colour and COD removals. With both of the coagulants a COD level below 100 mg/l can be obtained. An alum dosage of 75 mg/l or a bentonite dosage of 500 mg/l is monitored to be adequate in achieving the required COD removal efficiencies and good sludge characteristics. Schematic flow diagrams of chemical treatment alternatives employing alum and bentonite are illustrated in Figures 9.6 and 9.7.

Appropriate treatment that would apply to end-of-pipe effluent depends upon discharge alternatives of either to sewers or to receiving water. As indicated previously the Wastewater A (the effluent generated without any in-plant control) is weak in character and on the average, its quality can satisfy the discharge to sewer conditions without any treatment. For discharging directly to the receiving waters, a high rate biological treatment (sludge age lower than 4 days) can be prescribed as activated sludge systems are stated to yield often more reliable results than chemical treatment when effluents generated in textile finishing mills are considered. A schematic diagram of such a treatment facility is demonstrated in Figure 9.8. This configuration would be equally appropriate for Wastewater B (the effluent generated with water conservation), although a conventional activated sludge system (sludge age between 4 to 8 days) would be more reliable to treat a COD level of around 1200 mg/l down to acceptable limits.

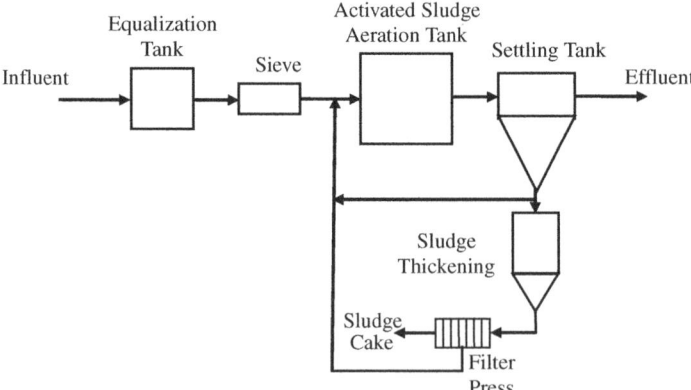

Figure 9.8. Schematic diagram of the activated sludge treatment applied to end-of-pipe effluent

The only alternative for treating Wastewater C (the remaining wastewater generated after water conservation and segregation of reusable streams) is biological treatment typically applicable to textile wastewaters with a sludge age higher than 8 days and a hydraulic retention time in the range of 15 to 20 hours.

According to the current understanding of the environmental biotechnology, the biodegradability of the effluents directly affects the performance of biological treatment. Thus the end-of-pipe wastewaters must be evaluated in terms of their COD fractions before prescribing a biological type of treatment. In this sense, the experimentally assessed COD fractions of Wastewater A and Wastewater C are tabulated in Table 9.25 (Dulkadiroglu et al., 2002).

As can be seen from the table, the biodegradable COD accounts for 85% of both Wastewater A (the effluent generated without any in-plant control) and Wastewater C (the remaining wastewater generated after water conservation and segregation of reusable streams). Both of the investigated wastewaters have the same ratio of initial soluble inert COD and initial particulate inert COD, 5% and 10%, respectively. The percentages of hydrolysable fractions are also similar for both of them. It can be concluded that although the application of in-plant control measures elevates the initial soluble inert COD concentration over 100%, practically it has no effect on the COD fractionation.

9.2.2.4 Feasibility analysis

The investigated textile mill obtains the water from the municipality at a price of US$ 3.00/m^3. Apart from the fresh water price, this rate also includes the sewer charge, without considering whether the sewer system is used for discharge or not. Thus the feasibility analysis should cover the water charge and the cost of treatment associated with each water management option defined with the

Table 9.25. COD fractions of wastewater A and wastewater C

Influent COD fractions	Wastewater A (mg/l)	(%)	Wastewater C (mg/l)	(%)
Total biodegradable COD, C_{S1}	583	85	1244	85
Readily biodegradable COD, S_{S1}	220	32	485	33
Rapidly hydrolysable COD, S_{H1}	203	30	418	29
Slowly hydrolysable COD, X_{S1}	160	23	341	23
Total inert COD, C_{I1}	104	15	216	15
Soluble inert COD, S_{I1}	32	5	67	5
Particulate inert COD, X_{I1}	72	10	149	10
Total COD, C_{T1}	687		1460	

340   Industrial Wastewater Treatment by Activated Sludge

relevant in-plant control strategy. Each water management option must be evaluated by considering the operating and investment costs involved for the treatment. Therefore different treatment configurations described according to pertinent in-plant control alternatives are designed in a detailed manner. Important features of the mentioned design exercise can be enumerated as (Erdogan et al., 2004):

  (i) an equalization tank with 1/3 the volume of the daily wastewater flow appropriate for the 3 shift (24 hours/day);
  (ii) mixing of the equalization tank with a submersible pump-ejector system;
  (iii) a wedge wire type of a static sieve at the outlet of equalization for the removal of fibres before biological treatment;
  (iv) appropriate selection of the sludge age for activated sludge alternative;
  (v) mechanical surface aeration of the aeration tank, controlled by dissolved oxygen sensors;
  (vi) diammonium phosphate feeding in the aeration tank to establish the necessary COD/N/P balance in biological treatment;
  (vii) a filter press system with a sludge holding tank and lime feeding to ensure 35% dry weight in dewatering; three charges per day for the press filter operation;
  (viii) a single joint sludge dewatering system for the chemical treatment of the reusable stream and biological treatment of the remaining wastewater.

The cost breakdown of each treatment scheme defined is tabulated in Table 9.26 (Erdogan et al., 2004). The appraisal of the costs is performed by considering the discharge media i.e. discharge to sewers or to receiving water. From a practical point of view an effluent discharge to receiving waters requires a higher treatment efficiency than can be achieved with biological treatment. In this context, the evaluation basically compares three different management options that can be expressed as:

  (a) *do nothing*
  (b) *only conserve water*
  (c) *conserve water and reuse a fraction of the wastewater*

by considering two different end-of-pipe discharge media of:

  (a) to sewers
  (b) to receiving water.

Table 9.26. Cost breakdown of different treatment schemes

Type of WW	Treatment process	Discharge media	Investment costs (USD)				Operation costs (USD/month)			
			Construction	Electro-mechanic	Other	Total	Man-power	Chemical	Energy	Total
A	Biological	Receiving water	57,526	57,929	12,147	127,602	1,874	29	765	**2,668**
A	Chemical	Sewer	36,910	49,000	8,147	**94,057**	1,874	585	360	**2,819**
B	Biological	Receiving water	46,326	56,769	12,147	**115,242**	1,874	41	765	**2,680**
B	Chemical	Sewer	25,258	49,000	8,147	**82,405**	1,874	646	360	**2,880**
C	Biological	Receiving water or sewer	45,415	53,769	12,147	**111,331**	1,874	41	765	**2,680**
Reusable	Chemical (Alum)	—	9,500	19,811	6,650	**37,961**	1,874	101	270	**2,245**
Reusable	Chemical (Bent.)	—	7,500	18,595	5,650	**32,109**	1,874	220	270	**2,364**

Cost implications of all the options are given in Table 9.27 (Erdogan et al., 2004). The cost analysis has been done by calculating the net present value for each of the alternatives. In the analysis, the discount factor for the operating costs has been determined using the monthly LIBOR rate. The depreciation expense which has been deducted as part of the running costs has been determined on a monthly basis using the straight line method. The borrowing rate for the total investment has been assumed to be the equivalent of the 20-year LIBOR swap rate plus 3% credit spread (which at the time of the analysis was equal to 8.2%). The discount factor for the investment component of the costs has been calculated using the 20-year LIBOR swap curve. Table 9.28 shows the net present values of the investments calculated on the basis of the assumptions presented above for both discharges to sewer and to receiving water (Erdogan et al., 2004).

In conclusion, discharge to sewers under present conditions involving the treatment of the 444 m^3 of daily wastewater generated requires a capital investment of US$ 94,057 and a monthly running cost of US$ 42,668 that corresponds to the operation cost of the wastewater treatment plant together with the water charge. While the price of water consumed representing 94% of the monthly running expenses is the main contributor of the monthly expenditure, the capital investment for wastewater treatment forms a negligible fraction of the total costs. Assuming a service duration of 20 years for the treatment facility, the capital investment represents less than 1% of the overall costs. Using the cost structure described herein, the overall cost per unit for discharge to sewers under present conditions is calculated to be US$ 262.16 per ton of fabric processed.

Water conservation leads to a significant reduction in capital investment as well as the running costs bringing the overall cost per unit down to US$ 180.72 per ton of fabric processed. The mentioned US$ 81.44 reduction in overall costs per ton of fabric processed corresponds to a saving of 31.1%.

Given the complex nature of the remaining wastewater and the necessity of additional treatment for the reusable stream, the capital investment requirement of the water reuse alternative is higher than that of the water conservation option by US$ 61,035. Nevertheless, the monthly savings of US$ 6,836 achieved in running costs by implementing the water reuse option (instead of only water conservation) will compensate for the incrementally higher capital expenditure as such savings will allow the investment differential to be paid back in less than ten months. In conclusion, the water reuse alternative proves to be the most cost effective option as it brings the overall cost per unit down to US$ 140.78 per ton of fabric processed from US$ 262.16 obtained for not applying any in-plant control measures (current conditions) and US$ 180.72 for

Table 9.27. Cost implications of the options

	Water usage & wastewater generation	Type of wastewater treatment	Investment cost for WW treatment (USD)	Running costs			Total running costs (USD/month)
				Operation cost for WW treatment (USD/month)	Water charge (USD/month)		

	Water usage & wastewater generation	Type of wastewater treatment	Investment cost for WW treatment (USD)	Operation cost for WW treatment (USD/month)	Water charge (USD/month)	Total running costs (USD/month)
Discharge to sewer						
Present use (Wastewater A)	444					
Total		Chemical settling*	94,057	2,819	39,960	**42,668**
Water conservation (Wastewater B)	294					
Total		Chemical settling*	82,405	2,880	26,440	**29,340**
Water conservation and WW reuse (Wastewater C)	194	Biological	*111,331*	*2,680*	*17,460*	*20,140*
(Reusable stream)	100	Chemical settling*	*32,109*	*2,364*	*–*	*2,364*
Total			**143,440**	**5,044**	**17,460**	**22,504**
Discharge to receiving water						
Present use (Wastewater A)	444					
Total		Biological	127,602	2,668	39,960	**42,668**
Water conservation (Wastewater B)	294					
Total		Biological	115,242	2,680	26,440	**29,120**
Water conservation and WW reuse (Wastewater C)	194	Biological	*111,331*	*2,680*	*17,460*	*20,140*
(Reusable stream)	100	Chemical settling*	*32,109*	*2,364*	*–*	*2,364*
Total			**143,440**	**5,044**	**17,460**	**22,504**

* with bentonite

Table 9.28. Net present values of the options

	Investment cost (USD)	Operation cost (USD/month)	Amortization cost (USD/year)	Amortization cost (USD/month)	NPV of investment (USD)	NPV of investment (USD/kg fabric)
Discharge to sewer						
Present use	94,057	42,668	4,703	392	9,239,668	0.23
Water conservation	82,405	29,340	4,120	343	6,397,860	0.16
Water conservation and WW reuse	143,440	22,504	7,172	598	5,107,894	0.13
Discharge to receiving water						
Present use	127,602	42,668	6,380	532	9,323,573	0.23
Water conservation	115,242	29,120	5,762	480	6,433,566	0.16
Water conservation and WW reuse	143,440	22,504	7,172	598	5,107,894	0.13

water conservation alone. Similar results in cost structure are observed for discharge to receiving water.

The case presented here can be regarded as a concrete example of obtaining considerable positive financial outputs due to adopting a sound environmental management strategy covering in-plant control measures prior to end-of pipe treatment.

9.3 REFERENCES

Bredereck, K. (1978) Neue Möglichkeiten der Mercerization von Baumwolltextilien durch Flussigammoniak-Behandlung und modifizierte NaOH Mercerisation; Grundlagen und praktische Aspekte. *Textilverdlung*, **13**, 498–506.

Dogruel, S., Dulkadiroglu, H., Eremektar, G., Germirli Babuna, F. and Orhon, D. (2002) Water conservation and wastewater recovery and reuse for a wool finishing textile industry. *Proceedings of ISWA 2002, Appropriate Environmental and Solid Waste Management and Technologies for Developing Countries*, **3**, 1915–1922, July, Istanbul, Turkey.

Dogruel, S. (2000) *The Effect of Ozonation on COD Fractions - A Case Study for Cotton Finishing ill.* M.Sc. Thesis, Institute of Science and Technology, Istanbul Technical University, Istanbul (in Turkish).

Dulkadiroglu, H., Eremektar, G., Dogruel, S., Uner, H., Germirli Babuna, F. and Orhon, D. (2002) In-plant control applications and their effect on treatability of a textile mill wastewater. *Water Sci. Technol.* **45**(12), 287–295.

Erdogan, A.O., Orhon, H.F., Dulkadiroglu, H., Dogruel, S., Eremektar, G., Germirli Babuna, F. and Orhon, D. (2004) Feasibility analysis of in-plant control for water minimization and wastewater reuse in a wool finishing textile mill. *J. Environ. Sci. Health Part A.* **39** (7), 1819–1832.

EPA. (1997) *Profile of the Textile Industry*. EPA Office of Compliance Sector Notebook Project, EPA/310-R-97-009, September.

European Council Directive (1996) *Integrated Pollution Prevention and Control (IPPC)*. 96/61/EC.

European Commission, (2003) *Integrated Pollution Prevention and Control (IPPC) Reference Document on Best Available Techniques for the Textiles Industry*, 586 pages, Brussels.

Germirli Babuna, F., Orhon, D., Ubay Cokgor, E., Insel, G. and Yaprakli, B. (1998a) Modelling of activated sludge for textile wastewaters. *Water Sci. Technol.* **38**(4–5), 9–17.

Germirli Babuna, F., Eremektar, G., and Yaprakli, B. (1998b) Inert COD Fractions of Various Textile Dyeing Wastewaters. *Fresen. Environ. Bull.* **7**, 959–966.

Germirli Babuna, F., Soyhan, B., Eremektar, G. and Orhon, D. (1999) Evaluation of treatability for two textile mill effluents. *Water Sci. Technol.* **40**(1), 145–152.

Hao, O.J., Kim, H. and Chiang, P.C. (2000) Decolorization of wastewater. *Crit. Rev. Env. Sci. Tec.* **30**(4), 449–505.

Hoehn, W. (1998) *Textile Wastewater-Methods to Minimize and Reuse*. Textilveredlung, reuse standards, Thies-Handbuch für den Garnfaerber.

Kabdasli, N.I., Gurel, M. and Tunay, O. (2000) Characterization and treatment of textile printing wastewaters. *Environ. Technol.* **21,** 1147–1155.

Li, X.Z. and Zhao, Y.G. (1999) Advanced treatment of dyeing wastewater for reuse, *Water Sci. Technol.* **39** (10–11), 245–255.

Orhon, D., Sozen, S. and Tasli, R. (1996) *Investigation of the wastewaters and evaluation of treatment scheme in Plant Vakko.* Report, Istanbul Technical University, 29 pages (in Turkish).

Orhon, D., Germirli Babuna, F., Kabdasli, I., Sozen, S., Karahan, O., Insel, G., Dulkadiroglu, H. and Dogruel, S. (2000a) *Appropriate Technologies for the Minimization of Environmental Impact from Industrial Wastewaters – Textile Industry, A Case Study.* Final Report, Technical University of Istanbul Environmental Engineering Department / GSF – National Research Center for Environmental and Health Institute of Ecological Chemistry Technical University of Munich, Chair of Ecological Chemistry, VW Foundation.

Orhon, D., Sozen, S., Kabdasli, I., Germirli Babuna, F., Karahan, O., Insel, G., Dulkadiroglu, H., Dogruel, S., Kiran, N., Baban, A. and Kemerdere Kaya, N. (2000b) Recovery and reuse in the textile industry – A case study at a wool and blends finishing mill. In: *Chemical Water and Wastewater Treatment VI,* ed. H. H. Hahn, E. Hoffman and H. Odegaard, 305–315, Springer Verlag, Berlin.

Orhon, D., Germirli Babuna, F. and Insel, G. (2001a) Characterization and modelling of denim processing wastewaters for activated sludge. *J. Chem. Technol. Biot.* **76**(1), 1–13.

Orhon, D., Germirli Babuna, F., Kabdasli, I., Insel, G., Karahan, O., Dulkadiroglu, H., Dogruel, S., Sevimli, F. and Yediler, A. (2001b) A scientific approach to wastewater recovery and reuse in the textile industry. *Water Sci. Technol.* **43**(11), 223–231.

Orhon, D., Kabdasli, I., Germirli Babuna, F., Sozen, S., Dulkadiroglu, H. Dogruel, S., Karahan-Gul, O. and Insel, G. (2003) Wastewater reuse for the minimization of fresh water demand in coastal areas - selected cases from the textile finishing industry. *J. Environ. Sci. Heal. A.* **38**(8), 1641–1657.

Rieger, P.G., Meier H.M., Gerle, M., Vogt, U., Groth, T. and Knackmuss, H.J. (2002) Xenobiotics in the environment: present and future strategies to obviate the problem of biological persistence. *J. Biotechnol.* **94**, 101–123.

Sewekow, U. (1995) *Ullmann's Encyclopedia of Industrial Chemistry.* **A26**, Chapter 14, VCH Verlagsgesellschaft, Germany.

Smith, B. (1986) *Identification and Reduction of Pollution Sources in Textile Wet Processing.* Office of Waste Reduction, North Carolina Department of Environment, Health and Natural Resources, 129 pages.

Smith, B. (1988) *A Workbook for Pollution Prevention by Source Reduction in Textile Wet Processing.* Office of Waste Reduction, Department of Environment, Health and Natural Resources (DEHNR), 69 pages.

The World Bank. (1997) *Pollution Prevention and Abatement Handbook- Part 3: Textile Industry.* Environment Department, September.

UNEP IE. (1994) *The Textile Industry and the Environment.* Technical Report, No. 16.

Uner, H., Dogruel, S., Arslan Alaton, I., Germirli Babuna, F. and Orhon, D. (2006) Evaluation of coagulation-flocculation on a COD-based molecular size distribution for a textile finishing mill effluent. *J. Environ. Sci. Heal. A.* **41**, 1899–1908.

US EPA Pollution Prevention Act of 1990.

Van Veldhuisen, D.R. (1991) *Technical and Economical Aspects of Measures to Reduce Water Pollution from the Textile Finishing Industry.* Comission des Communautes Europeennes, Direction Generale, Environment Securite Nucleaire et Protection Civile, 175 pages.

Yildiz, G., Insel, G., Cokgor, E.U. and Orhon, D. (2007) Respirometric assessment of biodegradation for acrylic fiber based carpet finishing wastewaters. *Water Sci. Technol.* **55**(10), 99–106.

Yildiz, G., Insel, G., Cokgor, E. U. and Orhon, D. (2008) Biodegradation kinetics of slowly biodegradable substrate in polyamide carpet finishing wastewater, *J. Chem. Technol. Biot.* **83**, 34–40.

10
Management of tannery wastewaters

10.1 GENERAL ASPECTS

The tannery industry can be defined as the processing of sheep and goat skins and bovine hides to produce leather that will be further processed for the production of different goods. The operations involve chemical reactions and mechanical processes in order to convert raw hide and skin subject to fast deterioration to the stable and useful product of leather which displays physical properties such as softness, plasticity, elasticity, resistance towards tear, abrasion and pressure, heat and water. While the skins of other animals like pig, camel, horse, reptiles, etc. are also manufactured, these make up a very small portion of the whole industrial activity in the sector (UNIDO, 1991).

The tannery industry, with its liquid, solid and gaseous waste streams has been one of the major concerns with respect to the environmental impacts (UNIDO, 1991). Wastewaters mainly originate from the wet processes carried out in the beamhouse, the tanyard and the post-tanning steps. Tannery effluents

© 2009 IWA Publishing. *Industrial Wastewater Treatment by Activated Sludge*, by Derin Orhon, Fatos Germirli Babuna and Ozlem Karahan. ISBN: 9781789065282. Published by IWA Publishing, London, UK.

are characterized by high pollution loads of conventional pollutants and also by the specific pollutants like biocides, surfactants and organic solvents. One of the most important pollutants is the trivalent chromium used for tanning which has severe adverse effects both on the environment and on treatment processes.

10.2 THE TANNING PROCESS

Raw skins and hides are generally treated with salt at the slaughterhouses and abattoirs prior to their transport to tanneries. The curing process with salt is applied for the preservation of the raw material in good condition but brings an additional burden of salt to tannery effluents. Chilling is also applied for preservation but only when the transportation period is short enough such that raw skins and hides are processed within 5–8 days. Fresh (unsalted) hides and skins can also be processed in the tanneries but they have to be processed within 8–12 hours after slaughter if they are to be kept without chilling (European Commission, 2003). The processing of fresh hides and skins needs an integration and synchronization between the slaughterhouse and tannery operations and thus most of the tanneries use salted raw material available in the global market (UNIDO, 1991).

The manufacturing processes of tanneries can be grouped in 4 main categories. These are:

(1) Beamhouse operations
(2) Tanyard processes
(3) Post-tanning processes
(4) Finishing operations

10.2.1 Beamhouse operations

Beamhouse operations are the operations carried out to mainly prepare the raw skins and hides for tanning through cleaning the undesired inter-fibrillary material and conditioning. The beamhouse operations are common in all tanning operations irrespective of the type of the subsequent tanning process (UNIDO, 1991). Typically applied beamhouse operations are, soaking, liming and unhairing, fleshing and splitting.

Soaking: The operation is carried out in vessels, such as mixers, drums, paddles, pits, or raceways for re-hydrating the hides and skins, for removing the salt and other additives applied in the curing process and for cleaning the raw material from dirt, dung, blood and inter-fibrillary material. The operation can also involve the use of additives like sodium hydroxide, sodium hypochlorite, sulphide, wetting agents, emulsifiers, surfactants, enzyme preparations and bactericides.

Liming and Unhairing: The process is applied for removing hair, interfibrillary components and epidermis. The fibre structure of the hides and skins are opened up. If wool from sheepskin is to be collected with minimum damage paint (sulphide and lime) is applied on the flesh side of the skin and the wool can then be pulled out of the skin manually or mechanically. Unhairing of hides is also performed with lime and sodium sulphide and the process is carried out in process vessels such as drums, paddles, mixers, or pits. The duration of the process may vary from 8 hours (drum) to 7 days (pit).

Fleshing: The process is the mechanical scraping off of the organic material from hides and skins. Fleshing can be performed prior to soaking, after soaking, after liming or after pickling.

Splitting: The process is applied to regulate the thickness of hides and skins using mechanical band knifes. Splitting can be applied after liming or tanning, however more accurate splitting is carried out if it is applied in tanned condition, which is done in most cases.

10.2.2 Tanyard processes

The tanyard processes are the processes performed to permanently preserve the cleaned hides and skins and to obtain semi-finished products (wet-blue) that will be subject to post-tanning operations. The products of tanning are non-putrescible and therefore tradable materials (European Commission, 2003). Typical tanyard processes are, deliming, bating, pickling, and tanning. Degreasing may be also applied to sheep skins before or after pickling or after tanning.

Deliming: The process is applied to remove residual lime from the pelt in order to prevent interference with the tanning process. Deliming is applied through washing for gradual lowering of the pH (with the addition of weak acidic salts like ammonium chlorides and ammonium sulphate). Neutralization and removal of residual chemicals and degraded skin components are achieved. The process is generally carried out in a processing vessel such as a drum, mixer or paddle.

Bating: The process is conducted by the application of selective enzymes to remove globular and inter-fibrilliary proteins so that the grain of the pelts is improved.

Pickling: The process is carried out in order to reduce the pH (pH 2 to 3) of the pelts mostly with sodium chloride and sulphuric acid. Pickling sterilizes the pelts, ends the activity of bating enzymes and improves the penetration into the pelt prior to tanning. Pickled sheep skins are tradable materials if they are also treated with fungicides.

Degreasing: The process is applied to remove excess grease mostly from sheep and pig skins in order to prevent the formation of insoluble chrome-soaps during the tanning step. Degreasing can be done with organic solvents or non-ionic surfactants.

Tanning: Tanning is performed in order to stabilize the collagen fibre so that the hide would not go through putrefaction or rotting and the resistance to mechanical action and heat would increase. Stabilization is achieved by the cross-linking action of the natural or synthetic tanning agents.

Three main tanning operations or their combinations are applied to produce leather:

(i) mineral tanning (chrome tanning)
(ii) vegetable tanning
(iii) alternative tanning agents, such as synthetic tannins (syntans), aldehydes and oils.

Chromium and vegetable tanning agents are the most commonly used tanning agents.

10.2.3 Post-tanning operations

Post-tanning operations are composed of retanning and wet finishing processes which in the end produce leather with the desired physical and aesthetic properties (UNIDO, 1991). Leather produced after retanning and wet finishing processes is called "crust" and is marketable as an intermediate product (European Commission, 2003). Post-tanning operations typically include samming, setting, splitting, shaving, retanning, dyeing, fatliquoring and drying processes.

Draining, samming and setting: Tanning is followed by subsequent draining and washing. Samming is applied to reduce the moisture content prior to further processes like splitting and shaving. Setting is performed mechanically to stretch out the leather.

Shaving: The process is applied to achieve uniform thickness and shaving is applied as fine adjustment or when splitting is not possible.

Neutralization: The process is applied to tanned leather for obtaining the suitable pH value for subsequent processes of retanning, dyeing and fatliquoring.

Bleaching: The process may be applied to remove stains, or to reduce the colouring in the hair, wool, or leather prior to retanning and dyeing for tanned skins with wool or hair.

Retanning: The process is applied for improving the feel and handle, for filling looser and softer parts of the leather to produce even properties, for improving resistance to alkali and perspiration, for the production of corrected grain leathers and for improving the wetting back property.

Different tanning agents like vegetable tanning extracts, synthetic tannins, aldehydes, mineral tanning agents and resins can be used for the retanning.

Dyeing: The process is applied to produce colours over the whole surface of each hide and skin mostly with water-based acid dyes and less commonly with basic and reactive dyes.

Fatliquoring: Since leather should be lubricated after loosing the fat content during tanning processes, fatliquoring is applied using animal or vegetable oils or synthetic oils.

Drying: Various techniques are used for drying and thus optimizing the quality and area yield of the leather. Different techniques have different influences on the characteristics of the leather.

10.2.4 Finishing operations

The finishing operations are mechanical treatment and coating steps performed to obtain the final product depending on the area of use and specifications of the finished leather. Finishing operations involve mechanical processes and surface coating, like conditioning, staking, buffing, applying a finish, milling, plating and embossing. Finishing is applied to provide desired characteristics such as colour, gloss, handle, flex, adhesion, rub fastness, extensibility, break, light and perspiration fastness, water vapour permeability and water resistance.

10.3 SUB-CATEGORIZATION IN TANNERY INDUSTRY

Tanneries house a variety of operations depending on the raw material and the desired products. Table 10.1 shows the matrix of tannery operations categorized according to raw materials, types of leather manufacturing processes and types of finished products (European Commission, 2003).

The variety in the manufacturing processes performed in the tannery industry leads to a variety of wastewater streams with different pollution characteristics. It is therefore necessary to group different types of leather production with different effluent characteristics to be able to provide a sound wastewater management system for the tannery industry. In the "Development Document for Effluent Limitations Guidelines and Standard Leather Tanning and Finishings", US EPA (EPA, 1979) has suggested that the tannery industry should be investigated in 7 subcategories as given in Table 10.2, according to the manufacturing processes performed in different tanneries.

Table 10.1. Categorization of tannery operations (European Commission, 2003)

Raw material	Type of leather manufacture	Finished products
Cattle	Fellmongery	Shoe upper
Sheep	Raw to wet-blue	Shoe lining
Goat	Raw to crust	Sole leather
Pig	Raw to finished	Upholstery leather for furniture
Buffalo	Wet-blue to finished	Upholstery leather for automotive use
	Crust to finished	Clothing
		Protective clothing (fire resistant, water resistant)
		Fancy goods
		Gloving
		Bookbinding
		Chamois
		Saddlery
		Belting

Subcategorization of the tannery industry may differ in different countries depending on the raw material and the processes applied and also on the final product. An example of this is presented by Tunay et al. (1995), who have systematically determined six different sub-categories in terms of wastewater generation and pollution characteristics in the tannery industry, in Turkey. The six different categories identified in the study of Tunay et al. (1995) are presented in Table 10.3, together with the comparison of the subcategories defined by US EPA (1979).

As Table 10.3 suggests it is possible that some subcategories different than the ones defined by US EPA (1979) may also be present depending on the regional properties of the industry and it is also possible that not all of the manufacturing subcategories exist in all regions.

Table 10.2. Subcategorization of tannery industry in terms of manufacturing process schemes

Subcategory	Description
A	Hair pulp/chrome tan/retan-wet finish
B	Hair save/chrome tan/retan-wet finish
C	Nonchrome tan/retan-wet finish
D	No beamhouse
E	Retan-wet finish
F	Through-the-blue
G	Shearling

Table 10.3. Subcategorization of tannery industry in terms of wastewater characteristics

Subcategories in Turkey	Subcategories defined by US EPA (1979)
Cattle raw hide-finished chrome	(A) Hair pulp/chrome tan/retan-wet finish
Cattle raw hide- finished vegetable	–
Cattle wet blue-finished	(E) Retan-wet finish
Sheep raw to finished	–
Sheep pickle to finished	(D) No beamhouse
Sheep fur-suede	(G) Shearling
–	(B) Hair save/chrome tan/retan-wet finish
–	(C) Nonchrome tan/retan-wet finish
–	(F) Through-the-blue

Although different subcategories are defined based on raw materials, applied processes and final products most operations are similar in different subcategories. However, the presence or absence of some operations extremely affects both effluent characteristics and environmental impacts. If sheep fur-suede sub-category is considered for instance, sulphide in the effluents will not be a problem since very low amounts of sulphide will be applied because unhairing is not needed. On the contrary, the sub-categories with hair-destroying unhairing processes using lime and sulphide will produce effluents highly loaded with sulphides and organics due to dissolving of hair. In addition to that, if hair saving methods are applied during unhairing operations 50–60% sulphide reduction will be achieved. Thus, sub-categorization can be effectively used as a waste management tool although similar unit operations are present in most tanneries.

10.4 WASTEWATER GENERATION AND CHARACTERISTICS

The tannery industry uses high amounts of water and chemicals. Tannery effluents are characterized by high organic carbon, suspended solids, chrome and salt content. The amount of water used for processing 1 tone of raw hide is reported to range between 15–80 m^3 (UNIDO, 1991; European Commission, 2003). Approximately 250 kg of leather is produced per tone of raw hide together with 15–80 m^3 of effluent carrying nearly 230–250 kg COD, 100 kg BOD, 150 kg TSS, 5–6 kg chrome and 10 kg sulphide (European Commission, 2003). Table 10.4 shows the illustrative amounts of major chemicals used in the tanning process.

Tanning process requires tanning agents and auxiliaries. The specific agents and auxiliary chemicals used for the tanning step are presented in Table 10.5 according to the type of tannage applied.

Table 10.4. Major chemicals used in the tanning process (UNIDO, 1991)

Chemicals	Formula	Heavy leather	Light leather
		kg per 100 kg of raw hides	
General purpose chemicals			
Sodium sulphide	Na_2S	3.0	3.0
Calcium hydroxide	$Ca(OH)_2$	4.5	4.5
Hydrochloric acid (conc.)	HCl	0.3	0.3
Ammonium sulphate	$(NH4)_2SO_4$	2.0	2.0
Sodium bisulphate	$NaHSO_3$	1.5	1.5
Sodium chloride	NaCl	10.0	10.0
Calcium formate	$Ca(COOH)_2$		2.0
Sulphuric acid (conc. 96%)	H_2SO_4	4.0	4.0
Sodium carbonate	Na_2CO_3		2.0
Sodium sulphite	Na_2SO_3		2.0
Basic tanning materials			
Chrome salts	$Cr_2(SO4)_3$		10.0
Vegetable tanning materials		12.0	3.0
Performance chemicals			
Bates		0.8	0.8
Bactericides		0.3	0.3
Syntans			3.0
Fat liquors			4.0
		kg per 100 kg of shaved weight	
Dyeing auxiliaries			3.8
Dyes			0.6
Finishes			4.0

10.4.1 Process and pollution profiles

Tannery industry wastewaters are characterized by highly concentrated pollutants such as solids determined as TSS, VSS and TDS; organic carbon identified in terms of COD and BOD_5; nitrogen compounds expressed as TKN and NH_3-N; phosphorus containing compounds determined as TP or PO_4^--P; sulphides (S^{2-}); sulphates (SO_4^{2-}); chloride (Cl^-) and total chromium (Cr).

Different operations applied in the tanning process release different pollutants depending on the type of operation and the chemicals used. Figure 10.1 shows the general process and pollution profile for the tannery industry.

10.4.2 Conventional wastewater characterization

The pollution loads generated by each operation in the tanning process varies according to the chemicals used for the operation. The pollutants are generated

Table 10.5. Specific tanning agents and auxiliary chemicals used for tanning (European Commission, 2003)

Type of tannage	Tanning agents used	Auxiliaries used
Chrome tannage	Basic sulphate complex of trivalent chrome	Salt, basifying agents (magnesium oxide, sodium carbonate, or sodium bicarbonate), fungicides, masking agents (e.g. formic acid, sodium diphthalate, oxalic acid, sodium sulphite), fatliquors, syntans, resins
Other mineral tannages	Aluminium, zirconium, and titanium salts	Masking agents, basifying agents, fatliquors, salts, syntans, resins, etc.
Vegetable tannage	Polyphenolic compounds leached from vegetable material (e.g. quebracho, mimosa, oak)	Pre-tanning agents, bleaching and sequestering agents, fatliquors, formic acid, syntans, resins etc.

by proteinaceous matter, hair, tissue, blood, unfixed chemicals, tanning agents, extracts, dyes, pigments, dirt, grit and manure. The pollution loads for unit raw material in the main operations conducted in the tannery industry are presented in Table 10.6.

The wastewater characteristics of tannery effluents in terms of conventional pollution parameters presented in different studies is given in Table 10.7. The table shows that the character of tannery effluents may vary depending on the wastewater generating processes.

10.5 IN-PLANT CONTROL MEASURES FOR TANNERY INDUSTRY

Despite its highly polluting characteristics, the environmental impacts of tannery industries can be minimized by appropriate management and application of in-plant control measures. The major principles of European Council Integrated Pollution Prevention and Control Directive (IPPC, 96/61/EC) and US EPA Pollution Prevention Act of 1990 (EPA, 1990) include efficient raw material and energy use, optimum process chemical utilization, recovering and recycling of waste and substitution of harmful substances in manufacturing

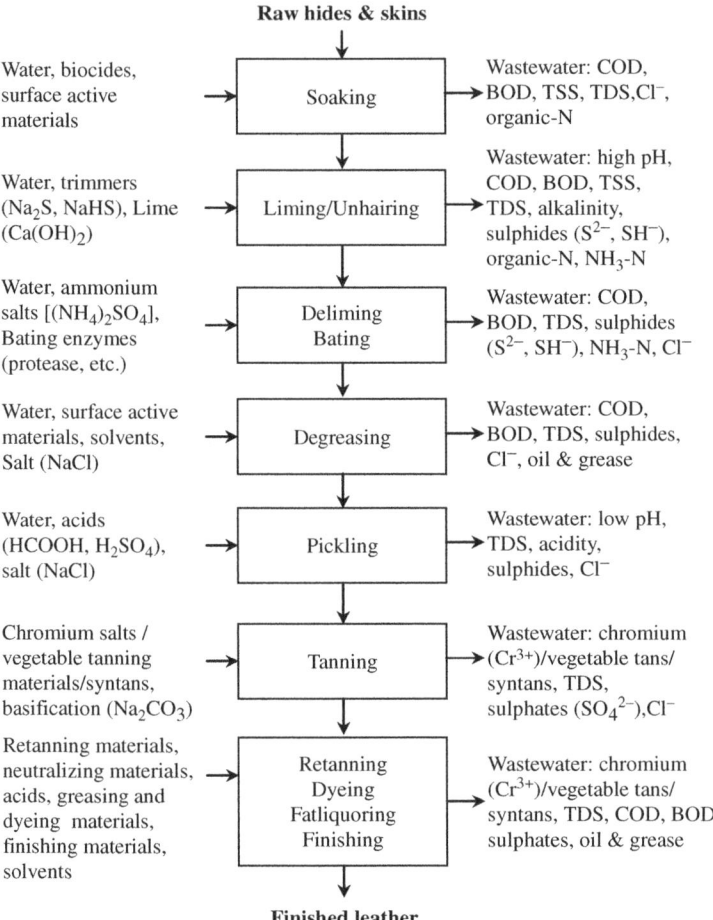

Figure 10.1. General process and pollution profile of tannery industry

processes that have significant impact on the environment, such as the tannery industry. It is possible to reduce the environmental impacts of tanneries mainly by:

(i) increasing efficiency of chemical utilization and substitution of harmful substances with environmentally friendly ones,
(ii) reducing water and energy consumption, and
(iii) recovery and reuse of water and chemicals.

Table 10.6. Average pollutant loads (kg/tones of hide/skin) (UNIDO, 1991)

Parameter	Soaking operations	Beamhouse operations	Tanyard processes	Post-tanning and wet finishing	Total pollution load
COD	23	94	10	27	154
BOD$_5$	12	44	3	12	71
TSS	28	65	9	6	108
TDS	160	85	130	44	419
Total Chromium (Cr)	–	–	5	1.5	6.5
Sulphides (S^{2-})	–	4	–	–	4
Sulphates (SO$_4^{2-}$)	8	9	45	22	84
Chlorides (Cl$^-$)	107	25	54	6	192

Table 10.7. Concentrations of conventional parameters in tannery effluents (mg/l)

COD	BOD$_5$	TKN	NH$_3$-N	TSS	S^{2-}	Total Cr	References
–	1000	250	–	2500	280	140	Huber and Jones, 1974
–	1323	180–445	–	1530	–	0.17	Nemerow et al., 1978
1320	440	–	–	710	–	47	Bishop, 1978
–	3734	–	–	4806	300	82	Huber and Doane, 1980
1500–4500	600–1600	–	–	1000–3000	100–300	20–80	Kalender, 1981
536	460	–	–	1630	224	–	Cheda et al., 1984
3600	3100	–	–	1500	–	–	Cheda et al., 1984
3087	930	–	–	952	200	–	Cheda et al., 1984
609	380	–	–	640	95	104	Cheda et al., 1984
5000	100	–	–	360	300	23.5	Kaul et al., 1993
4775–6400	–	993–1280	560–568	1232–2993	59–158	367–1329	Kabdasli et al., 1993
4230	1500	–	–	3270	134	61	Sengul and Gurel, 1993
10380–26100	–	–	–	–	20	192	Kabdasli et al., 1993
3250	1844	–	–	6840	49.8	209	Sengul and Gurel, 1993
3500–11200	2100–5400	24–106	–	780–2030	150	5–168	Talinli, 1994
7450–9635	–	1068–1180	900–906	–	10	10–110	Tunay et al., 1994
11145	–	–	–	–	278	76	Tunay et al., 1994
3235–7420	600–2600	112–640	48–245	1470–3474	17–110	58–213	Ates et al., 1997
2525	–	488	438	1593	–	0.23	Moussa et al., 2001

10.5.1 Material substitution and reuse

There is no efficient chemical that could be used for the substitution of sulphide in the unhairing/liming process. However, lime/sulphide application may be replaced by sulphide/caustic soda liquors or the volumes can be reduced. Reuse of spent lime/sulphide liquors after filtering may also be applied. Stripping sulphide after acidification as H_2S and collecting the gas in caustic soda to be reused in liming/unhairing applications is a more sophisticated way of recovery. In addition to sulphide reduction, hair save techniques can also be applied to remove the high COD, TKN and sulphide loads in the effluents.

When chrome tanning is applied the amount of chrome used can be minimized by high chrome exhaustion techniques, by recycling and by chrome recovery. High chrome exhaustion is applied by increasing the ratio of chrome fixation with short float, increasing temperature, increasing reaction time for tanning, increasing basification using dicarboxylic acids and their salts, and decreasing neutral salts. Recycling of used chrome liquors is possible to tannage or to pickle processes. Chrome recovery is applied after the precipitation of chrome with alkalis and the precipitated hydroxides can be reused for tanning. Precipitation and reuse of chrome is more cost efficient than high exhaustion techniques. Salinity of the effluents will also be reduced if chrome recovery is applied. There is a possibility that chromium used in the tanning process can be substituted with aluminium, titanium or zirconium if the market will be ready to accept slightly lower quality final products.

Halogenated organic compounds are used as biocides in the curing, soaking, pickling, tanning and post-tanning processes. It is possible to substitute halogenated organics with sodium or potassium salt of dimethyl-dithiocarbamate which is less toxic and less persistent in the aquatic environment.

Surface active reagents used in various processes such as soaking, liming, degreasing, tanning, dyeing, etc. are mostly composed of nonyl phenol etoxylates (NPEs), which can be replaced with alcohol ethoxylates for less fatty hides and skins. EDDS and MGDA can be applied instead of EDTA and NTA used as sequestering agents.

Deliming process carried out with ammonium salts can also be applied using carbon dioxide for a portion or all of ammonia using automated processes. Ammonium salts used in deliming can be replaced with weak organic acids, such as lactic, formic and acetic acid. The ammonia levels in the waste water will be reduced, but these agents increase the COD load.

The salinity of the effluents can be reduced by processing fresh (unsalted) hides. However, the hides should be processed within 8–12 hours after slaughter, if the hides are not to be refrigerated. The hides and skins would be preserved for 5–8 days with refrigeration. The manufacturing of unsalted fresh hides and skins with or without chilling can only be possible if an integrated system of slaughterhouses and tanneries are established.

10.5.2 Water use reduction

The amount of water used in the tannery industry can be reduced by several methods as given below. Although water reduction practices would also result in reduction of the amounts of some of the chemicals used, it should be noted that water reduction applications would generate more polluted effluents since the used water will be reduced more than the pollution loads, leading to more concentrated wastewaters.

Increasing volume control in the operations: A great amount of water is spent for running water washes, overflowing vessels, continuously running pipes and too frequent washing of drums and floors. Water could be saved with careful control measures, additional equipment like water meters and valves, and through the training of employers.

Using batch washes instead of running washes: Uncontrolled running washes cause extreme amounts of water loss. Preferring batch wash sequences not only saves water but also yields a more uniform product quality.

Low floatation: It is possible to reduce floatation ratios up to the point the equipment can tolerate. Replacement of paddles and pits with drums that would be appropriate for low floating would provide significant water savings.

Reuse of spent water: Wastewaters from relatively clean rinsing and washing operations can be reused in other less critical processes. Soaking process might be a good point of reusing lime washings. Many other recycling opportunities would be possible depending on the applied processes within tanning.

Recycling of individual liquors: Individual effluents can be collected separately and reused with the addition of spent chemicals in the same process. Reuse of lime/sulphide liquors and chrome tanning liquors are the most used recycling applications as explained above.

10.6 TREATMENT OF TANNERY EFFLUENTS

10.6.1 Treatment requirements

Tannery industry wastewaters include high amounts of organic and inorganic pollutants. Tanneries generally involve batch operations and a number of raw materials and chemicals which lead to complex effluents with variable characteristics depending on the applied processes. Tannery industry wastewaters need to be treated before they are discharged to sewers or receiving water bodies.

The extent of required treatment is set by the character of the wastewater streams generated within the industrial establishment and the discharge limits to the receiving body set by the authorities. Thus, each environmental authority would set the discharge limits dependent on the sub-categories of the tanneries present in their region, the tolerance of the receiving body for discharge and the appropriate treatment technologies. The achievable degree of treatment with appropriate technologies should be carefully determined while setting the discharge limits otherwise too stringent standards would not be compliable with possible appropriate treatment schemes.

10.6.2 Treatment schemes

Due to the variable characteristics of different tannery effluents it is not easy to generalize the required treatment schemes for tannery wastewaters. However, the wastewater treatment strategies employed can be simply described in four major steps such as;

(1) Physical treatment (Mechanical pre-treatment)
(2) Physico-chemical treatment
(3) Biological treatment
(4) Sludge handling

Tannery effluents are first subjected to mechanical pre-treatment operations such as screening, filtering, grit and grease removal, gravity settling, etc. The physico-chemical treatment steps are basically required for sulphur and chromium removal and they are mainly applied to segregated streams carrying most of the sulphide from beamhouse effluents and chrome from tanning and post-tanning operations. These operations include oxidation, precipitation, and flotation. The sulphide and chromium removal operations are followed by equalization, neutralization and settling prior to biological treatment.

The pre-treatment steps, i.e., primary physical treatment and physicochemical treatment steps are necessary for the treatment of tannery effluents since compounds such as Cr^{3+}, S^{2-} and other toxic or inhibitory chemicals that would adversely effect the following biological treatment steps are present. The pre-treatment is basically applied in two different ways:

(1) Effluents containing chromium and sulphide can be segregated and treated separately using oxidation, chemical precipitation, etc. These pre-treated streams are then combined with general wastewaters and undergo plain settling with little chemical polishing before biological treatment.

(2) A more straight forward way is the application of chemical settling for the generated wastewaters as a whole prior to biological treatment.

The biological treatment after pre-treatment involves carbon and nitrogen removal via the application of nitrification/denitrification processes. Biological oxidation of sulphides would also be possible in the aeration tank of the activated sludge systems.

Generally sludge from the primary settler and the sludge wasted from activated sludge process are combined and homogenized. Mechanical thickening and dewatering operations are applied and the dried sludge cakes are subject to final disposal.

Treatment of tannery wastewaters may also involve recycling of solvents that could be reused in manufacturing operations.

10.6.3 Pre-treatment applications

Tannery effluents need to be subjected to pre-treatment prior to biological treatment since the effluents possess highly toxic pollutants like chromium and sulphide as well as high amounts of nitrogen. Oxidation and chemical precipitation has been mostly applied as pre-treatment steps as given in the case studies presented below. Ozonation has recently been used as a new pre- or post-treatment alternative that would enhance biological treatment (Dogruel et al., 2004).

10.6.3.1 Case study-1

The presented case study was conducted on three different tannery effluents of organized industrial districts in Turkey, where chemical precipitation has been conducted using alum and iron salts, pH adjustment sustained with lime (Ates et al., 1997). Tuzla Organized Leather Tanning Industrial District housed 107 tanneries processing both cattle hide and sheepskin; Corlu Leather Tanning Industrial District housed 87 tanneries processing predominantly sheepskin; and Biga Tanneries District with 46 tanneries processing sheepskin. The results of pre-treatment with chemical precipitation presented in Table 10.8, have revealed that COD removal of up to 80% could be achieved and the resulting effluent practically had no suspended solids and chromium, presenting ideal wastewater characteristics for biological treatment.

10.6.3.2 Case study-2

Another case study presents the effect of pre-treatment on the wastewater characteristics of combined tannery wastewater streams where stream

Table 10.8. Pre-treatment of tannery effluents from different tannery districts (Ates et al., 1997)

Parameter	Organized leather district (Tuzla)		Organized leather district (Corlu)		Organized leather district (Biga)	
	Raw wastewater	Chemical precipitation effluent	Raw wastewater	Chemical precipitation effluent	Raw wastewater	Chemical precipitation effluent
COD (mg/l) TOTAL	5756	1050	4705	1108	4180	1120
FILTERED	1170	ND	1600	ND	1495	ND
TKN (mg/l)	363	208	ND	ND	250	175
TSS (mg/l)	2640	248	2300	128	2070	205
pH	8.1		8.4		8.3	
Sulphide (mg S^{2-}/l)	78	24	42	10.6	68	16
Total Cr (mg Cr/l)	42	<0.5	167	1.9	65	<0.5

ND: Not Determined

segregation was not possible (Kabdasli et al., 1999). The study was conducted with the effluents generated in a leather tanning district (Biga, Turkey) which houses 46 tanneries mostly processing sheepskin from raw hide to finished leather.

The pre-treatment scheme was determined as sulphide oxidation followed by chromium precipitation due to the high pH of the original wastewater. Sulphide was oxidized by catalytic air oxidation process at pH > 9 (sustained with lime addition where necessary) and using $MnSO_4$ as catalyst. Chemical precipitation of chromium salts were investigated with $FeCl_3$, $FeSO_4$ and $Al_2(SO_4)_3$ both on raw and on sulphide oxidized wastewaters.

Table 10.9 shows the experimental results obtained after sulphide oxidation and chromium precipitation, together with the conventional characterization of the two samples taken at different times. The results show that 57–75% sulphide oxidation was achieved and the oxidation also removed some of the organic carbon and nitrogen fractions. Chemical precipitation applied alone and after sulphide oxidation proved that, chromium could be removed totally with all coagulants and additional COD and TKN removals are also achieved. However, the results have shown that application of chemical precipitation after sulphide oxidation proved to be more efficient in terms of organic carbon and nitrogen removal with a smaller amount of chemical sludge production. Thus the application of a wise pre-treatment strategy plays an important role in the treatment of tannery effluents.

10.7 BIODEGRADABILITY CHARACTERISTICS

10.7.1 Biodegradability based characterization

Tannery effluents are highly complex wastewaters that should be characterized in terms of contemporary approach for biodegradability. In this sense it is necessary to characterize tannery effluents to identify the COD fractionations. Research efforts have been conducted to understand the biodegradable and inert fractions of COD present in tannery wastewaters. The COD fractions however are determined by the pre-treatment operations applied to tannery effluents.

Pre-treatment not only reduces pollutants like chromium and sulphide together with organic carbon and nitrogen fractions but also changes the COD fractionation and the biodegradation characteristics as shown by Orhon et al. (1999c). Table 10.10 presents the effect of chemical precipitation on the conventional pollutants present in tannery effluents. The analytical data in Table 10.10 shows that the effluent has a strong and complex character, with high COD and TKN amounts and very low phosphorus.

Table 10.9. Different alternatives of pre-treatment for tannery effluents (Kabdasli et al., 1999)

| | Untreated wastewater | | Removal efficiency | | | | | |
| | | | Sulphide oxidation | | Chemical precipitation | | Chemical precipitation after sulphide oxidation | |
Parameter	Sample I	Sample II	Sample I	Sample II	Sample I	Sample II	Sample I	Sample II
COD (mg/l) TOTAL	5235	4480	6%	33%	55–68%	55–71%	78–82%	69–74%
SETTLED		3200		18%				
FILTERED	2470	1830		18%				
TKN (mg/l)	325	295	43%	15%		7–36%	59–65%	31–48%
NH_3-N (mgN/l)	185	135		22%				
TSS (mg/l)	2275	2070						
pH	8.35	9.74						
Oil and Grease	430	550		57%				
Sulphide (mg S^{2-}/l)	100	56	75%		>99%	>99%	>99%	>99%
Total Cr (mg Cr/l)	60	65						

Table 10.10. Effect of chemical pre-treatment on conventional wastewater characterization (Orhon et al., 1999c)

Parameter		Raw tannery Effluent	Plain settled tannery effluent	Chemically settled tannery effluent
COD (mg/l)	TOTAL	4180	2300	1100
	FILTERED	1495	1300	1100
TKN (mg/l)	TOTAL	250	215	197
	FILTERED		180	175
NH_3-N (mgN/l)		160	164	160
TP (mgP/l)			5.9	1.8
TSS (mg/l)		2070	768	157
VSS (mg/l)			467	55
Alkalinity (mg $CaCO_3$/l)			1417	1140
pH			7.9	8.4
Sulphide (mg S^{2-}/l)			37	10
Total Cr (mg Cr/l)			40	<0.5

Different COD fractions present in plain and chemically settled tannery effluents have been identified as given in Table 10.11. Chemical pre-treatment strongly effects the distribution of COD fractions present in the effluents, while decreasing the total organic load. Chemical pre-treatment removes the particulate COD fractions and increases the ratio of the soluble biodegradable COD fractions in the tannery wastewater, hence enhancing the biodegradability of the wastewater. Chemically treated effluents have a totally different composition in terms of COD fraction with 16% of soluble inert, 35% of readily biodegradable and 49% of rapidly hydrolysable and almost no particulate fractions (Figure 10.2).

10.7.2 Activated sludge modelling for tannery effluents

Process kinetics and stoichiometry for tannery wastewaters should be evaluated by using a multi-component model based on the concept of endogenous decay, including initial soluble inert matter and residual microbial products as separate model components (Orhon and Artan, 1994).

The stoichiometric coefficients for microbial product generation obtained for tannery effluents in different studies are given in Table 10.12. All of the stoichiometric coefficients involved in the modelling of tannery effluents, except for the heterotrophic yield coefficient, Y_H, are estimated in the inert COD experiments conducted as a part of COD fractionation studies.

The stoichiometric coefficients shown in Table 10.12 indicate that activated sludge treatment of tannery effluents may generate high amounts of particulate

Table 10.11. COD fractionation of tannery effluents

Wastewater	C_{T1} (mg/l)	C_{S1} (mg/l)	C_{I1} (mg/l)	S_{T1} (mg/l)	X_{T1} (mg/l)	S_{S1} (mg/l)	S_{H1} (mg/l)	X_{S1} (mg/l)	X_{I1} (mg/l)	S_{I1} (mg/l)	References
Plain settled effluent	1500									323	Kabdasli et al., 1994
Plain settled effluent	2050	1525	525	1125	925				285	240	Orhon et al., 1999a
Plain settled effluent	1980	1500	480	1345	635			1170			Orhon et al., 1999b
Plain settled effluent	2410	2005	405	1210	1200	330	770	935	265	140	Orhon et al., 1999c
Plain settled effluent	2300	1347	485	1300	1000	300	650	730	270	215	Orhon et al., 1999c
Plain settled effluent	9385	8590	795	5170	4215	435	4245	3490	725	70	Tunay et al., 2004
Chemically settled effluent	1100	915	175	1100	–	855	540	–	–	175	Orhon et al., 1999b
MAP precipitated effluent	5130	5065	65	5130	–	385	2725	–	–	65	Tunay et al., 2004

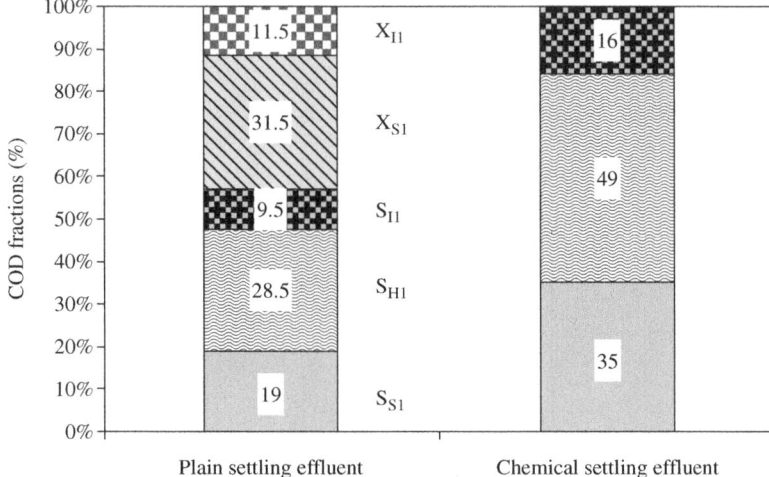

Figure 10.2. Effect of chemical pre-treatment on the COD fractions of tannery effluents (Orhon *et al.*, 1999c)

residual microbial products up to 25% of the biodegradable COD that have to be disposed with excess sludge.

Although the soluble microbial products generated are around 5% of the total biodegradable COD, at a comparable level to that of domestic sewage, the effluent microbial products concentration (S_P) will be high due to the high concentrations of biodegradable COD (C_{S1}) present in the wastewater. Thus, the achievable effluent COD concentrations with biological treatment will be challenging since the discharge concentration will be set by the concentrations of soluble inert COD (S_{I1}), present in the influent and the soluble residual products (S_P) generated. It is evident from the S_{I1} concentrations around 200 mg COD/l and the S_P concentrations that will be generated around 50–100 mg COD/l, as presented in Table 10.12 that tannery effluents can not comply with the effluent COD limits of 150–250 mg COD/l after biological treatement. Therefore it is crucial that the effluent standards should be set in accordance with the achievable discharge concentrations by the best appropriate techniques (BAT) which is defined as biological treatment for the case of tannery industry wastewaters.

The kinetic values obtained by experimental studies (Orhon *et al.*, 1999c) are presented in Table 10.13. The values of the kinetic coefficients show that biological degradation of tannery effluents are harder than that of domestic sewage since lower growth and hydrolysis rates are obtained for plain settled wastewaters.

Table 10.12. Stoichiometric coefficients for metabolic product generation for tannery effluents

Wastewater	Y_H	Y_{SP}	Y_{XP}	f_{ES}	f_{EX}	References
		(g COD/g COD)				
Plain settled effluent	0.04					Kabdasli et al., 1994
Plain settled effluent	0.64	0.07	0.25	0.11	0.40	Orhon et al., 1999a
Plain settled effluent (filtered through 0.45 μm)	0.64	0.03		0.05		Orhon et al., 1999a
Plain settled effluent	0.06	0.06		0.10	0.20	Orhon et al., 1999c
Chemically settled effluent	0.06	0.06		0.10	0.20	Orhon et al., 1999b
Plain settled effluent	0.64	0.047	0.10	0.073	0.16	Tunay et al., 2004
MAP precipitated effluent	0.71	0.046	0.085	0.064	0.12	Tunay et al., 2004

Pre-treatment studies with alum, $FeCl_3$ and $FeSO_4$ have shown that (Table 10.13), chemical precipitation enhances biodegradability as implied by the higher growth and hydrolysis rates achieved with chemically treated effluents (Orhon et al., 1999c). Magnesium ammonium phosphate (MAP) has also been examined as a new pre-treatment method that could be used in the chemical precipitation of tannery wastewaters. The study conducted by Tunay et al. (2004) has shown that formation of MAP actively removed ammonia present in the effluents and the precipitation of MAP flocs adsorbed the particulate portion of the organic matter. MAP precipitation is reported to be very effective for the enhancement of biodegradability (Tunay et al., 2004).

Table 10.13. Kinetic coefficients for tannery effluents

Wastewater	Y_H (g cell COD/ g COD)	$\hat{\mu}_H$ (day^{-1})	b_H (day^{-1})	K_S (mg/l)	k_h (day^{-1})	K_X (day^{-1})	References
Plain settled effluent	0.64	2.2	0.12	28	0.9	0.3	Orhon et al., 1999c
Chemically treated effluent	0.64	4.3	0.12	26	2	0.08	Orhon et al., 1999c
Plain settled effluent	0.64	2.6	0.20	5	1.6	0.03	Tunay et al., 2004
MAP Precipitated effluent	0.71	2.8	0.20	6	1.8	0.03	Tunay et al., 2004

Model simulation is a useful tool both for the assessment of the kinetic constants and the prediction of process performance. Modelling with respirometric methods generates reliable and consistent kinetic information. The respirometric studies reveal results which are independent of the hydraulic conditions of the system and sludge age. This feature is extremely important for the slowly biodegradable COD that needs to be taken into account in the design of biological treatment for strong wastewaters such as tannery effluents to meet stringent effluent COD limitations.

Modelling studies with tannery effluents have shown that a single-hydrolysis mechanism does not satisfactorily simulate the respirometric behaviour obtained in the experimental studies. In this respect, a dual hydrolysis mechanism containing both soluble and particulate slowly biodegradable COD as separate model components provided better results than single-hydrolysis mechanism in the interpretation for the simulations of respirometric response of tannery wastewaters (Orhon *et al.*, 1999b). A similar study conducted on the effluents of a tannery industry manufacturing leather for shoe-production has also shown that dual hydrolysis was able to predict the hydrolysis kinetics markedly better (Rojas *et al.*, 2001). Table 10.14 presents the kinetic coefficients estimated for the hydrolysis processes. Incorporation of a two-stage hydrolysis mechanism where soluble and particulate slowly biodegradable COD fractions are separately accounted, provides substantial improvement in the model simulation results.

Chemical precipitation removes the particulate slowly biodegradable COD together with the particulate inerts. Thus, chemical precipitation increases the ratio of biodegradable organic fraction as given in Figure 10.2. Chemical precipitation also decreases inhibition exerted on biochemical processes

Table 10.14. Dual hydrolysis kinetics for tannery effluents

	Dual hydrolysis model				Single hydrolysis model		
	Rapidly hydrolysable COD, S_{H1}		Slowly hydrolysable COD, X_{S1}				
Wastewater	k_{hs} (day^{-1})	K_{XS} (g COD/ g COD)	k_{hx} (day^{-1})	K_{XX} (g COD/ g COD)	k_h (day^{-1})	K_X (g COD/ g COD)	References
Plain settled effluent	1.1	0.2	0.3	0.2	0.9	0.2	Orhon *et al.*, 1999b
Chemically settled effluent					1.1	0.2	Orhon *et al.*, 1999b
Plain settled effluent	1.1	0.2	0.3	0.2			Rojas *et al.*, 2001

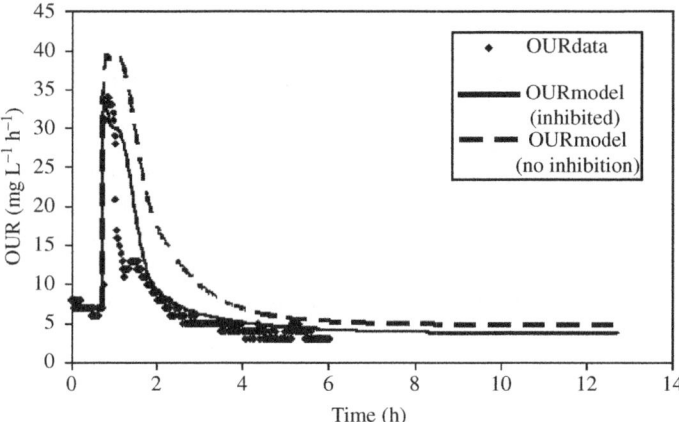

Figure 10.3. Effect of inhibition imposed by the particulate fraction of tannery effluents on the OUR response of the activated sludge system (Karahan *etal.*, 2008)

occurring in the activated sludge system. The presence of trivalent chromium and sulphide are the major pollutants imposing inhibitory effects on microbial growth and since most of the trivalent chromium present in tannery effluents are in particulate forms, the removal of particulate material by chemical precipitation enhances the biodegradation characteristics of the wastewaters. It is also observed that the rate of hydrolysis is improved by the removal of the more slowly biodegradable particulate COD (Orhon *et al.*, 1999b).

In the experimental study conducted by Karahan *et al.* (2008) the particle size distribution of tannery effluents yielded that 60% of the total COD was at the particulate range, 25% was at the soluble range and the remaining 15% was well distributed among the colloidal range. The respirometric experiments conducted with samples filtered through different size ranges between 2 and 1600 nm could be calibrated with the same set of stoichiometric and kinetic coefficients and the extent of the biodegradability of different particle size distribution of all major COD fractions were obtained. The results have revealed that the majority of S_H was in the soluble range (<2 nm), while S_I was well distributed among the colloidal range and therefore removable by physical entrapment and adsorption.

The respirometric data obtained in the study of Karahan *et al.* (2008) confirmed that the biodegradation of X_S was hindered for plain settled wastewater samples and a significant portion of that COD fraction appeared to be inert. Model calibration results also revealed rapid hydrolysis and direct growth rates being 10% lower than the values obtained for the filtered portions. These results showed that the particulate portion of the tannery effluent imposed inhibitory effects on activated sludge and the inhibition was observed as a decrease in the amount of biodegradable COD (C_S) and decrease in the rates of hydrolysis, growth and decay related processes, indicating a non-competitive inhibition where a portion of the available enzymes is irreversibly blocked by the inhibitor. This inhibition, among other possible complex organic compounds, has been stated to be mostly attributed to the presence of trivalent chromium in particulate form, probably entrapped in the particulate organic matter. The inhibitory effects imposed on the biochemical reactions is shown in Figure 10.3, by modelling the respirometric response for the inhibited real case and for the non-inhibitory conditions, with the set of parameters obtained for the filtered effluents.

REFERENCES

Ates, E., Orhon, D. and Tunay, O. (1997) Characterization of tannery wastewaters for pre-treatment — Selected case studies. *Water Sci. Technol.* **36**(2–3), 217–223.

Bishop, P.L. (1978) Physicochemical Pre-treatment of Wastes from a secondary tannery. *Proceedings of the 33rd Industrial Waste Conference.* 64–72, 9–11 May, Purdue University.

Cheda, P.V., Mandlekar, U.V., Handa, B.K. and Khanna, P. (1984) Joint wastewater management for a cluster of Tanneries at Kanpur. *Proceedings of the 39th Industrial Waste Conference.* Purdue University.

Dogruel S., Genceli E.A., Germirli Babuna F. and Orhon, D. (2004) Ozonation of nonbiodegradable organics in tannery wastewater. *J. Environ. Sci. Heal. A.* **39**(7), 1705–1715.

EPA (1979) *Development Document for Effluent Limitations Guidelines and Standard Leather Tanning and Finishings.* United States Environmental Protection Agency, Office of Water and Waste Management, Effluent Guidelines Division, Washington, DC. EPA-44011-79-016.

EPA (1990) US EPA Pollution Prevention Act of 1990.

European Commission (2003) *Integrated Pollution Prevention and Control (IPPC) Reference Document on Best Available Techniques for the Tanning of Hides and Skins.* Brussels, Belgium.

Huber, C.V. and Doane, T. (1980) A case study — Tannery meets EPA Pre-treatment Standards. *Proceedings of the 35th Industrial Waste Conference.* 7–9 May, Purdue University.

Huber, C.V. and Jones, G.E. (1974) Combined treatment of leather tanning and municipal waste at Grand Haven, Michigan. *Proceedings of the 29th Industrial Waste Conference.* 7–9 May, Purdue University.

IPPC, (1996) European Council Directive 96/61/EC, *Integrated Pollution Prevention and Control.*

Kabdasli, I., Tunay, O. and Orhon, D. (1993) The treatability of chromium tannery wastewaters. *Water Sci. Technol.* **28**(2), 97–105.

Kabdasli, I., Tunay, O. and Orhon, D. (1994) The treatability of chromium tannery wastes. *Water Sci. Technol.* **28**(2), 97–106.

Kabdasli, I., Tunay, O. and Orhon, D. (1999) Wastewater control and management in a leather tanning district. *Water Sci. Technol.* **40**(1), 261–267.

Kalender, A. (1981) A study on the kinetics of activated sludge treatment of tannery industry wastewaters. Ph.D. Thesis, Istanbul Technical University, Turkey. (*in Turkish*)

Kaul, S.N. and Szpyrkowicz, L. (1996) Biological processes as a pre-treatment for electrochemical oxidation — A case study for Clusterof Tanneries from a common effluent treatment plant. *2nd Specialized Conference on Pre-treatment of Industrial Wastewaters.* October 16–18, Athens, Greece.

Karahan, O., Dogruel, S., Dulekgurgen, E. and Orhon, D., (2008) COD fractionation of tannery wastewaters — particle size distribution, biodegradability and modelling. *Water Res.* **42**(4–5), 1083–1092.

Moussa, M.S., Rojas A.R., Hooijmans, C.M., Van Loosdrecht, M.C.M. and Gijzen, H.J. (2002) Model-based evaluation of the upgrading of a full-scale industrial wastewater treatment plant. *Proceedings of The 2nd Biennial Conference on Management of Wastewaters.* 295–304, 15–17 April. 2, Edinburgh, Scotland.

Nemerow, N.L., Warne, D. and Falk, F. (1978) A new and effective solution for treatment of tannery wastewaters. *Proceedings of the 33rd Industrial Waste Conference.* 596–603. 9–11 May, Purdue University.

Orhon, D. and Artan, N. (1994) *Modelling of Activated Sludge Systems.* Technomic Press, Lancaster, PA.

Orhon, D., Karahan, O. and Sozen, S. (1999a). The effect of residual microbial products on the experimental assessment of the particulate inert cod in wastewaters. *Water Res.* **33**(14), 3191–3203.

Orhon, D., Ubay Cokgor, E. and Sozen, S. (1999b) Experimental basis for the hydrolysis of slowly biodegradable substrate in different wastewaters. *Water Sci. Technol.* **39**(1), 87–95.

Orhon, D., Ates Genceli, E. and Ubay Cokgor, E. (1999c) Characterization and modelling of activated sludge for tannery wastewater. *Water Environ. Res.* **71**(1), 50–63.

Rojas, A.R. (2001) Computer modelling of an activated sludge wastewater treatment plant treating tannery wastewater. M.Sc. thesis, Delft University, The Netherlands.

Sengul, F. and Gurel, O. (1993) Pollution profile of leather industries waste characterization and pre-treatment of pollutants. *Water Sci. Technol.* **28**(2), 87–97.

Talinli, I. (1994) Pre-treatment of tannery wastewaters. *Water Sci. Technol.* **29**(9), 175–178.

Tunay, O., Orhon, D. and Kabdasli, I. (1994) Pre-treatment requirements for leather tanning industry wastewaters. *Water Sci. Technol.* **29**(9), 121–128.

Tunay, O., Kabdasli, I., Orhon, D. and Ates, E. (1995) Characterization and pollution profile of leather tanning industry in Turkey. *Water Sci. Technol.* **32**(12), 1–9.

Tunay, O., Zengin, G.E., Kabdasli, I. and Karahan, O. (2004) Performance of magnesium ammonium phosphate precipitation and its effect on biological treatability of leather tanning industry wastewaters. *J. Environ. Sci. Heal. A.* **39**(7), 1891–1902.

UNIDO (1991) *Tanneries and the environment: a technical guide to reducing the environmental impact of tannery operations.* Authors: UNEP, IEO, UNIDO. Technical report series / United Nations Environment Programme, Industry and Environment Office; no. 4. ISBN 92-807-1276-4.

Index

A
Acinetobacter 98
acrylic 311
acute toxicity tests 175
aerobic conditions 166
aerobic respiration test 106–109
aerobic sludge age
  EBPR 250
  nitrification 220–221
alkalinity
  buffer capacity 3
  nitrification 223
  textile wastewater 336
alum 337
aluminium 357
ammonia
  oxidation 16
  pH 313–314
  pollution 313–314
anabolic reactions 12–13
anaerobic conditions 166–167

anaerobic sludge age 250
anoxic respiration 89
anoxic volume 89
ASM1
  death regeneration 65
  organic carbon removal 65
  OUR profile 70, 110, 141–142
ASM2 100
ASM3
  microbial products 68
  NUR profile 111
  OUR profile 70, 111
assessment
  coefficients 126–135
  inhibition 135–140
  models 140–146
  respirometry 104–112
  toxicity 135–140
ATP 14
autotrophs 15–16
  biomass 222

nitrification 222–223
nitrogen removal 79
oxygen demand 222–223
auxilliary chemicals 171–173
  activated sludge inhibition 175
  acute toxicity tests 175

B
batch flow 258
batch reactors 39, 45
batting 351
beamhouse operations 155, 350–351
biochemical oxygen demand *see* BOD
biocides 171–175
biodegradation
  biological treatment 7
  characteristics 112–126
  COD hydrolysis 94–95
  coefficients 126–135
  denitrification 94–95
  fractions 124–126
  inert fraction assessment 113–124
  inhibition 135–140
  models 140–146
  respirometry 104–112
  toxicity 135–140
biological nitrogen removal 219–247
  dual anoxic phases 284–286
  intermittent aeration 286
  nitrogen balance 281–283
  pre-denitrification 284
  process options 283–286
  SBR technology 280–296
  step feeding 284
  see also nitrogen removal
biological phosphorus removal 248–253
  SBR technology 296–299
  see also phosphorus removal
biological reactors
  fill-and-draw 258
biomass 16, 18
  active 19
  assessment 18–20
  BOD_5 27–29
  COD 30–31, 40, 42
  heterotrophic growth 49
  VSS 26–29, 31–35, 42–43

biosynthesis 17
bleaching 352
BOD 20–21
  biomass 27–29
  model 41–45
  pollution 157–158, 161–164
  substrate 27–29, 32–35
  tannery effluent 360
buffer capacity 3
  alkalinity 3
  pH 3
butyl benzoate 171

C
carbon
  total organic carbon (TOC) 20
  see also organic carbon
carbon sources 15
  external 234
catabolic reactions 12
chemicals
  auxilliary 171–173, 175
  biological treatment 6–7
  OUR profile 71
  pre-treatment 320, 368, 370
  substitution 152
  tanneries 356–357
  textile wastewater 337
chlorine 321
chrome 357
chromium 135, 139
  EC_{50} 139
  tanneries 360
clean technologies 2
COD 3–4, 19–22
  biomass 30–31
  characterization 3
  COD/P ratio 249, 297
  COD/TKN ratio 249
  pollution 157–158, 161–165, 313–321
  removal and nitrification 225
  size distribution 211
  substrate 30–32
  sugar recovery 171
  tannery effluent 360
  textile wastewater 336

Index 379

COD fractionation 212
  continuous flow process 212–213
  domestic sewage 212
  polyamide carpet finishing wastewater 73
  pre-treatment 210–211, 320, 370
  tannery wastewater 369–370
  textile wastewater 322
COD fractions 5, 168–169
  confectionery wastewater 169
  corn wet mill wastewater 169
  dairy processing wastewater 168
  drink manufacturing wastewater 168
  meat processing wastewater 169
  textile wastewaters 168, 339
  total biodegradable C_{S1} 46–47
  total C_{T1} 46
  total inert C_{I1} 46
  total inert particulate X_{I1} 46
  total inert soluble S_{I1} 46
colour
  pollution 321
  textile wastewater 336
complex industrial waste models 67–71
conceptual design 211–219
  continuous flow technology 211–219
  process 211–219
conductivity 336
confectionery industry
  endogenous decay coefficient 57
  heterotrophic growth 51
  hydrolysis 53
  kinetic coefficients 52
  pollution parameters 164
  wastewater 160, 168–169
continuous filling 286
continuous flow technology 181–208
  functions 186–209
  organic carbon removal 209–219
  parameters 182–186
  reactors 39
control volume 44
corn wet mill wastewater
  characterization 160
  COD fractions 168–169
  endogenous decay coefficient 57
  hydrolysis 53

cotton 311–312
cycle frequency 261–263

D
dairy industry wastewater
  characterization 160
  endogenous decay coefficient 57
  heterotrophic growth 51
  hydrolysis 53
  kinetic coefficients 52
  pollution 164
Daphnia magna 174–175
death regeneration models 64–65
decay 49
degreasing 352
deliming 351
denitrification
  anoxic respiration 89
  anoxic volume 89
  biodegradable COD hydrolysis 94–95
  denitrifier decay 95–96
  denitrifier growth 93–94
  models 89–96
  nitrate dissimilation 89–90
  nitrate reductase 89–90
  nitrogen removal 236–237
  nitrogen utilization rate (NUR) 96
  pre-denitrification 233
  process kinetics 96–97
  stoichiometry 90–93, 97
  yield 94
design algorithms
  nitrification 224
  nutrient removal 287–288
  organic carbon removal 214, 224, 276
  SBR 276, 287–288
direct discharge standards 2
dissolved oxygen, S_O 40, 107
dodecyl benzene 171
domestic sewage 218
  COD fractionation 212
  COD fractions 168–169
  design 218
  dual hydrolysis kinetics 53–54, 136
  endogenous decay coefficient 54, 57

heterotrophic growth kinetics 50
hydrolysis 53
OUR 136
pollution 164
soluble inert microbial product 60
dual anoxic phases
  biological nitrogen removal 284–286
  SBR operation 285
dual anoxic zones, nitrogen removal 234
dual hydrolysis
  domestic sewage 54, 136
  kinetics 54–55, 371
  OUR 136
  poultry wastewater 54
  slaughterhouse wastewater 54
  tannery wastewater 54, 371
dyeing 352

E
Enhanced Biological Phosphorus Removal, EBPR 16, 96–101, 248–253
  acetate 297
  aerobic sludge age 250
  anaerobic sludge age 250
  carbon sources 297
  COD/P ratio 249, 297
  COD/TKN ratio 249
  nitrogen removal 251–253, 298–299
  PAOs 250–251, 297–298
  phosphorus removal 298–299
  process kinetics 100
economic concepts 2
effluent
  BOD 360
  chromium 360
  COD 171, 360
  discharge limitations 3
  residual 166
  sugar recovery 171
  tanneries 360, 362–366, 365, 367
  TKN 360
  treatment 362–366
  TSS 360
electron acceptors 14–15
electron transport chain 14

end-of-pipe effluent
  inspections 2
  textile wastewater 309–310, 338
  treatment 6–7, 152, 309–310
endogenous decay 18, 49
  model 62–64
  OUR 133
  rate 133
endogenous decay coefficient 54
  assessment 130–133
  confectionary industry wastewater 57
  corn wet mill wastewater 57
  dairy industry wastewater 57
  domestic sewage 54, 57
  meat processing wastewater 57
  tannery wastewater 57
  textile industry wastewater 57
endogenous respiration 17, 40, 55–57
energy drink manufacturing wastewater 170
enhanced biological phosphorus removal see EBPR
environmental factors 250–251
Escherichia coli 144
ether 171

F
fat liquoring 352
feasibility analysis
  in-plant control 332–335
  textile wastewater 339–345
fermentable substrate 98
fermented substrate 98
fill-and-draw 258
finishing operations 352–353
fleshing 351
front-of-pipe waste prevention 152

G
glucose 143–144
glycogen 143
gravity settling 8
growth 17
  heterotrophic 50–52
  microbial 40
  objective function 145

Index

H
hazardous waste 5–7
heterotrophic growth 50–52
  biomass 18, 49, 183–185
  coefficient assessment 133–134
  confectionary 51
  dairy 51
  domestic sewage 50
  kinetics 50–51
  meat processing 52
  pharmaceutical 52
  tannery 51
  textiles 51
  yield 18, 126–127, 183–185
heterotrophs 15–16
  energy utilization 17
  nitrogen removal 79
  substrate utilization 17
hexavalent chromium 139
hydraulic retention time 186
  SBR 263–264
hydrolysis 49, 52–55
  confectionary industry wastewater 53
  corn wet mill wastewater 53
  dairy integrated processing 53
  domestic sewage 53
  kinetics 53, 318
  meat processing 53
  objective function 145
  pharmaceutical 53
  tannery wastewater 53
  textile wastewater 53, 318

I
in-plant control 2
  effluent 325–332
  feasibility analysis 332–335
  material reclamation 167–171
  process modifications 308
  processes 325–332
  substitution of chemicals 307
  tannery wastewater 357–362
  textile wastewater 305–309
  wastewater reuse 329–332
  water conservation 329
inert fraction assessment 122–124

inhibition
  auxilliary chemicals 175
  biodegradation 135–140
integrated dairy processing wastewater 168
intermittent aeration 286
isobutanol 171

K
kinetic coefficients 213
  confectionary industry wastewater 52
  dairy industry wastewater 52
  meat processing wastewater 52
  pharmaceutical industry wastewater 52
  tannery wastewater 52, 371
  textile industry wastewater 52
kinetics
  hydrolysis 318
  reaction kinetics 44
  textile wastewater 318

L
lignin sulphonate 172–175
liming 351
liquor 19

M
management
  resources 167–176
  textile wastewater 303–345
  tools 153–167
mass balance 44–45
materials
  reclamation 167–171
  resource management 167–171
  reuse 152, 167–171
matrix modelling
  endogenous decay 62
  organic carbon removal 39–43
meat processing wastewater 160
  COD fractions 168–169
  endogenous decay coefficient 57
  heterotrophic growth 52
  hydrolysis 53
  kinetic coefficients 52
  pollution parameters 164

mechanistic models
  chemical industry effluent 72
  particulate COD 49
  soluble COD 49
membrane bioreactors 8
membrane filtration 8
metabolic functions 16
metabolic products 371
microbes
  ASM3 68
  biosynthesis 12
  energy 12, 17
  endergonic and exergonic reactions 12
  growth 17, 40
  metabolism 11
  nutrition 14–16
  PAOs 80
  ratio 185–186
  residual products 58–61
  substrate utilization 17
  yield 17
micropollutants 5–7
mills 335
  see also textile wastewater
mixed liquor
  dissolved oxygen 107
  OUR 109
  suspended solids 19
models
  biodegradation 126–135, 140–146
  coefficients 126–135
  complex industrial waste 67–71
  components 40
  death regeneration concept 64–65
  denitrification 89–96
  endogenous decay model 62–64
  mechanistic 38–39
  nitrification 81–89
  organic carbon removal 39–45, 61–71
  physical 38
  substrate storage concept 65–67

N
naphthalene sulphonate 171, 173–175
net yield coefficient 18
neutralization 352
nickel 135, 138–139

nitrate
  denitrification 89–90
  reductase 89–90
  utilization rate (NUR) 105
nitrification
  aerobic sludge age 220–221
  algorithm 224
  alkalinity consumption 223
  autotrophic biomass 222
  autotrophic oxygen demand 222–223
  capacity 221–222
  COD removal 225
  denitrification, models 89–96
  design 219–232, 223–232, 231
  models 81–89
  pre-denitrification 233
  process kinetics 88
  process stoichiometry 88
  specific nitrogen removal 221–222
nitrifiers 15
nitrogen
  organic carbon removal 205
  reaction kinetics 205
  total Kjeldahl nitrogen see TKN
  utilization rate (NUR) 96, 111
nitrogen removal 16, 232–249
  autotrophs 79
  balance 281–283
  biological 219–247
  denitrification potential 236–237
  design 232–249
  dual anoxic zones 234
  heterotrophs 79
  oxygen requirement 237
  and phosphorus removal 298–299
  pre-denitrification 238–248
  process 232–235
  SBR operation 298
  sludge age 235–236
  system design 251–253
  see also biological nitrogen removal
nutrient balance 3, 186, 204–209
  COD/N and COD/P ratios 3
  industrial waste 3
  nutrient requirements 206–209
nutrient removal 80–96, 204–206
  denitrification 89

models 79–101
nitrification 81
nitrogen 79
phosphorus 96–101
SBR 287–288

O
organic carbon, total (TOC) 20
organic carbon removal 16, 187,
    267–269
  aeration 273–275
  ASM1 65
  continuous flow technology 209–219
  design 213–218, 224, 231, 276
  excess sludge production rate 270–271
  models 37–74
    characterization 45–61
    mass balance 44–45
    matrix representation 39–43
    multicomponent 61–71
    system design 71–74
  nitrogen reaction kinetics 205
  phosphorus reaction kinetics 205
  process design 209–218, 209–219,
    275–280
  reactor volume 271–273
  SBR 267–280, 276
  sludge age 269–270
organic chemical removal 187
organic content 173
oxygen uptake rate, OUR 104–112
  acetate storage 130
  aerobic respiration test 106–109
  area 130
  ASM1 70, 141–142
  ASM3 70, 111
  chemical industry effluent 71
  cumulative 129
  domestic sewage 136
  dual hydrolysis model 136
  endogenous decay rate 133
  Escherichia coli 144
  nickel addition 139
  readily biodegradable substrate 110
  single hydrolysis model 136
  starch utilization 143
  starch-acetate mixture 139

tannery wastewater 373
utilization, S_{SI} 125
Y_H 127
oxidation-reduction system 13–14
oxidised nitrogen removal 16
ozonation tests 336
ozone flux 336

P
PAOs 80, 96–99
  EBPR 250–251, 297–298
  temperature effect on 250–251
particulate COD 49
particulate fraction inhibition 373
peptone-meat extract substrate 138–139
pH
  buffer capacity 3
  drink manufacturing wastewater 170
  textile wastewater 336
Phaeodactylum tricornutum 174–175
pharmaceutics
  heterotrophic growth 52
  hydrolysis 53
  kinetic coefficients 52
phosphorus
  accumulating organisms *see* PAOs
  ammonia 313–314
  COD/P ratio 249
  drink manufacturing 170
  EBPR 96, 248–253
  organic carbon removal 205
phosphorus removal 16
  environmental factors 250–251
  SBR operation 298
  system parameters 250
  wastewater characteristics 249
  see also biological phosphorus removal
physico-chemical treatability 319
pickling 351
plant operation
  pollution profile 211–212
  wool 323–325
pollution 2
  ammonia 313–314
  BOD 157–158, 161–164
  chlorine 321
  COD 157–158, 161–165, 313–321

colour 321
heterotrophs 157
parameters 156–158, 163–164
phosphorus 313–314
profiles
    pollution profile 153
    process profile 153
polyamide carpet finishing wastewater 73
post denitrification 234
post-tanning operations, tanning process 352–353
poultry industry wastewater 54
pre-denitrification 233, 240–241
    nitrogen removal 238–248, 284
    recommended design 240–241
    SBR 284, 286–296
    schematic process 233
pre-treatment 3
    COD fractionation 210–211
    COD size distribution 211
    limitations 1
    process 209–211
    standards 1
    tannery wastewater 365, 367
precipitation 372
process kinetics
    denitrification 96–97
    EBPR 100
    nitrification 88
process profiles 153–156
processes
    continuous flow technology 209–219
    design 209–218, 275–280
    in-plant control 325–332
    nitrification 88
    nitrogen removal 232–235, 283–286
    organic carbon removal 209–219, 275–280
    phase periods 264
    polyamide carpet finishing wastewater 73
    pre-treatment 209–211
    SBR 264
    stoichiometry 73, 88
    tanneries 358
pyridine nucleotides 14

R
rayon 311
reaction kinetics 44
reaction rate 39
reactors
    batch 39, 45
    continuous flow 39
    hydrolytics 38–39, 44
    kinetics 38
    organic carbon removal 271–273
    see also SBR
readily biodegradable substrate 110
reclamation
    materials 152
    wastewater 152
recycle ratio 186, 202–204
removal rate 39
resource management 167–171
respiration 17, 40, 55–57
respirometer 105

S
samming 352
SBR
    continuous filling 286
    design 260–261
    dual anoxic phases 285
    intermittent aeration 286
    nitrogen removal 280–296, 298
    nutrient removal 287–288
    organic carbon removal 267–280, 276
    performance 297
    phosphorus removal 296–299, 298
    pre-denitrification 284, 286–296
    process 262–267
    step feeding 285
    technology 259–260
segregated effluent discharge 173–174
segregated streams
    pollution profile 170
    textile wastewater 321
sequencing batch reactor see SBR
setting 352
sewer discharge 160–161
shaving 352
shower rinsings 331
silk 311

Index

single hydrolysis model 136
single sludge 234
slaughterhouses 54
sludge
  COD 42, 73
  excess 17, 186, 194–198, 270–271
  organic carbon removal 270–271
  retention time 265
sludge age 18, 182–183, 197
  anaerobic 250
  nitrogen removal 235–236
  organic carbon removal 269–270
soaking 350
solids
  mixed liquor suspended (MLSS) 19
  total suspended (TSS) 19
  volatile suspended (VSS) 19
soluble COD
soluble inert microbial product 60
source
  management 7
  pollution profiles 153–156
  process profiles 153–156
specific nitrogen removal 221–222
splitting 351
sulfur, S^{2-} 360
standards
  direct discharge 2
  pre-treatment 1
starch utilization 143
step feeding 284–285
stoichiometry
  coefficients 213, 371
  denitrification 90–93, 97
  substrate removal 30–35
substrate 18, 40
  BOD_5 27–29, 32–35, 43
  COD 30–32, 40, 42
  fermentable 98
  fermented 98
  process rate 31
  removal energetics 11–16
  removal stoichiometry 30–35
  removal yield 16–30
  storage concept 65–67
  utilization 17
  yield coefficient 128–130

sugar recovery
  effluent COD 171
  wastewater generation 171
sustainable environments 2

T
tannage
  aluminium 357
  chrome 357
  titanium 357
  vegetables 357
  zirconium 357
tannery wastewater 160, 349–350
  biodegradability 366–374
  chemicals 356–357, 368
  COD 169, 360, 369–370
  dual hydrolysis kinetics 54, 371
  endogenous decay coefficient 57
  generation 355–357
  heterotrophic growth 51
  hydrolysis 53
  in-plant control 357–362
  kinetic coefficients 52, 371
  management 349–374
  metabolic products 371
  operation categorization 354–355
  OUR response 373
  parameters 360
  particulate fraction inhibition 373
  pollution 164, 358–359
  pre-treatment 365, 367
  precipitation 372
  processes 358
  stoichiometric coefficients 371
  subcategorization 353–355
  treatment 362–366
tanning 350–353
  beamhouse operations 350–351
  finishing operations 352–353
  materials 361–362
  post-tanning operations 352–353
  water use reduction 362
tannins 171–175
total dissolved solids, TDS 321
temperature 250–251
tetra propylene benzene sulphonate 171
textile mills 335

textile production 324
textile wastewater
  alkalinity 336
  alum treatment 337
  chemical pre-treatment 320
  chemical treatability 337
  COD 168, 316, 320, 322, 331, 336, 339
  colour 336
  conductivity 336
  discharge standards 335
  end-of-pipe 309–310, 338
  endogenous decay coefficient 57
  feasibility analysis 339–345
  flow rates 335
  heterotrophic growth 51
  hydrolysis 53, 318
  in-plant control 305–309, 332, 334
  kinetic coefficients 52, 317
  kinetics 318
  management 303–345
  ozonation tests 336
  ozone flux 336
  pH 336
  physico-chemical treatability 319
  pollution 163, 310–322
  process specifications 311–312
  schematic processes 326–328
  segregated streams 321
  shower rinsings 331
  stoichiometric coefficients 317
  treatability 337
  treatment 338, 341, 343–344
  water conservation 323–345
  water recovery 323–345
  water reuse 321, 323–345, 335–337
titanium 357
TKN
  COD/TKN ratio 249
  drink manufacturing wastewater 170
  tannery effluent 360
total Kjeldahl nitrogen *see* TKN
total organic carbon (TOC) 20
  pollution 157
total suspended solids 19, 26, 162, 196, 271
toxicity 135–140

treatability
  biological 7
  textile wastewater 337
treatment 8
  costs 341, 343
  end-of-pipe 309
  gravity settling 8
  membrane bioreactors 8
  membrane filtration 8
  net present values 344
  plant performance 167
  reusable streams 152, 315, 323
total suspended solids, TSS 19
  pollution 313–314, 321
  tannery effluent 360

U
unhairing 351
unoxidized nitrogen fractions 82

V
vegetables 357
volatile suspended solids (VSS) 19, 26–29, 31–35, 41–44, 313–314, 321

W
waste
  chemical substitution 152
  end-of-pipe treatment 152
  front-of-pipe prevention 152
  minimization 2
  recovery 2
  reuse 2
  technological change 152
  water conservation 152
wastewater
  COD fractions 46, 48, 165–167
  confectionery 160
  corn wet mill 160
  dairy processing 160
  energy drink manufacture 170
  meat processing 160
  phosphorus removal 249
  pollutant parameters 156–158
  reclamation 152
  reliability 158–161
  reuse 152, 329–332

sewer discharge 160
sugar recovery 171
tannery 160
textile 159
unoxidized nitrogen fractions 82
water conservation
  in-plant control 329
  textile wastewater 323–345
  waste 152
water recovery 323–345
water reuse 321, 323–345
water use reduction 362
wool finishing
  plant operation 323–325
  pollution 311–312

X
xenobiotics 175–176
sludge age 193

Y
yield 17
  definition 22–25
  denitrification 94
  endogenous respiration 17
  expression 26–29
  microbial growth 17
  substrate removal 16–30
  true yield cofficient 17–18
yield coefficient assessment 126–130
  heterotrophic growth yield 126–127
  substrate storage yield 128–130
  OUR profile 127
  soluble COD profile 127

Z
zero discharge concept 2
zirconium 357

IWA Publishing's authorised EU representative for General Product Safety Regulations is Diane D'Arras, 15 rue Duret, 75116 Paris, France, e-mail: safety@iwap.co.uk.

Printed and bound by CPI Group (UK) Ltd, Croydon, CR0 4YY
27/03/2026
02079974-0004